Probabilistic Machine Learning for Civil Engineers

# Probabilistic Machine Learning for Civil Engineers

James-A. Goulet

The MIT Press
Cambridge, Massachusetts
London, England

This book was set in LATEX by the author. Printed and bound in the United States of America.

Library of Congress Cataloging-in-Publication Data

Names: Goulet, James-A., author.
Title: Probabilistic machine learning for civil engineers / James-A. Goulet.
Description: Cambridge, Massachusetts : The MIT Press, 2020. | Includes bibliographical references and index.
Identifiers: LCCN 2019027152 | ISBN 9780262538701 (paperback)
Subjects: LCSH: Machine learning. | Probabilities.
Classification: LCC Q325.5 .G68 2020 | DDC 006.3/1—dc23
LC record available at https://lccn.loc.gov/2019027152

10 9 8 7 6 5 4 3 2 1

# Contents

# List of Figures

# List of Algorithms

# *Acknowledgments*

I would like to acknowledge the help I have received from colleagues, students, and friends in the process of writing this book. I especially thank Jonathan Jalbert, Shervin Khazaeli, Saeid Amiri, Sébastien Le Digabel, Marco Broccardo, Luong-Ha Nguyen, Rocio Lilen Segura, and Gerd Brandstetter for helping me with reviewing the manuscript. I also acknowledge the feedback I have received from many students who attended my CIV6540 graduate class at Polytechnique Montreal. I want to thank Michel Goulet, who contributed to the book with several of his photos; as well as Daphne and Daniel Dethier, who provided helpful insights during the design of the book cover. I thank the anonymous reviewers who took the time to read the manuscript and whose comments were integrated into the final version.

In a broader perspective, I would like to thank my family for their inexhaustible support over the years. Also, I want to recognize the constructive influence from my advisers, who guided me since the beginning of my university years: André Picard, Mario Fafard, Ian F.C. Smith, and last but certainly not least, Armen Der Kiureghian, who was an inspiration through his lectures and mentoring.

Finally, I want to acknowledge the exceptional freedom I was allowed to have as a young professor by my host institution, Polytechnique Montreal. Without it, I do not see how this book would have been possible.

# Nomenclature & Abbreviations

## General Mathematical Symbols

| | |
|---:|:---|
| $\mathbb{R}$ | Real domain $\equiv (-\infty, \infty)$ |
| $\mathbb{R}^+$ | Real positive domain $\equiv (0, \infty)$ |
| $\mathbb{Z}$ | Real integer domain $\equiv \{\cdots, -1, 0, 1, 2, \cdots\}$ |
| $(0, 1)$ | Continuous close interval between 0 and 1, which includes 0 and 1 |
| $(0, 1]$ | Continuous open interval between 0 and 1, which includes 0 and not 1 |
| $\lim\limits_{n \to \infty}$ | A limit when $n$ tends to infinity |
| $\forall$ | For all |
| $:$ | Such that |
| $\check{x}$ | The true value for $x$ |
| $\hat{x}$ | An approximation of $x$ |
| $\sum$ | Sum operation |
| $\neg$ | The negation symbol |
| $\prod$ | Product operation |
| $\int dx$ | Integral operation with respect to $x$ |
| $\frac{dv(x)}{dx} \equiv \nabla_x v$ | Derivative or gradient of $v(x)$ with respect to $x$ |
| $\frac{\partial v(x,y,z)}{\partial x}$ | Partial derivative of $v(x, y, z)$ with respect to $x$ |
| $|x|$ | Absolute values of $x$ |
| $\approx$ | Approximately equal |
| $\propto$ | Proportional to |
| $\equiv$ | Equivalent |
| $\ln(x) \equiv \log_e(x)$ | Natural logarithm of $x$ $\ln(\exp(x)) = x$ |
| $\exp(x) \equiv e^x$ | Exponential function of $x$, $= 2.71828^x$, $\exp(\ln(x)) = x$ |
| $\Delta x$ | An infinitesimal interval for $x$ |
| $A \Leftrightarrow B$ | $A$ implies $B$ and $B$ implies $A$ |

## Linear Algebra

| | |
|---|---|
| $x$ | A scalar variable |
| $\mathbf{x}$ | A column vector, $\mathbf{x} = [x_1\ x_2\ \cdots\ x_{\mathsf{X}}]^{\mathsf{T}}$ |
| $\mathbf{X}$ | A matrix |
| $x_i \equiv [\mathbf{x}]_i$ | $i^{\text{th}}$ element of a vector |
| $x_{ij} \equiv [\mathbf{X}]_{ij}$ | $\{i,j\}^{\text{th}}$ element of a matrix |
| $\mathbf{X} = \text{diag}(\mathbf{x})$ | Square matrix $\mathbf{X}$ where the terms on the main diagonal are the elements of $\mathbf{x}$ and 0 elsewhere |
| $\mathbf{x} = \text{diag}(\mathbf{X})$ | Vector $\mathbf{x}$ consisting in the main diagonal terms of a matrix $\mathbf{X}$ |
| $\mathbf{I}$ | The identity matrix, i.e., a square matrix with 1 on the main diagonal and 0 elsewhere |
| $\text{blkdiag}(\mathbf{A}, \mathbf{B})$ | Block diagonal matrix where matrices $\mathbf{A}$ and $\mathbf{B}$ are concatenated on the main diagonal of a single matrix |
| $\mathsf{T}$ | Transposition operator : $[\mathbf{X}]_{ij} = [\mathbf{X}^{\mathsf{T}}]_{ji}$ |
| $\cdot$ | Scalar product |
| $\times$ | Matrix multiplication |
| $\odot$ | Hadamar (element-wise) product |
| $\|\mathbf{x}\|_p$ | $L^p$-norm of a vector $\mathbf{x}$ |
| $\det(\mathbf{A}) \equiv |\mathbf{A}|$ | Determinant of a Matrix $\mathbf{A}$ |
| $\text{tr}(\mathbf{A})$ | Sum of the elements on the main diagonal of $\mathbf{A}$ |
| $\rightarrow$ | A transformation from a space to another |
| $\mathbf{J}_{\mathbf{y},\mathbf{x}}$ | The Jacobian matrix so that $[\mathbf{J}_{\mathbf{y},\mathbf{x}}]_{k,l} = \frac{\partial y_k}{\partial x_l}$ |
| $\frac{\partial g(\mathbf{x})}{\partial x_i}$ | Partial derivative of $g(\mathbf{x})$ with respect to the $i^{\text{th}}$ variable $x_i$ |
| $\nabla g(\mathbf{x})$ | A gradient vector, $= \left[\frac{\partial g(\mathbf{x})}{\partial x_1}\ \cdots\ \frac{\partial g(\mathbf{x})}{\partial x_n}\right]$ |
| $\mathbf{H}$ | The Hessian matrix containing second-order partial derivatives, $[\mathbf{H}[g(\mathbf{x})]]_{ij} = \frac{\partial^2 g(\mathbf{x})}{\partial x_i \partial x_j}$ |

## Probability Theory and Random Variables

| | |
|---|---|
| $\mathcal{A} = \{E_1, E_2, \cdots\}$ | A set is described by a *calligraphic* letter |
| $\#\mathcal{A}$ | Number of elements in a set $\mathcal{A}$ |
| $\mathcal{S}$ | Universe/sampling space, i.e., the ensemble of possible results |
| $x$ | An elementary event, $x \in \mathcal{S}$ |
| $x \in \mathcal{S}$ | $x$ belongs to the sampling space $\mathcal{S}$ |
| $E$ | An ensemble of elementary events |
| $E \subset \mathcal{S}$ | E is a subset of $\mathcal{S}$ |
| $E \subseteq \mathcal{S}$ | E is a subset or is equal to $\mathcal{S}$ |
| $E = \mathcal{S}$ | A certain event |
| $E = \emptyset$ | An impossible event |
| $\overline{E}$ | The complement of the event $E$ |
| $\text{Pr}(\cdot)$ | The probability of an event ($\in (0,1)$) |

| | |
|---|---|
| $\cup$ | Union operation for events, i.e., "or" |
| $\cap$ | Intersection operation for events, i.e., "and" |
| $X$ | A random variable |
| $\mathbf{X}$ | A vector of random variables |
| $f(x) \equiv f_X(x)$ | Probability density function of a random variable $X$ |
| $X \sim f(x)$ | $X$ is distributed as described by its marginal probability density function $f(x)$ |
| $\mathbf{X} \sim f(\mathbf{x})$ | $\mathbf{X}$ is distributed as described by its joint probability density function $f(\mathbf{x})$ |
| $\xrightarrow{d}$ | Converges in distribution |
| $x, x_i$ | Realization of a random variable $x : X \sim f(x)$ |
| $F(x) \equiv F_X(x)$ | Cumulative distribution (or mass) function of a random variable $X$ |
| $\Phi(x)$ | Cumulative distribution function of a standard Normal random variable with mean equal to 0 and variance equal to 1 |
| $p(x) \equiv p_X(x)$ | Probability mass function of a random variable $X$ |
| $X \perp\!\!\!\perp Y$ | The random variables $X$ and $Y$ are statistically independent |
| $X \perp\!\!\!\perp Y|z$ | The random variables $X$ and $Y$ are conditionally independent given z |
| $X|y$ | The random variable $X$ is conditionally dependent on $y$ |
| $\mathbb{E}[X]$ | Expectation operation for a random variable $X$ |
| $\mathrm{var}[X]$ | Variance operation for a random variable $X$ |
| $\mathrm{cov}(X, Y)$ | Covariance operation for a pair of random variables $X, Y$ |
| $\delta_X$ | Coefficient of variation of a random variable $X$ |
| $\mu_X$ | The mean of a random variable $X$ |
| $\sigma_X^2$ | The variance of a random variable $X$ |
| $\rho$ | The correlation coefficient |
| $\boldsymbol{\mu}_{\mathbf{X}}$ | The mean values for a vector of random variables $\mathbf{X}$ |
| $\boldsymbol{\Sigma}_{\mathbf{X}}$ | A covariance matrix for a vector of random variables $\mathbf{X}$ |
| $\mathbf{R}_{\mathbf{X}}$ | A correlation matrix for a vector of random variables $\mathbf{X}$ |
| $\mathbf{D}_{\mathbf{X}}$ | A standard deviation matrix for a vector of random variables $\mathbf{X}$ |
| $\mathcal{N}(x; \mu, \sigma^2)$ | The probability density function of a univariate Normal random variable $X$, parameterized by its mean and variance |
| $\mathcal{N}(\mathbf{x}; \boldsymbol{\mu}_{\mathbf{X}}, \boldsymbol{\Sigma}_{\mathbf{X}})$ | The joint probability density function of a multivariate Normal random variable $\mathbf{X}$, parameterized by its mean vector and covariance matrix |
| $\ln \mathcal{N}(x; \lambda, \zeta)$ | The probability density function of a log-normal random variable $X$, parameterized by its mean and standard deviation defined in the log space |
| $\mathcal{B}(x; \alpha, \beta)$ | The probability density function of a Beta random variable $X$, parameterized by $\alpha$ and $\beta$ |
| $\mathcal{U}(x; 0, 1)$ | The uniform probability density function for a random variable $X$ defined for the interval $(0, 1)$ |
| $\delta(x)$ | The dirac-delta function |

$$f(y|x) \equiv \mathcal{L}(x|y)$$   The likelihood, i.e., the prior probability density of observing $Y = y$ given $x$

## Optimization

$$\tilde{f}(\boldsymbol{\theta})$$   Target function we want maximize

$$\tilde{f}'(\theta) = \frac{d\tilde{f}(\theta)}{d\theta}$$   First derivative of a function

$$\tilde{f}''(\theta) = \frac{d^2\tilde{f}(\theta)}{d\theta^2}$$   Second derivative of a function

$$\arg\max_{\boldsymbol{\theta}} \tilde{f}(\boldsymbol{\theta})$$   The values of $\boldsymbol{\theta}$ that maximize $\tilde{f}(\boldsymbol{\theta})$

$$\max_{\boldsymbol{\theta}} \tilde{f}(\boldsymbol{\theta})$$   The maximal value of $\tilde{f}(\boldsymbol{\theta})$

$$\boldsymbol{\theta}^*$$   An optimal value

$$d$$   Search direction

$$\lambda$$   Scale factor for the search direction

$$\mathbb{I}(i)$$   An indicator vector for which all values are equal to 0, except the $i^{\text{th}}$, which is equal to one

## Sampling

$$\boldsymbol{\theta}$$   Vector of parameters

$$q(\boldsymbol{\theta}'|\boldsymbol{\theta})$$   Proposal distribution describing the probability of moving to $\boldsymbol{\theta}'$ from $\boldsymbol{\theta}$

$$\tilde{f}(\boldsymbol{\theta})$$   Target distribution from which we want to draw samples

$$\alpha$$   Acceptance probability

$$\beta$$   Acceptance rate

$$\hat{R}$$   Estimated potential scale reduction

## Utility & Decisions

$$\mathcal{A} = \{a_1, \cdots, a_A\}$$   A set of possible actions

$$x \in \mathbb{X} \subseteq \mathbb{Z}$$   An outcome from a set of discrete states

$$L$$   A lottery

$$\mathbb{U}(a, x)$$   Utility given a state $x$ and an action $a$

$$\overline{\mathbb{U}}(a) \equiv \mathbb{E}[\mathbb{U}(a, X)]$$   Expected utility conditional on an action $a$

$$\overline{\mathbb{U}}(s) \equiv \overline{\mathbb{U}}(s, \pi) \equiv \mathbb{E}[\mathbb{U}(s, \pi)]$$   Long-term expected utility conditional on a current state $s$ and that a policy $\pi$ is followed

$$\pi(s)$$   A policy defining an action $a$ to take, conditional on a state $s$

$$r(s, a, s')$$   The reward for being in state $s$, taking the action $a$ and ending in state $s'$

$$\mathbb{Q}(s, a)$$   Action-utility function

$$\gamma$$   Discount factor

$$\alpha$$   The learning rate

$\epsilon$    The probability that an agent takes a random action rather than following the policy $\pi(s)$

$\leftarrow$    An assignment in a recurrent equation, e.g., $x \leftarrow 2x$

## *Abbreviations*

| | |
|---|---|
| AI | Artificial intelligence |
| BN | Bayesian network |
| CDF | Cumulative distribution function |
| CLT | Central limit theorem |
| CMF | Cumulative mass function |
| CPT | Conditional probability table |
| CV | Cross-validation |
| DAG | Directed acyclic graph |
| DBN | Dynamic Bayesian network |
| e.g. | For example |
| EM | Expectation maximization |
| EPSR | Estimated potential scale reduction |
| FCNN | Fully connected feedforward neural network |
| GA | Gradient ascent |
| GMM | Gaussian mixture model |
| GP | Gaussian process |
| GPC | Gaussian process classification |
| GPR | Gaussian process regression |
| GPU | Graphical processing unit |
| HMM | Hidden Markov model |
| i.e. | That is |
| iid | Independent identically distributed |
| KF | Kalman filter |
| KS | Kalman smoother |
| LDA | Linear discriminant analysis |
| LLOCV | Leave-one-out cross-validation |
| MAP | Maximum a-posteriori |
| MCMC | Markov chain Monte Carlo |
| MDP | Markov decision process |
| ML | Machine learning |
| MLE | Maximum likelihood estimate |
| NB | Naïve Bayes |
| NN | Neural network |
| NR | Newton-Raphson |
| PCA | Principal component analysis |
| PDF | Probability density function |

| | |
|---|---|
| PMF | Probability mass function |
| POMDP | Partially observable Markov decision process |
| PSD | Positive semi-definite |
| QDA | Quadratic discriminant analysis |
| RL | Reinforcement learning |
| R.V. | Random variable |
| SKF | Switching Kalman filter |
| SSM | State-space model |
| sym. | Symmetric |
| TD | Temporal difference |
| VOI | Value of information |
| VPI | Value of perfect information |

# 1
# *Introduction*

*Machine learning* (ML) describes a family of methods that allows learning from data what relationships exist between quantities of interest. The goal of learning relationships between quantities of interest is to gain information about how a system works and to make predictions for unobserved quantities. This new knowledge can then be employed to support decision making.

Learning from data is not new in the field of engineering. As depicted in figure 1.1a, without machine learning, the task of learning is typically the role of the engineer, who has to figure out the relationships between the variables in a data set, then build a hard-coded model of the system, and then return predictions by evaluating this model with the help of a computer. The issue with *hard-coding* rules and equations is that it becomes increasingly difficult as the number of factors considered increases. Figure 1.1b depicts how, with machine learning, the role of the human is now shifted from learning to programming the computer so

**Civil engineering examples**

*Geotechnics*: From a set of discrete observations of soil resistances measured across space, we want to predict the resistance for other unobserved locations.

*Environment*: For a lake, we want to build a model linking the effect of temperature and the usage of fertilizers with the prevalence of cyanobacteria, fish mortality, and water color.

*Transportation*: We want to predict the demand for public transportation services from multiple heterogeneous data sources such as surveys and transit card usage.

*Structures*: From observations made on the displacement of a structure over time, we want to detect the presence of anomalies in its behavior.

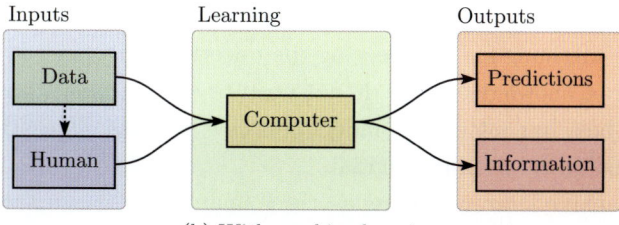

(a) <u>Without</u> machine learning

(b) <u>With</u> machine learning

Figure 1.1: Comparison of how, in the context of engineering, relationships are learned from data with and without using machine learning.

**Note:** For a generic model $y = g(x; \theta)$, $x$ and $y$ are the input and output *variables* of interest that are evaluated by the model. $\theta$ is a *parameter* characterizing the model behavior.

it can learn. The question is, How do we program a *computer to learn*? The quick answer is this: by defining a generic model that can be adapted to a wide array of problems by changing its parameters and variables. For most machine learning methods, *learning* consists in inferring these *parameters* and *variables* from data. Notwithstanding the specificities for reinforcement learning that we will cover later, the big picture is the following: learning about parameters and variables is done by quantifying, for a given set of their values, how good the model is at predicting observed data. This task typically leads to nonunique solutions, so we have to make a choice: either pick a single set of values (that is, those leading to the best predictions) or consider all possible values that are compatible with data. In figure 1.1b, the dashed arrow represents the practical reality where ML typically relies on a human decision for the selection of a particular mathematical method adapted to a given data set and problem.

One term often employed together with machine learning is *artificial intelligence* (AI). ML and AI are closely interconnected, yet they are not synonyms. In its most general form, artificial intelligence consists in the reproduction of an intelligent behavior by a machine. AI typically involves a system that can perceive and interact dynamically with its environment through the process of making *rational decisions*. Note that such an AI system does not have to take a physical form; in most cases it is actually only a computer program. In AI systems, decisions are typically not hard-coded or learned by imitation; instead, the AI system chooses actions with the goal of maximizing an objective function that is given to it. Behind such an AI system that interacts with its environment are machine learning methods that allow extracting information from observations, predict future system responses, and choose the optimal action to take. As depicted in the Venn diagram in figure 1.2, machine learning is part of the field of artificial intelligence.

Now, going back to machine learning, why opt for a *probabilistic* approach to it? It is because the task of learning is intrinsically uncertain. Uncertainties arise from the imperfect models employed to learn from data, because data itself often involves imperfect observations. Therefore, probabilities are at the core of machine learning methods for representing the uncertainty associated with our lack of knowledge. If we do not want to use machine learning as a black box, but instead to understand how it works, going through the probabilistic path is essential.

In machine learning, there are three main subfields: supervised

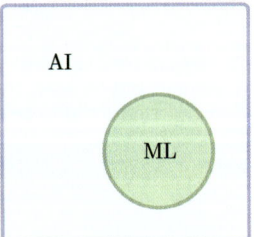

Figure 1.2: The field of artificial intelligence (AI) includes the field of machine learning (ML).

learning, unsupervised learning, and reinforcement learning. *Supervised learning* applies in the context where we want to build a model describing the relationships between the characteristics of a *system* defined by *covariates* and observed system responses that are typically either continuous values or categories. With *unsupervised learning*, the objective is to discover structures, patterns, subgroups, or even anomalies without knowing what the right answer is because the target outputs are not observed. The third subfield is *reinforcement learning*, which involves more abstract concepts than supervised and unsupervised learning. Reinforcement learning deals with sequential decision problems where the goal is to learn the optimal action to choose, given the knowledge that a system is in a particular state. Take the example of infrastructure maintenance, where, given the state of a structure today, we must choose between performing maintenance or doing nothing. The key is that there is no data to train on with respect to the decision-making behavior that the computer should reproduce. With reinforcement learning, the goal is to identify a *policy* describing the optimal action to perform for each possible state of a system in order to maximize the long-term accumulation of rewards. Note that the classification of machine learning methods within supervised, unsupervised, and reinforcement learning has limitations. For many methods, the frontiers are blurred because there is an overlap between more than one ML subfield with respect to the mathematical formulations employed as well as the applications.

*This book is intended to help making machine learning concepts accessible to civil engineers who do not have a specialized background in statistics or in computer science.* The goal is to dissect and simplify, through a step-by-step review, a selection of key machine learning concepts and methods. At the end, the reader should have acquired sufficient knowledge to understand dedicated machine learning literature from which this book borrows and thus expand on advanced methods that are beyond the scope of this introductory work.

The diagram in figure 1.3 depicts the organization of this book, where arrows represent the dependencies between different chapters. Colored regions indicate to which machine learning subfield each chapter belongs. Before introducing the fundamentals associated with each machine learning subfield in Parts II–V, Part I covers the *background* knowledge required to understand machine learning. This background knowledge includes *linear algebra* (chapter 2), where we review how to harness the potential of matrices to describe systems; *probability theory* (chapter 3) and *probability*

**Note:** We employ the generic term *system* to refer to either a single object or many interconnected objects that we want to study. We use the term *covariates* for variables describing the characteristics or the properties of a system.

Figure 1.3: Each square node describes a chapter, and arrows represent the dependencies between chapters. Shaded regions group chapters into the five parts of the book. Note that at the beginning of each part, this diagram is broken down into subparts in order to better visualize the dependencies between the current and previous chapters.

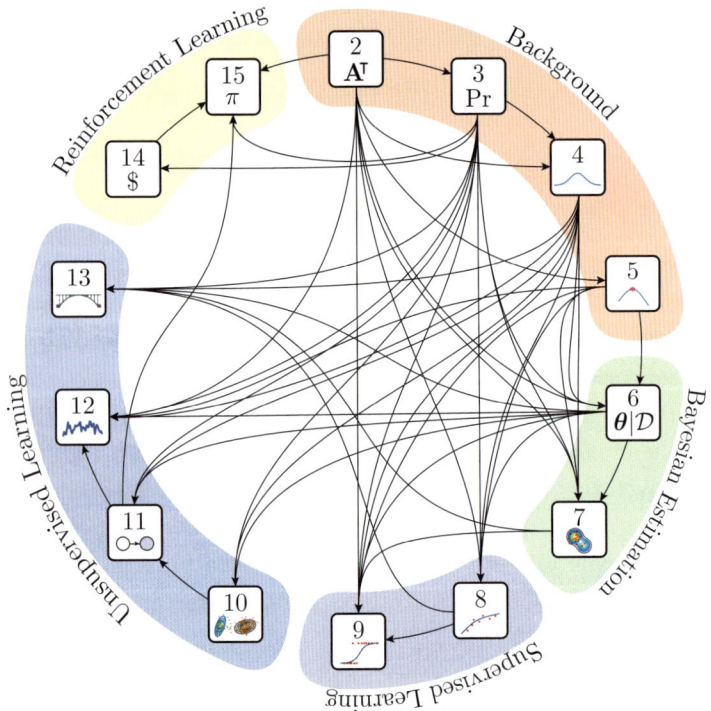

*distributions* (chapter 4) to describe our incomplete knowledge; and *convex optimization* (chapter 5) as a first method for allowing a computer to learn about model parameters.

Part II first covers *Bayesian estimation* (chapter 6), which is behind the formulation of supervised and unsupervised learning problems. Second, it covers *Markov chain Monte Carlo* (MCMC) methods (chapter 7), allowing one to perform Bayesian estimation in complex cases for which no analytical solution is available.

Part III explores methods and concepts associated with supervised learning. Chapter 8 covers regression methods, where the goal is to build models describing continuous-valued system responses as a function of covariates. Chapter 9 presents classification methods, which are analogous to regression except that the system responses are categories rather than continuous values.

Part IV introduces the notions associated with unsupervised learning, where the task is to build models that can extract the underlying structure present in data without having access to direct observations of what this underlying structure should be. In chapter 10, we first approach unsupervised learning through *clustering* and *dimension reduction*. For clustering, the task is to identify subgroups within a set of observed covariates for which we do not have access to the subgroup labels. The role of dimension reduction is, as its name implies, to reduce the number of dimensions required to represent data while minimizing the loss of information. Chapter 11 presents *Bayesian networks*, which are graph-based probabilistic methods for modeling dependencies within and between systems through their joint probability. Chapter 12 presents *state-space models*, which allow creating probabilistic models for time-dependent systems using sequences of observations. Finally, chapter 13 presents how we can employ the concepts of probabilistic inference for the purpose of *model calibration*. Model calibration refers to the task of using observations to improve our knowledge associated with hard-coded mathematical models that are commonly employed in engineering to describe systems. This application is classified under the umbrella of *unsupervised learning* because, as we will see, the main task consists in inferring hidden-state variables and parameters, for which observations are not available.

Part V presents the fundamental notions necessary to define reinforcement learning problems. First, chapter 14 presents how rational decisions are made in uncertain contexts using the *utility theory*. Chapter 15 presents how to extend rational decision making to a sequential context using the *Markov decision process* (MDP). Finally, building on the MDP theory, we introduce the fundamental concepts of *reinforcement learning*, where a virtual agent learns how to take optimal decisions through trial and error while interacting with its environment.

# Part I

# Background

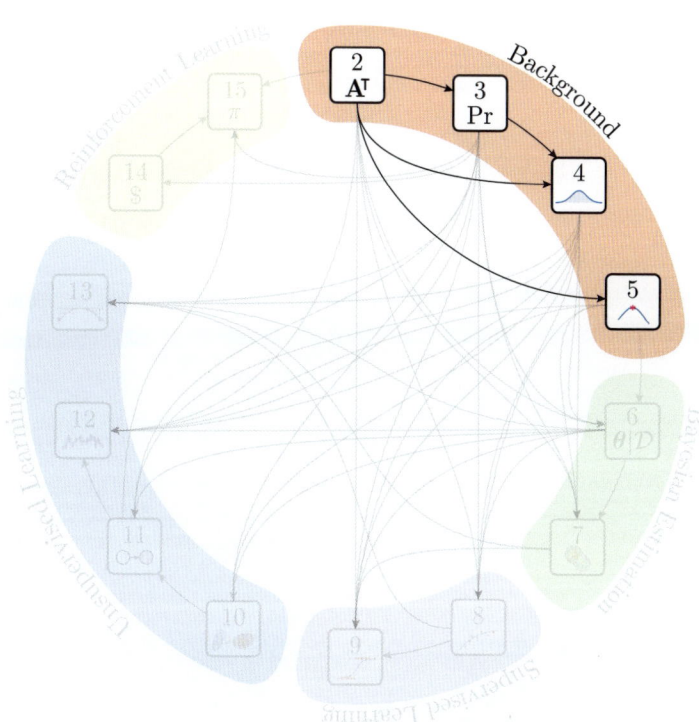

# 2
# Linear Algebra

*Linear algebra* is employed in a majority of machine learning methods and algorithms. Before going further, it is essential to understand the mathematical notation and basic operations.

## 2.1  Notation

We employ lowercase letters $x, s, v, \cdots$ in order to describe variables that can lie in specific domains such as real numbers $\mathbb{R}$, real positive $\mathbb{R}^+$, integers $\mathbb{Z}$, closed intervals $[\cdot, \cdot]$, open intervals $(\cdot, \cdot)$, and so on. Often, the problems studied involve multiple variables that can be regrouped in arrays. A 1-D array or *vector* containing scalars is represented as

$$\mathbf{x} = \begin{bmatrix} x_1 \\ x_2 \\ \vdots \\ x_n \end{bmatrix}.$$

By convention, a vector $\mathbf{x}$ implicitly refers to a $n \times 1$ column vector. For example, if each element $x_i \equiv [\mathbf{x}]_i$ is a real number $[\mathbf{x}]_i \in \mathbb{R}$ for all $i$ from 1 to $n$, then the vector belongs to the $n$-dimensional real domain $\mathbb{R}^n$. This last statement can be expressed mathematically as $[\mathbf{x}]_i \in \mathbb{R}, \forall i \in \{1 : n\} \rightarrow \mathbf{x} \in \mathbb{R}^n$. In machine learning, it is common to have 2-D arrays or *matrices*,

$$\mathbf{X} = \begin{bmatrix} x_{11} & x_{12} & \cdots & x_{1n} \\ x_{21} & x_{22} & \cdots & x_{2n} \\ \vdots & \vdots & \ddots & \vdots \\ x_{m1} & x_{m2} & \cdots & x_{mn} \end{bmatrix},$$

where, for example, if each $x_{ij} \equiv [\mathbf{X}]_{ij} \in \mathbb{R}, \forall i \in \{1 : m\}, j \in \{1 : n\} \rightarrow \mathbf{X} \in \mathbb{R}^{m \times n}$. Arrays beyond two dimensions are referred to as *tensors*. Although tensors are widely employed in the field of *neural networks*, they will not be treated in this book.

| | | |
|---|---|---|
| $x$ | : | scalar variable |
| $\mathbf{x}$ | : | column vector |
| $\mathbf{X}$ | : | matrix |
| $x_i \equiv [\mathbf{x}]_i$ | : | $i^{\text{th}}$ element of a vector |
| $x_{ij} \equiv [\mathbf{X}]_{ij}$ | : | $\{i,j\}^{\text{th}}$ element of a matrix |

Examples of variables $x$ belonging to different domains

$$\begin{aligned} x &\in \mathbb{R} &&\equiv (-\infty, \infty) \\ &\in \mathbb{R}^+ &&\equiv (0, \infty) \\ &\in \mathbb{Z} &&\equiv \{\cdots, -1, 0, 1, 2, \cdots\} \end{aligned}$$

There are several matrices with specific properties: A diagonal matrix is square and has only terms on its main diagonal,

$$\mathbf{Y} = \mathrm{diag}(\mathbf{x}) = \begin{bmatrix} x_1 & 0 & \cdots & 0 \\ 0 & x_2 & \cdots & 0 \\ \vdots & \vdots & \ddots & \vdots \\ 0 & 0 & \cdots & x_n \end{bmatrix}_{n \times n}.$$

An identity matrix $\mathbf{I}$ is similar to a diagonal matrix except that elements on the main diagonal are 1, and 0 everywhere else,

$$\mathbf{I} = \begin{bmatrix} 1 & 0 & \cdots & 0 \\ 0 & 1 & \cdots & 0 \\ \vdots & \vdots & \ddots & \vdots \\ 0 & 0 & \cdots & 1 \end{bmatrix}_{n \times n}.$$

A *block diagonal* matrix concatenates several matrices on the main diagonal of a single matrix,

$$\mathrm{blkdiag}(\mathbf{A}, \mathbf{B}) = \begin{bmatrix} \mathbf{A} & \mathbf{0} \\ \mathbf{0} & \mathbf{B} \end{bmatrix}.$$

We can manipulate the dimensions of matrices using the *transposition* operation so that indices are permuted $[\mathbf{X}^\mathsf{T}]_{ij} = [\mathbf{X}]_{ji}$. For example,

$$\mathbf{X} = \begin{bmatrix} x_{11} & x_{12} & x_{13} \\ x_{21} & x_{22} & x_{23} \end{bmatrix} \rightarrow \mathbf{X}^\mathsf{T} = \begin{bmatrix} x_{11} & x_{21} \\ x_{12} & x_{22} \\ x_{13} & x_{23} \end{bmatrix}.$$

The *trace* of a square matrix $\mathbf{X}$ corresponds to the sum of the elements on its main diagonal,

$$\mathrm{tr}(\mathbf{X}) = \sum_{i=1}^{n} x_{ii}.$$

## 2.2   Operations

In the context of machine learning, linear algebra is employed because of its capacity to model *linear systems of equations* in a format that is compact and well suited for computer calculations. In a 1-D case, such as the one represented in figure 2.1, the $x$ space is *mapped* into the $y$ space, $\mathbb{R} \rightarrow \mathbb{R}$, through a linear (i.e., *affine*) function. Figure 2.2 presents an example of a 2-D linear function where the $\mathbf{x}$ space is *mapped* into the $y$ space, $\mathbb{R}^2 \rightarrow \mathbb{R}$. This can be generalized to linear systems $\mathbf{y} = \mathbf{Ax} + \mathbf{b}$, defining a mapping so that $\mathbb{R}^n \rightarrow \mathbb{R}^m$, where $\mathbf{x}$ and $\mathbf{y}$ are respectively $n \times 1$ and $m \times 1$

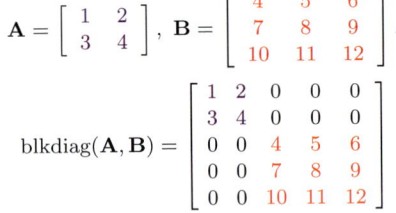

$$\mathbf{A} = \begin{bmatrix} 1 & 2 \\ 3 & 4 \end{bmatrix}, \mathbf{B} = \begin{bmatrix} 4 & 5 & 6 \\ 7 & 8 & 9 \\ 10 & 11 & 12 \end{bmatrix}.$$

$$\mathrm{blkdiag}(\mathbf{A}, \mathbf{B}) = \begin{bmatrix} 1 & 2 & 0 & 0 & 0 \\ 3 & 4 & 0 & 0 & 0 \\ 0 & 0 & 4 & 5 & 6 \\ 0 & 0 & 7 & 8 & 9 \\ 0 & 0 & 10 & 11 & 12 \end{bmatrix}$$

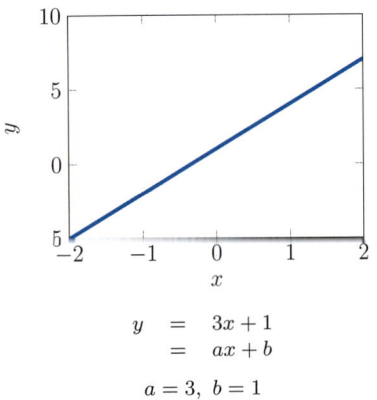

$$\begin{aligned} y &= 3x + 1 \\ &= ax + b \end{aligned}$$

$$a = 3, \ b = 1$$

Figure 2.1: 1-D plot representing a linear system, $y = ax + b$.

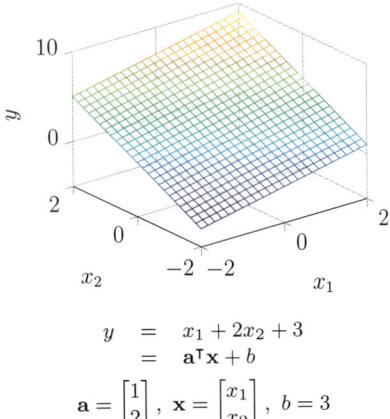

$$\begin{aligned} y &= x_1 + 2x_2 + 3 \\ &= \mathbf{a}^\mathsf{T}\mathbf{x} + b \end{aligned}$$

$$\mathbf{a} = \begin{bmatrix} 1 \\ 2 \end{bmatrix}, \ \mathbf{x} = \begin{bmatrix} x_1 \\ x_2 \end{bmatrix}, \ b = 3$$

Figure 2.2: 2-D plot representing a linear system, $y = \mathbf{a}^\mathsf{T}\mathbf{x} + b$.

vectors. The product of the matrix $\mathbf{A}$ with the vector $\mathbf{x}$ is defined as $[\mathbf{Ax}]_i = \sum_j [\mathbf{A}]_{ij} \cdot [\mathbf{x}]_j$.

In more general cases, linear algebra is employed to multiply a matrix $\mathbf{A}$ of size $n \times k$ with another matrix $\mathbf{B}$ of size $k \times m$, so the result is a $n \times m$ matrix,

$$\begin{aligned} \mathbf{C} &= \mathbf{AB} \\ &= \mathbf{A} \times \mathbf{B}. \end{aligned}$$

The *matrix multiplication* operation follows $[\mathbf{C}]_{ij} = \sum_k [\mathbf{A}]_{ik} \cdot [\mathbf{B}]_{kj}$, as illustrated in figure 2.3. Following the requirement on the size of the matrices multiplied, this operation is not generally commutative so that $\mathbf{AB} \neq \mathbf{BA}$. Matrix multiplication follows several properties such as the following:

$$\begin{aligned} \text{Distributivity} && \mathbf{A}(\mathbf{B} + \mathbf{C}) &= \mathbf{AB} + \mathbf{AC} \\ \text{Associativity} && \mathbf{A}(\mathbf{BC}) &= (\mathbf{AB})\mathbf{C} \\ \text{Conjugate transposability} && (\mathbf{AB})^{\mathsf{T}} &= \mathbf{B}^{\mathsf{T}}\mathbf{A}^{\mathsf{T}}. \end{aligned}$$

When the matrix multiplication operator is applied to $n \times 1$ vectors, it reduces to the *inner product*,

$$\begin{aligned} \mathbf{x}^{\mathsf{T}}\mathbf{y} &\equiv \mathbf{x} \cdot \mathbf{y} \\ &= [x_1 \ \cdots \ x_n] \times \begin{bmatrix} y_1 \\ \vdots \\ y_n \end{bmatrix} \\ &= \sum_{i=1}^{n} x_i y_i. \end{aligned}$$

Another common operation is the *Hadamar product* or *element-wise product*, which is represented by the symbol $\odot$. It consists in multiplying each term from matrices $\mathbf{A}_{m \times n}$ and $\mathbf{B}_{m \times n}$ in order to obtain $\mathbf{C}_{m \times n}$,

$$\begin{aligned} \mathbf{C} &= \mathbf{A} \odot \mathbf{B} \\ [\mathbf{C}]_{ij} &= [\mathbf{A}]_{ij} \cdot [\mathbf{B}]_{ij}. \end{aligned}$$

The element-wise product is seldom employed to define mathematical equations; however, it is extensively employed when implementing these equations in a computer language. Matrix addition is by definition an element-wise operation that applies only to matrices of same dimensions,

$$\begin{aligned} \mathbf{C} &= \mathbf{A} + \mathbf{B} \\ [\mathbf{C}]_{ij} &= [\mathbf{A}]_{ij} + [\mathbf{B}]_{ij}. \end{aligned}$$

One last key operation is the *matrix inversion* $\mathbf{A}^{-1}$. In order to be invertible, a matrix must be square and must not have *linearly dependent* rows or columns. The product of a matrix with its in-

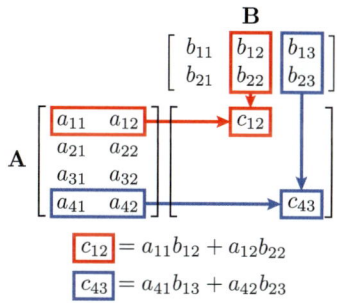

$$c_{12} = a_{11}b_{12} + a_{12}b_{22}$$
$$c_{43} = a_{41}b_{13} + a_{42}b_{23}$$

Figure 2.3: Example of matrix multiplication operation $\mathbf{C} = \mathbf{AB}$.

**Linearly dependent vectors**

Vectors $\mathbf{x}_1 \in \mathbf{R}^n$ and $\mathbf{x}_2 \in \mathbf{R}^n$ are *linearly dependent* if a nonzero vector $\mathbf{y} \in \mathbf{R}^2$ exists, such that $y_1\mathbf{x}_1 + y_2\mathbf{x}_2 = \mathbf{0}$.

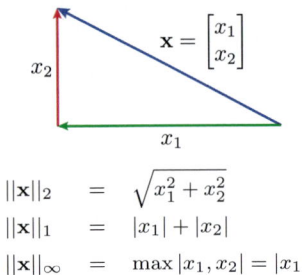

$$\begin{aligned}
||\mathbf{x}||_2 &= \sqrt{x_1^2 + x_2^2} \\
||\mathbf{x}||_1 &= |x_1| + |x_2| \\
||\mathbf{x}||_\infty &= \max |x_1, x_2| = |x_1|
\end{aligned}$$

Figure 2.4: Examples of applications of different norms for computing the length of a vector $\mathbf{x}$.

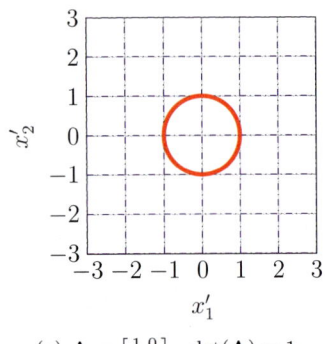

(a) $\mathbf{A} = \begin{bmatrix} 1 & 0 \\ 0 & 1 \end{bmatrix}$, $\det(\mathbf{A}) = 1$

(b) $\mathbf{A} = \begin{bmatrix} 1 & 0 \\ 0 & 2 \end{bmatrix}$, $\det(\mathbf{A}) = 2$

(c) $\mathbf{A} = \begin{bmatrix} 1.5 & 0 \\ 0 & 1 \end{bmatrix}$, $\det(\mathbf{A}) = 1.5$

Figure 2.5: Examples of linear transformations $\mathbf{x}' = \mathbf{A}\mathbf{x}$.

verse is equal to the identity matrix $\mathbf{A}^{-1}\mathbf{A} = \mathbf{I}$. Matrix inversion is particularly useful for solving linear systems of equations,

$$\begin{aligned}
\mathbf{A}\mathbf{x} &= \mathbf{b}, \\
\mathbf{A}^{-1}\mathbf{A}\mathbf{x} &= \mathbf{A}^{-1}\mathbf{b}, \\
\mathbf{I}\mathbf{x} &= \mathbf{A}^{-1}\mathbf{b}, \\
\mathbf{x} &= \mathbf{A}^{-1}\mathbf{b}.
\end{aligned}$$

## 2.3   Norms

Norms measure how large a vector is. In a generic way, the $L^p$-norm is defined as

$$||\mathbf{x}||_p = \left( \sum_i |[\mathbf{x}]_i|^p \right)^{1/p}.$$

Special cases of interest are

$$\begin{aligned}
||\mathbf{x}||_2 &= \sqrt{\sum_i [\mathbf{x}]_i^2} \equiv \sqrt{\mathbf{x}^\mathsf{T}\mathbf{x}} &&\text{(Euclidian norm)} \\
||\mathbf{x}||_1 &= \sum_i |[\mathbf{x}]_i| &&\text{(Manhattan norm)} \\
||\mathbf{x}||_\infty &= \max_i |[\mathbf{x}]_i|. &&\text{(Max norm)}
\end{aligned}$$

These cases are illustrated in figure 2.4. Among all cases, the $L^2$-norm (Euclidian distance) is the most common. For example, §8.1.1 presents for the context of linear regression how choosing a Euclidian norm to measure the distance between observations and model predictions allows solving the parameter estimation problem analytically.

## 2.4   Transformations

Machine learning involves transformations from one space to another. In the context of linear algebra, we are interested in the special case of linear transformations.

### 2.4.1   Linear Transformations

Figure 2.1 presented an example for a $\mathbb{R} \to \mathbb{R}$ linear transformation. More generally, a $n \times n$ square matrix can be employed to perform a $\mathbb{R}^n \to \mathbb{R}^n$ linear transformation through multiplication. Figures 2.5a–c illustrate how a matrix $\mathbf{A}$ transforms a space $\mathbf{x}$ into another $\mathbf{x}'$ using the matrix product operation $\mathbf{x}' = \mathbf{A}\mathbf{x}$. The deformation of the circle and the underlying grid (see (a)) show the effect of various transformations. Note that the terms on the main diagonal of $\mathbf{A}$ control the transformations along the $x_1'$ and $x_2'$ axes, and the nondiagonal terms control the transformation dependency between both axes, (see, for example, figure 2.6).

The *determinant* of a square matrix $\mathbf{A}$ measures how much the transformation contracts or expands the space:

- $\det(\mathbf{A}) = 1$: preserves the space/volume

- $\det(\mathbf{A}) = 0$: collapses the space/volume along a subset of dimensions, for example, 2-D space $\rightarrow$ 1-D space (see figure 2.7)

In the examples presented in figure 2.5a–c, the determinant quantifies how much the area/volume is changed in the transformed space; for the circle, it corresponds to the change of area caused by the transformation. As shown in figure 2.5a, if $\mathbf{A} = \mathbf{I}$, the transformation has no effect so $\det(\mathbf{A}) = 1$. For a square matrix $[\mathbf{A}]_{n\times n}$, $\det(\mathbf{A}) : \mathbb{R}^{n\times n} \rightarrow \mathbb{R}$.

### 2.4.2   Eigen Decomposition

Linear transformations operate on several dimensions, such as in the case presented in figure 2.6 where the transformation introduces dependency between variables. *Eigen decomposition* enables finding a linear transformation that removes the dependency while preserving the area/volume. A square matrix $[\mathbf{A}]_{n\times n}$ can be decomposed in *eigenvectors* $\{\boldsymbol{\nu}_1, \cdots, \boldsymbol{\nu}_n\}$ and *eigenvalues* $\{\lambda_1, \cdots, \lambda_n\}$. In its matrix form,

$$\mathbf{A} = \mathbf{V}\mathrm{diag}(\boldsymbol{\lambda})\mathbf{V}^{-1},$$

where

$$\begin{aligned}\mathbf{V} &= [\boldsymbol{\nu}_1 \cdots \boldsymbol{\nu}_n]\\ \boldsymbol{\lambda} &= [\lambda_1 \cdots \lambda_n]^{\mathsf{T}}.\end{aligned}$$

Figure 2.6 presents the eigen decomposition of the transformation $\mathbf{x}' = \mathbf{A}\mathbf{x}$. Eigenvectors $\boldsymbol{\nu}_1$ and $\boldsymbol{\nu}_2$ describe the new referential into which the transformation is independently applied to each axis. Eigenvalues $\lambda_1$ and $\lambda_2$ describe the transformation magnitude along each eigenvector.

A matrix is *positive definite* if all eigenvalues $> 0$, and a matrix is *positive semidefinite* (PSD) if all eigenvalues $\geq 0$. The determinant of a matrix corresponds to the product of its eigenvalues. Therefore, in the case where one eigenvalue equals zero, it indicates that two or more dimensions are linearly dependent and have collapsed into a single one. The transformation matrix is then said to be *singular*. Figure 2.7 presents an example of a nearly singular transformation. For a positive semidefinite matrix $\mathbf{A}$ and for any

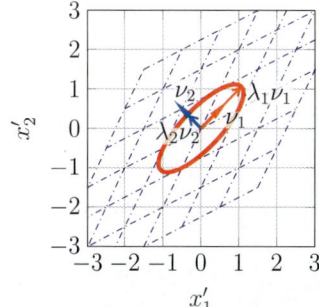

$$\begin{aligned}\mathbf{x}' &= \mathbf{A}\mathbf{x}\\ \mathbf{A} &= \begin{bmatrix} 1 & 0.5 \\ 0.5 & 1 \end{bmatrix}\\ \mathbf{V} &= [\boldsymbol{\nu}_1\ \boldsymbol{\nu}_2] = \begin{bmatrix} -0.71 & 0.71 \\ 0.71 & 0.71 \end{bmatrix}\\ \boldsymbol{\lambda} &= [0.5\ 1.5]^{\mathsf{T}}\end{aligned}$$

Figure 2.6: Example of eigen decomposition, $\mathbf{A} = \mathbf{V}\mathrm{diag}(\boldsymbol{\lambda})\mathbf{V}^{-1}$.

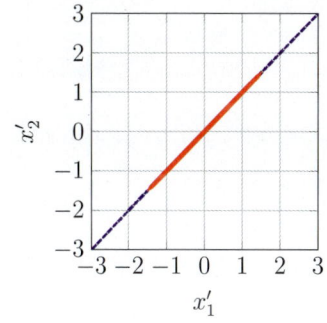

$$\mathbf{A} = \begin{bmatrix} 1 & 0.99 \\ 0.99 & 1 \end{bmatrix},\ \det(\mathbf{A}) = 0.02$$

Figure 2.7: Example of a nearly singular transformation.

vector $\mathbf{x}$, the following relation holds:

$$\mathbf{x}^{\mathsf{T}}\mathbf{A}\mathbf{x} \geq 0.$$

This property is employed in §3.3.5 to define the requirements for an admissible covariance matrix.

A more exhaustive review of linear algebra can be found in dedicated textbooks such as the one by Kreyszig.[1]

[1] Kreyszig, E. (2011). *Advanced engineering mathematics* (10th ed.). Wiley.

## Exercises

P2.1  Calculate the Euclidian, Manhattan, and max norm for the vector $\mathbf{x} = [1\ 4\ 2\ 7\ 4]^{\mathsf{T}}$.

P2.2  Reformulate the two equations $2x_1 + 6x_2 = 8$ and $5x_1 + x_2 = 0$ as a system of linear equations, and solve it for $[x_1\ x_2]^{\mathsf{T}}$ using linear algebra.

P2.3  Draw the result of transforming a unit radius circle through the function $\mathbf{x}' = \mathbf{A}\mathbf{x} + \mathbf{b}$, where $\mathbf{A} = \begin{bmatrix} 0.85 & -1.3 \\ -1.3 & 0.5 \end{bmatrix}$.

P2.4  For the transformation in P2.3, calculate the eigen decomposition for the matrix $\mathbf{A}$, and plot the deformed unit circle in a set of axes corresponding to the eigenvectors.

P2.5  Identify which matrix is positive semidefinite: $\mathbf{A} = \begin{bmatrix} 3 & 1 \\ 12 & 4 \end{bmatrix}$, $\mathbf{B} = \begin{bmatrix} 2 & 0 \\ -5 & 8 \end{bmatrix}$.

# 3
# *Probability Theory*

The interpretation of probability theory employed in this book follows Laplace's view of "*common sense reduced to calculus*." It means that probabilities describe our state of knowledge rather than intrinsically *aleatory* phenomena. In practice, few phenomena are actually intrinsically unpredictable. Take, for example, a coin as displayed in figure 3.1. Whether a coin toss results in either heads or tails has nothing to do with an inherently aleatory process. The outcome appears unpredictable because of the *lack of knowledge* about the coin's initial position, speed, and acceleration. If we could gather information about the coin's initial kinematic conditions, the outcome would become predictable. Devices that can throw coins with repeatable initial kinematic conditions will lead to repeatable outcomes.

Figure 3.2 presents another example where we consider the *elastic modulus*[1] $E$ at one specific location in a dam. Notwithstanding long-term effects such as *creep*,[2] at any given location, $E$ does not vary with time: $E$ is a deterministic, yet unknown constant. Probability is employed here as a tool to describe our incomplete knowledge of that constant.

There are two types of uncertainty: *aleatory* and *epistemic*. aleatory uncertainty is characterized by its *irreducibility*; no information can either reduce or alter it. Alternately, epistemic uncertainty refers to a lack of knowledge that can be altered by new information. In an engineering context, aleatory uncertainties arise when we are concerned with future realizations that have yet to occur. Epistemic uncertainty applies to any other case dealing with deterministic, yet unknown quantities.

This book approaches machine learning using probability theory because in many practical engineering problems, the number of observations available is limited, from a few to a few thousand. In such a context, the amount of information available is typically

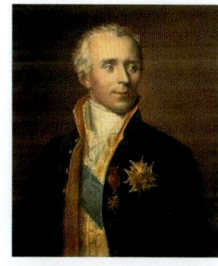

"La théorie des probabilités n'est, au fond, que le bon sens réduit au calcul; elle fait apprécier avec exactitude ce que les esprits justes sentent par une sorte d'instinct"
— Pierre-Simon, marquis de Laplace (1749–1827)

Figure 3.1: A coin toss illustrates the concept of epistemic uncertainty. (Photo: Michel Goulet)

[1] Elastic modulus relates the stress $\sigma$ and strains $\epsilon$ in Hooke's law $\sigma = \epsilon E$.
[2] Creep is the long-term (i.e., $\gg$years) deformation that occurs under constant stress.

Figure 3.2: The concrete elastic modulus $E$ at a given location is an example of deterministic, yet unknown quantity. The possible values for $E$ can be described using the probability theory.

**Set:** Ensemble of events or elements.

**Universe/sampling space** ($S$): Ensemble of all possible events.

**Elementary event** ($x$): A single event, $x \in S$.

**Event** (E): Ensemble of elementary events.

$E \subset S$ : Subset of $S$
$E = S$ : Certain event
$E = \emptyset$ : Impossible event
$\overline{E}$ : Complement of $E$

[3] Box, G. E. P. and G. C. Tiao (1992). *Bayesian inference in statistical analysis.* Wiley.

[4] Ang, A. H.-S. and W. H. Tang (1975). *Probability concepts in engineering planning and decision,* Volume 1—Basic Principles. John Wiley.

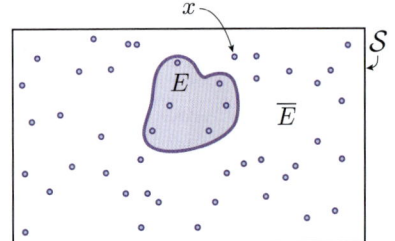

Figure 3.3: Venn diagram representing the sampling space $S$, an event $E$, its complement $\overline{E}$, and an elementary event $x$.

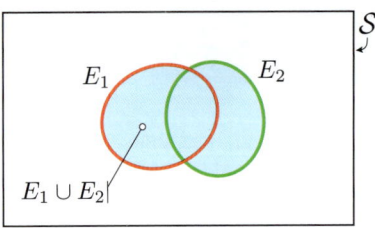

(a) Union operation

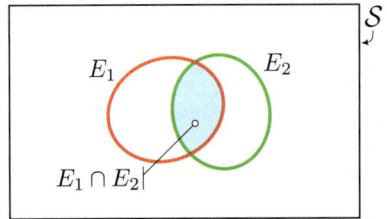

(b) Intersection operation

Figure 3.4: Venn diagrams representing the two basic operations.

[5] This example is adapted from Armen Der Kiureghian's course, CE229, at University of California, Berkeley.

insufficient to eliminate epistemic uncertainties. When large data sets are available, probabilistic and deterministic methods may lead to indistinguishable results; the opposite occurs when little data is available. Therefore, *the less we know about it, the stronger the argument for approaching a problem using probability theory.*

In this chapter, a review of *set theory* lays the foundation for *probability theory,* where the central part is the concept of *random variables.* Machine learning methods are built from an ensemble of functions organized in a clever way. Therefore, the last part of this chapter looks at what happens when random variables are introduced into deterministic functions.

For specific notions related to probability theory that are outside the scope of this chapter, the reader should refer to dedicated textbooks such as those by Box and Tiao;[3] Ang and Tang.[4]

### 3.1  Set Theory

A *set* describes an ensemble of *elements,* also referred to as *events.* An *elementary* event $x$ refers to a single event among a *sampling space* (or *universe*) denoted by the calligraphic letter $S$. By definition, a sampling space contains all the possible events, $E \subseteq S$. The special case where an event is equal to the sampling space, $E = S$, is called a *certain event.* The opposite, $E = \emptyset$, where an event is an empty set, is called a *null event.* $\overline{E}$ refers to the complement of a set, that is, all elements belonging to $S$ and not to $E$. Figure 3.3 illustrates these concepts using a *Venn diagram.*

Let us consider the example,[5] of the state of a structure following an earthquake, which is described by a sampling space,

$$S = \{\text{no damage}, \text{light damage}, \text{important damage}, \text{collapse}\}$$
$$= \{\mathsf{N}, \mathsf{L}, \mathsf{I}, \mathsf{C}\}.$$

In that context, an event $E_1 = \{\mathsf{N}, \mathsf{L}\}$ could contain the no damage and light damage events, and another event $E_2 = \{\mathsf{C}\}$ could contain only the collapsed state. The complements of these events are, respectively, $\overline{E_1} = \{\mathsf{I}, \mathsf{C}\}$ and $\overline{E_2} = \{\mathsf{N}, \mathsf{L}, \mathsf{I}\}$.

The two main operations for events, *union* and *intersection,* are illustrated in figure 3.4. A union is analogous to the "or" operator, where $E_1 \cup E_2$ holds if the event belongs to either $E_1$, $E_2$, or both. The intersection is analogous to the "and" operator, where $E_1 \cap E_2 \equiv E_1 E_2$ holds if the event belongs to both $E_1$ and $E_2$. As a convention, intersection has priority over union. Moreover, both operations are *commutative, associative,* and *distributive.*

Given a set of $n$ events $\{E_1, E_2, \cdots, E_n\} \in S$, $E_1, E_2, \cdots, E_n$, the events are *mutually exclusive* if $E_i E_j = \emptyset$, $\forall i \neq j$, that is, if the intersection for any pair of events is an empty set. Events $E_1, E_2, \cdots, E_n$ are *collectively exhaustive* if $\cup_{i=1}^{n} E_i = \mathcal{S}$, that is, the union of all events is the sampling space. Events $E_1, E_2, \cdots, E_n$ are *mutually exclusive and collectively exhaustive* if they satisfy both properties simultaneously. Figure 3.5 presents examples of mutually exclusive (3.5a), collectively exhaustive (3.5b), and mutually exclusive and collectively exhaustive (3.5c–d) events. Note that the difference between (b) and (c) is the absence of overlap in the latter.

## 3.2  Probability of Events

$\Pr(E_i)$ denotes the *probability* of the event $E_i$. There are two main interpretations for a probability: the *Frequentist* and the *Bayesian*. Frequentists interpret a probability as the number of occurrences of $E_i$ relative to the number of samples $s$, as $s$ goes to $\infty$,

$$\Pr(E_i) = \lim_{s \to \infty} \frac{\#\{E_i\}}{s}.$$

For Bayesians, a probability measures how likely is $E_i$ in comparison with other events in $\mathcal{S}$. This interpretation assumes that the nature of uncertainty is epistemic, that is, it describes our knowledge of a phenomenon. For instance, the probability depends on the available knowledge and can change when new information is obtained. Throughout this book we are adopting this Bayesian interpretation.

By definition, the probability of an event is a number between zero and one, $0 \leq \Pr(E_i) \leq 1$. At the ends of this spectrum, the probability of any event in $\mathcal{S}$ is one, $\Pr(\mathcal{S}) = 1$, and the probability of an empty set is zero, $\Pr(\emptyset) = 0$. If two events $E_1$ and $E_2$ are mutually exclusive, then the probability of the events' union is the sum of each event's probability. Because the union of an event and its complement are the sampling space, $E \cup \overline{E} = \mathcal{S}$ (see figure 3.5d), and because $\Pr(\mathcal{S}) = 1$, then the probability of the complement is $\Pr(\overline{E}) = 1 - \Pr(E)$.

When events are not mutually exclusive, the general addition rule for the probability of the union of two events is

$$\Pr(E_1 \cup E_2) = \Pr(E_1) + \Pr(E_2) - \Pr(E_1 E_2).$$

This general addition rule is illustrated in figure 3.6, where if we simply add the probability of each event without accounting for the subtraction of $\Pr(E_1 E_2)$, the probability of the intersection of both events will be counted twice.

**Union** ("or")
$E_1 \cup E_2$

**Intersection** ("and")
$E_1 \cap E_2 \equiv E_1 E_2$

**Commutativity**
$E_1 \cup E_2 = E_2 \cup E_1, \quad E_1 E_2 = E_2 E_1$
$\cup_{i=1}^{n} E_i = E_1 \cup E_2 \cup \cdots \cup E_n$
$\cap_{i=1}^{n} E_i = E_1 \cap E_2 \cap \cdots \cap E_n$

**Associativity**
$(E_1 \cup E_2) \cup E_3 = E_1 \cup (E_2 \cup E_3) = E_1 \cup E_2 \cup E_3$

**Distributivity**
$E_1 (E_2 \cup E_3) = (E_1 E_2 \cup E_1 E_3)$

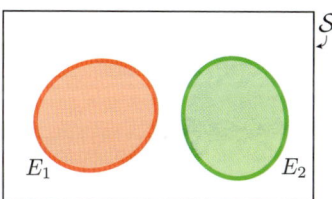

(a) Mutually exclusive

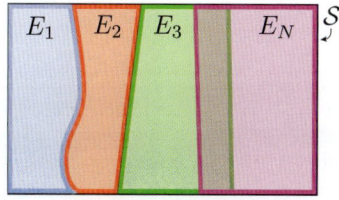

(b) Collectively exhaustive

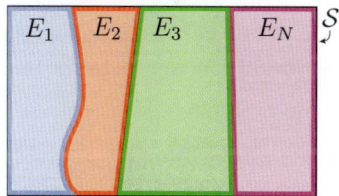

(c) Mutually exclusive and collectively exhaustive

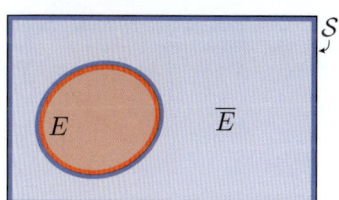

(d) Mutually exclusive and collectively exhaustive

Figure 3.5: Venn diagrams representing the concepts of mutual exclusivity and collective exhaustivity for events.

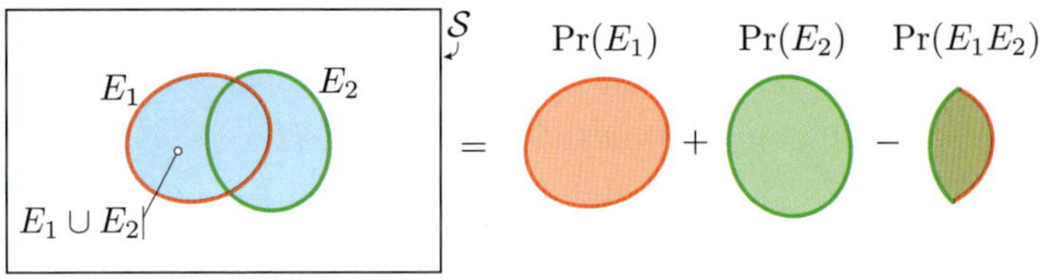

Figure 3.6: Venn diagram representing the addition rule for the probability of events.

$\Pr(E_1|E_2)$ denotes the probability of the event $E_1$ conditional on the realization of the event $E_2$. This conditional probability is defined as the joint probability for both events divided by the probability of $E_2$,

**Note:** $\Pr(E_2) \neq 0$ on the denominator because a division by 0 is not finite.

$$\Pr(E_1|E_2) = \frac{\overset{\text{Joint probability}}{\Pr(E_1 E_2)}}{\underset{\text{Marginal probability}}{\Pr(E_2)}}, \quad \Pr(E_2) \neq 0. \tag{3.1}$$

Conditional probability

The probability of a single event is referred to as a *marginal* probability. A *joint* probability designates the probability of the intersection of events. The terms in equation 3.1 can be rearranged to explicitly show that the joint probability of two events $\{E_1, E_2\}$ is the product of a conditional probability and its associated marginal,

$$\begin{aligned} \Pr(E_1 E_2) &= \Pr(E_1|E_2) \cdot \Pr(E_2) \\ &= \Pr(E_2|E_1) \cdot \Pr(E_1). \end{aligned}$$

In cases where $E_1$ and $E_2$ are *statistically independent*, $E_1 \perp\!\!\!\perp E_2$, conditional probabilities are equal to the marginal,

**Note:** Statistical independence ($\perp\!\!\!\perp$) between a pair of random variables implies that learning about one random variable does not modify our knowledge for the other.

$$E_1 \perp\!\!\!\perp E_2 \begin{cases} \Pr(E_1|E_2) &= \Pr(E_1) \\ \Pr(E_2|E_1) &= \Pr(E_2). \end{cases}$$

In the special case of statistically independent events, the joint probability reduces to the product of the marginals,

$$\Pr(E_1 E_2) = \Pr(E_1) \cdot \Pr(E_2).$$

The joint probability for $n$ events can be broken down into $n-1$ conditionals and one marginal probability using the *chain rule*,

$$\begin{aligned} \Pr(E_1 E_2 \cdots E_n) &= \Pr(E_1|E_2 \cdots E_n) \Pr(E_2 \cdots E_n) \\ &= \Pr(E_1|E_2 \cdots E_n) \Pr(E_2|E_3 \cdots E_n) \Pr(E_3 \cdots E_n) \\ &= \Pr(E_1|E_2 \cdots E_n) \Pr(E_2|E_3 \cdots E_n) \cdots \Pr(E_{n-1}|E_n) \Pr(E_n). \end{aligned}$$

Let us define $\{E_1, E_2, E_3, \cdots, E_n\} \in \mathcal{S}$, a set of mutually exclusive and collectively exhaustive events, that is, $E_i E_j = \emptyset$, $\forall i \neq j$, $\cup_{i=1}^{n} E_i = \mathcal{S}$ – and an event $A$ belonging to the same sampling space, that is, $A \in \mathcal{S}$. This context is illustrated using a Venn diagram in figure 3.7. The probability of the event $A$ can be obtained by summing the joint probability of $A$ and each event $E_i$,

$$\Pr(A) = \sum_{i=1}^{n} \underbrace{\Pr(A|E_i) \cdot \Pr(E_i)}_{\Pr(AE_i)}. \tag{3.2}$$

This operation of obtaining a marginal probability from a joint is called *marginalization*. The addition rule for the union of $E_1 \cup E_2$ conditional on $A$ is

$$\Pr(E_1 \cup E_2|A) = \Pr(E_1|A) + \Pr(E_2|A) - \Pr(E_1E_2|A),$$

and the intersection rule is

$$\Pr(E_1E_2|A) = \Pr(E_1|E_2, A) \cdot \Pr(E_2|A).$$

Using the definition of a conditional probability in equation 3.1, we can break $\Pr(AE_i)$ into two different products of a conditional and its associated marginal probability,

$$\begin{aligned} \Pr(AE_i) &= \Pr(A|E_i) \cdot \Pr(E_i) \\ &= \Pr(E_i|A) \cdot \Pr(A) \end{aligned} \tag{3.3}$$

$$\underbrace{\phantom{\Pr(E_i|A) \cdot \Pr(A) = \Pr(A|E_i) \cdot \Pr(E_i).}}$$
$$\Pr(E_i|A) \cdot \Pr(A) = \Pr(A|E_i) \cdot \Pr(E_i).$$

Reorganizing the right-hand terms of equation 3.3 leads to *Bayes rule*,

$$\Pr(E_i|A) = \frac{\overset{\text{Conditional probability}}{\Pr(A|E_i)} \cdot \overset{\text{Prior probability}}{\Pr(E_i)}}{\underset{\text{Evidence}}{\Pr(A)}}.$$

Posterior probability

On the left-hand side is the *posterior probability*: the probability of the event $E_i$ given the realization of the event $A$. On the numerator of the right-hand side is the product of the *conditional probability* of the event $A$ given the event $E_i$, times the *prior probability* of $E_i$. The term on the denominator is referred to as the *evidence* and acts as a normalization constant, which ensures that $\sum_i \Pr(E_i|A) = 1$. The normalization constant $\Pr(A)$ is obtained using the marginalization operation presented in equation 3.2. In practical applications, $\Pr(A)$ is typically difficult to estimate. Chapters 6 and 7 present analytic as well as numerical methods for tackling this challenge. Figure 3.8 illustrates the conditional occurrence of events in the context of Bayes rule.

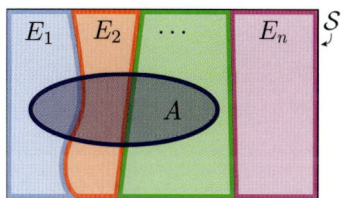

Figure 3.7: Venn diagram representing the conditional occurrence of events.

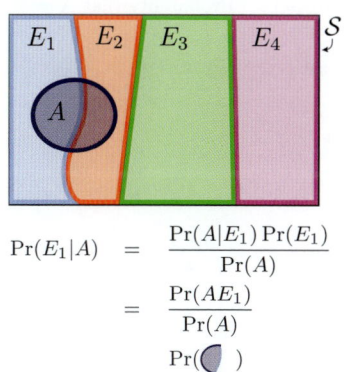

$$\begin{aligned} \Pr(E_1|A) &= \frac{\Pr(A|E_1)\Pr(E_1)}{\Pr(A)} \\ &= \frac{\Pr(AE_1)}{\Pr(A)} \\ &= \frac{\Pr(\text{⬤})}{\Pr(\text{◑})} \end{aligned}$$

Figure 3.8: Venn diagram representing the conditional occurrence of events in the context of Bayes rule.

## 3.3   Random Variables

Set theory is relevant for introducing the concepts related to probabilities. However, on its own, it has a limited applicability to practical problems that require defining the concept of *random variables*. A random variable is denoted by a capital letter $X$. Contrarily to what its name implies, a random variable is not intended to describe only intrinsically random events; in our case, it describes lack of knowledge. A random variable $X$ does not take any specific value. Instead, it takes any value in its valid sampling space $x \in \mathcal{S}$ and, as we will see shortly, the probability of occurrence of each value is typically not equal. Values of $x$ are either called *realizations* or *outcomes* and are elementary events that are mutually exclusive and collectively exhaustive. A sampling space $\mathcal{S}$ for a random variable can either be *discrete* or *continuous*. Continuous cases are always *infinite*, whereas discrete ones can either be *finite* or *infinite*. Figure 3.9 illustrates how the concepts of events and sampling space can be transposed from a Venn diagram representation to the domain of a random variable.

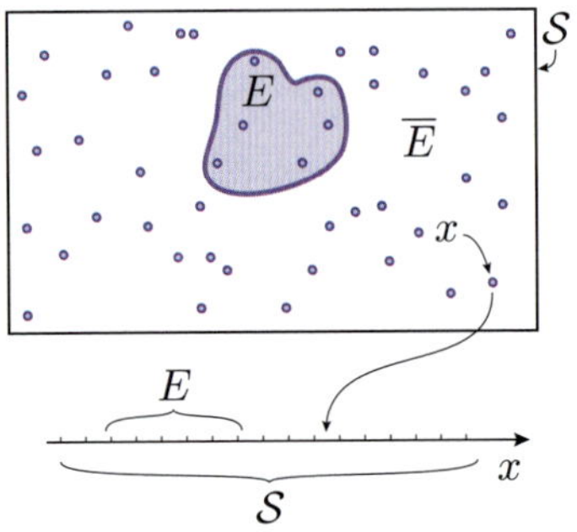

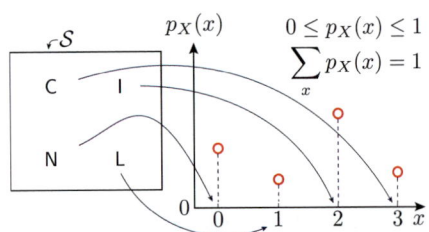

Figure 3.9: Parallel between a Venn diagram and a continuous domain to represent a random variable.

### 3.3.1   Discrete Random Variables

In the case where $\mathcal{S}$ is a discrete domain, the probability that $X = x$ is described by a *probability mass function* (PMF). In terms of notation $\Pr(X = x) \equiv p_X(x) \equiv p(x)$ are all equivalent. Moreover, we typically describe a random variable by defining its sampling space and its probability mass function so that $x : X \sim p_X(x)$. The symbol $\sim$ reads as *distributed like*. Analogously to the probability of events, the probability that $X = x$ must be

$$0 \leq p_X(x) \leq 1,$$

and the sum of the probability for all $x \in \mathcal{S}$ follows

$$\sum_x p_X(x) = 1.$$

For the post-earthquake structural safety example introduced in §3.1, where

$$\mathcal{S} = \left\{ \begin{array}{ll} \text{no damage} & \text{(N)} \\ \text{light damage} & \text{(L)} \\ \text{important damage} & \text{(I)} \\ \text{collapse} & \text{(C)} \end{array} \right\},$$

the sampling space along with the probability of each event can be represented by a probability mass function as depicted in figure 3.10.

**Notation**

$$
\begin{array}{rl}
X: & \text{Random variable} \\
x: & \text{Realization of } X \\
p_X(x): & \text{Probability that } X = x
\end{array}
$$

Figure 3.10: Representation of a sampling space for a discrete random variable.

The event corresponding to damages that are either light or important corresponds to $L \cup I \equiv \{1 \leq x \leq 2\}$. Because the events $x = 1$ and $x = 2$ are mutually exclusive, the probability

$$\begin{aligned} \Pr(L \cup I) &= \Pr(\{1 \leq X \leq 2\}) \\ &= p_X(x = 1) + p_X(x = 2). \end{aligned}$$

The probability that $X$ takes a value less than or equal to $x$ is described by a *cumulative mass function* (CMF),

$$\Pr(X \leq x) = F_X(x) = \sum_{x' \leq x} p_X(x').$$

Figure 3.11 presents on the same graph the *probability mass function* (PMF) and the cumulative mass function. As its name indicates, the CMF corresponds to the cumulative sum of the PMF. Inversely, the PMF can be obtained from the CMF following

$$p_X(x_i) = F_X(x_i) - F_X(x_{i-1}).$$

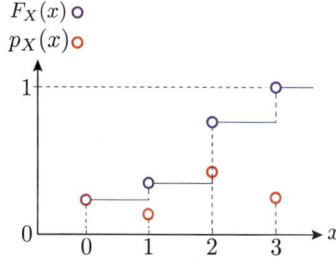

Figure 3.11: Comparison of a probability $p_X(x)$ and a cumulative $F_X(x)$ mass function.

### 3.3.2   Continuous Random Variables

The concepts presented for discrete sampling spaces can be extended for cases where $\mathcal{S}$ is a continuous domain. Because continuous domains are inevitably infinite, the probability that a random variable takes a specific value $X = x$ is zero,

$$\Pr(X = x) = 0.$$

For continuous random variables, the probability is only defined for intervals $x < X \leq x + \Delta x$,

$$\Pr(x < X \leq x + \Delta x) = f_X(x)\Delta x,$$

where $f_X(x) \equiv f(x)$ denotes a *probability density function* (PDF). A PDF must always be greater than or equal to zero $f_X(x) \geq 0$; however, unlike for the discrete case where $0 \leq p_X(x) \leq 1$, $f_X(x)$ can take values greater than one because it describes a *probability density* rather than a *probability*. In order to satisfy the property that $\Pr(\mathcal{S}) = 1$, the integral of $f_X(x)$ over all possible values of $x$ must be one,

$$\int_{-\infty}^{+\infty} f_X(x)dx = 1.$$

The probability that $\Pr(X \leq x)$ is given by the *cumulative density function* (CDF),

$$\Pr(X \leq x) = F_X(x) = \int_{-\infty}^{x} f_X(x')dx'.$$

**Note:** Here, the probability equal to zero does not mean that a specific value $x$ is impossible. Take, for example, a random variable defined in the interval $(0, 1)$, for which all the outcomes are equally probable. The probability that $X = 0.23642$ is only one out of an infinite number of possibilities in $(0, 1)$.

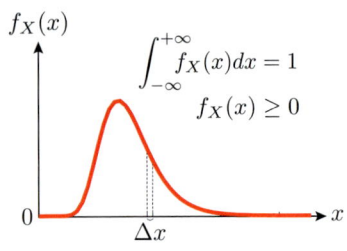

(a) Probability density function

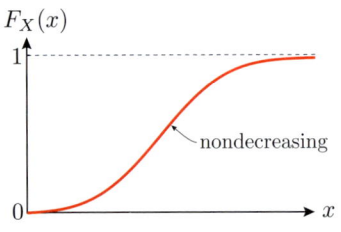

(b) Cumulative distribution function

Figure 3.12: Examples of PDF and CDF for a continuous random variable.

*Peak ground acceleration* is a metric quantifying the intensity of an earthquake using the maximal acceleration recorded during an event.

For a random variable $x \in \mathbb{R} : X \sim f_X(x)$, the CDF evaluated at the lower and upper bounds is, respectively, $F_X(-\infty) = 0$ and $F_X(+\infty) = 1$. Notice that the CDF is obtained by integrating the PDF, and inversely, the PDF is obtained by differentiating the CDF,

$$F_X(x) = \int_{-\infty}^{x} f_X(x')dx' \leftrightarrow f_X(x) = \frac{dF_X(x)}{dx}.$$

Moreover, because $F_X(x)$ is the integral of $f_X(x)$ and $f_X(x) \geq 0$, $F_X(x)$ is *nondecreasing*. Figure 3.12 presents examples of probability density and cumulative distribution function.

### 3.3.3   Conditional Probabilities

Conditional probabilities describe the probability of a random variable's outcomes, given the realization of another variable. The conditional notation for discrete random variables follows

$$X|y \sim p(x|y) \equiv p_{X|y}(x) \equiv \Pr(X = x|y) = \frac{p_{XY}(x,y)}{p_Y(y)},$$

and the conditional notation for continuous random variables follows,

$$X|y \sim f(x|y) \equiv f_{X|y}(x|y) = \frac{f_{XY}(x,y)}{f_Y(y)}.$$

Conditional probabilities are employed in Bayes rule to infer the posterior knowledge associated with a random variable, given the observations made for another.

Let us revisit the post-earthquake structural safety example introduced in §3.1, where the damage state $x \in \mathcal{S}$. If we measure the *peak ground acceleration* (PGA) after an earthquake $y \in \mathbb{R}^+$, we can employ the conditional probability of having structural damage given the PGA value to infer the structural state of a building that itself has not been observed. Figure 3.13 illustrates schematically how an observation of the peak ground acceleration $y$ can be employed to infer the structural state of a building $x$, using conditional probabilities. Because the structural state $X$ is a discrete random variable, $p(x|y)$ describes the posterior probability of each state $x \in \mathcal{S}$, given an observed value of PGA $y \in \mathbb{R}^+$. $f(y)$ is a normalization constant obtained by marginalizing $X$ from $f(x,y)$ and evaluating it for the particular observed value $y$,

$$f(y) = \sum_{x \in \mathcal{S}} f(y|x) \cdot p(x).$$

The posterior is obtained by multiplying the likelihood of observing the particular value of PGA $y$ given each of the structural states

$x$, times the prior probability of each structural state, and then dividing by the probability of the observation $y$ itself. Conditional probabilities can be employed to put in relation any combination of continuous and discrete random variables. Chapter 6 further explores Bayesian estimation with applied examples.

### 3.3.4  Multivariate Random Variables

It is common to study the joint occurrence of multiple phenomena. In the context of probability theory, it is done using multivariate random variables. $\mathbf{x} = [x_1\ x_2\ \cdots\ x_n]^{\mathsf{T}}$ is a vector (column) containing realizations for $n$ random variables $\mathbf{X} = [X_1\ X_2\ \cdots\ X_n]^{\mathsf{T}}$, $\mathbf{x} : \mathbf{X} \sim p_{\mathbf{X}}(\mathbf{x}) \equiv p(\mathbf{x})$, or $\mathbf{x} : \mathbf{X} \sim f_{\mathbf{X}}(\mathbf{x}) \equiv f(\mathbf{x})$. For the discrete case, the probability of the joint realization $\mathbf{x}$ is described by

$$p_{\mathbf{X}}(\mathbf{x}) = \Pr(X_1 = x_1 \cap X_2 = x_2 \cap \cdots \cap X_n = x_n),$$

where $0 \leq p_{\mathbf{X}}(\mathbf{x}) \leq 1$. For the continuous case, it is

$$f_{\mathbf{X}}(\mathbf{x})\Delta\mathbf{x} = \Pr(x_1 < X_1 \leq x_1 + \Delta x_1 \cap \cdots \cap x_n < X_n \leq x_n + \Delta x_n),$$

for $\Delta\mathbf{x} \to \mathbf{0}$. Note that $f_{\mathbf{X}}(\mathbf{x})$ can be $> 1$ because it describes a probability density. As mentioned earlier, two random variables $X_1$ and $X_2$ are statistically independent ($\perp\!\!\!\perp$) if

$$p_{X_1|x_2}(x_1|x_2) = p_{X_1}(x_1).$$

If $X_1 \perp\!\!\!\perp X_2 \perp\!\!\!\perp \cdots \perp\!\!\!\perp X_n$, the joint PMF is defined by the product of its marginals,

$$p_{X_1:X_n}(x_1, \cdots, x_n) = p_{X_1}(x_1)p_{X_2}(x_2)\cdots p_{X_n}(x_n).$$

For the general case where $X_1, X_2, \cdots, X_n$ are not statistically independent, their joint PMF can be defined using the *chain rule*,

$$
\begin{aligned}
p_{X_1:X_n}(x_1, \cdots, x_n) &= p_{X_1|X_2:X_n}(x_1|x_2, \cdots, x_n)\cdots \\
&\quad \cdot p_{X_{n-1}|X_n}(x_{n-1}|x_n) \cdot p_{X_n}(x_n).
\end{aligned}
$$

The same rules apply for continuous random variables except that $p_{\mathbf{X}}(\mathbf{x})$ is replaced by $f_{\mathbf{X}}(\mathbf{x})$. Figure 3.14 presents examples of marginals and a bivariate joint probability density function.

The multivariate cumulative distribution function describes the probability that a set of $n$ random variables is simultaneously lesser or equal to $\mathbf{x}$,

$$F_{\mathbf{X}}(\mathbf{x}) = \Pr(X_1 \leq x_1 \cap \cdots \cap X_n \leq x_n).$$

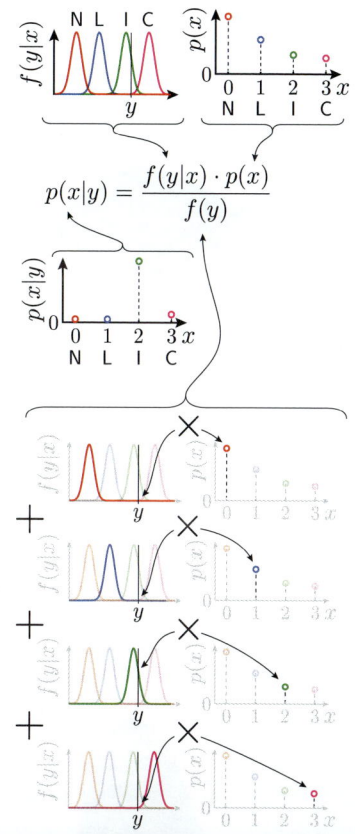

$$p(x|y) = \frac{f(y|x) \cdot p(x)}{f(y)}$$

Figure 3.13: Schematic example of how observations of the peak ground acceleration $y$ can be employed to infer the structural state of a building $x$ using conditional probabilities.

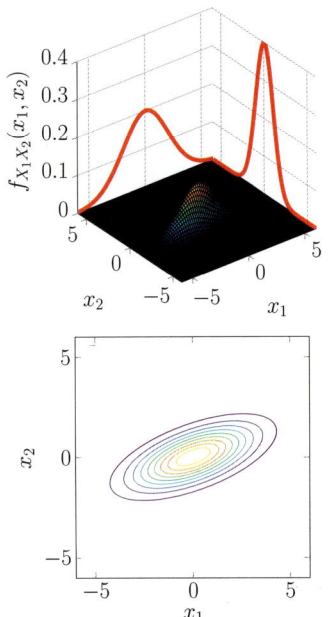

Figure 3.14: Examples of marginals and a bivariate probability density function.

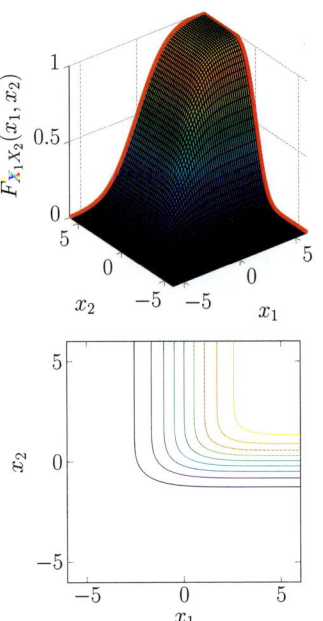

Figure 3.15: Examples of marginals and a bivariate cumulative distribution function.

**Marginalization**

$$
\begin{aligned}
\int_{-\infty}^{\infty} f_{X_1 X_2}(x_1, x_2) dx_2 &= f_{X_1}(x_1) \\
\sum_{x_2} p_{X_1 X_2}(x_1, x_2) &= p_{X_1}(x_1) \\
F_{X_1 X_2}(x_1, +\infty) &= F_{X_1}(x_1)
\end{aligned}
$$

The joint CDF is obtained by integrating the joint PDF over each dimension from its lower bound up to $\mathbf{x}$, and inversely, the joint PDF is obtained by differentiating the CDF,

$$
F_{\mathbf{X}}(\mathbf{x}) = \int_{-\infty}^{x_1} \cdots \int_{-\infty}^{x_n} f_{\mathbf{X}}(\mathbf{x}') d\mathbf{x}' \leftrightarrow f_{\mathbf{X}}(\mathbf{x}) d\mathbf{x} = \frac{\partial^n F_{\mathbf{X}}(\mathbf{x})}{\partial x_1 \cdots \partial x_n}.
$$

A multivariate CDF has values $0 \leq F_{\mathbf{X}}(\mathbf{x}) \leq 1$, and its value is zero at the lowest bound for any dimension and one at the upper bound for all dimensions,

$$
\begin{aligned}
F_{X_1:X_n}(x_1, \cdots, x_{n-1}, -\infty) &= 0 \\
F_{X_1:X_n}(+\infty, \cdots, +\infty, +\infty) &= 1.
\end{aligned}
$$

Figure 3.15 presents an example of marginals and a bivariate cumulative distribution function.

The operation consisting of removing a random variable from a joint set is called *marginalization*. For a set of $n$ joint random variables, we can remove the $i^{\text{th}}$ variable by summing over the $i^{\text{th}}$ dimension,

$$
\sum_{x_n} p_{X_1:X_n}(x_1, \cdots, x_n) = p_{X_1:X_{n-1}}(x_1, \cdots, x_{n-1}).
$$

If we marginalize all variables by summing over all dimensions, the result is

$$
\sum_{x_1} \cdots \sum_{x_n} p_{\mathbf{X}}(\mathbf{x}) = 1.
$$

For the example presented in figure 3.16, $X_1 \perp\!\!\!\perp X_2$ so the joint PMF is obtained by the product of its marginals. It is possible to obtain the marginal PMF for $x_1$ from the joint through marginalizing, $\sum_{i=1}^{3} p_{\mathbf{X}}(x_1, i)$:

$$
p_{\mathbf{X}}(x_1, x_2) = 
\begin{cases}
\end{cases}
$$

|  | $x_2 = 1$ | $x_2 = 2$ | $x_2 = 3$ | $\sum_{i=1}^{3} p_{\mathbf{X}}(x_1, i)$ |
|---|---|---|---|---|
| $x_1 = 1$ | 0.08 | 0.015 | 0.005 | 0.1 |
| $x_1 = 2$ | 0.4 | 0.075 | 0.025 | 0.5 |
| $x_1 = 3$ | 0.32 | 0.06 | 0.02 | 0.4 |

Marginalization applies to continuous random variables using integration,

$$
\int_{-\infty}^{\infty} f_{X_1:X_n}(x_1, \cdots, x_n) dx_n = f_{X_1:X_{n-1}}(x_1, \cdots, x_{n-1}),
$$

where again, if we integrate over all dimensions, the result is

$$
\int_{-\infty}^{\infty} \cdots \int_{-\infty}^{\infty} f_{\mathbf{X}}(\mathbf{x}) d\mathbf{x} = 1.
$$

For both continuous and discrete random variables, we can marginalize a random variable by evaluating its CDF at its upper bound,

$$F_{X_1:X_n}(x_1, \cdots, x_{n-1}, +\infty) = F_{X_1:X_{n-1}}(x_1, \cdots, x_{n-1}).$$

### 3.3.5  Moments and Expectation

The *moment* of order $m$, $\mathbb{E}[X^m]$ of a random variable $X$ is defined as

$$
\begin{aligned}
\mathbb{E}[X^m] &= \int x^m \cdot f_X(x)dx \quad \text{(continuous)} \\
&= \sum_i x_i^m \cdot p_X(x_i) \quad \text{(discrete)},
\end{aligned}
$$

where $\mathbb{E}[\cdot]$ denotes the *expectation* operation. For $m = 1$, $\mathbb{E}[X] = \mu_X$ is a measure of position for the centroid of the probability density or mass function. This centroid is analogous to the concept of *center of gravity* for a solid body or cross section. An *expected value* refers to the sum of all possible values weighted by their probability of occurrence. A key property of the expectation is that it is a *linear operation* so that

$$\mathbb{E}[X + Y] = \mathbb{E}[X] + \mathbb{E}[Y].$$

The notion of expectation can be extended for any function of random variables $g(X)$,

$$\mathbb{E}[g(X)] = \int g(x) \cdot f_X(x)dx.$$

The expectation of the function $g(X) = (X - \mu_X)^m$ is referred to as *centered moment of order* $m$,

$$\mathbb{E}[(X - \mu_X)^m] = \int (x - \mu_X)^m \cdot f_X(x)dx.$$

For the special cases where $m = 1$, $\mathbb{E}[(X - \mu_X)^1] = 0$, and for $m = 2$,

$$
\begin{aligned}
\mathbb{E}[(X - \mu_X)^2] &= \sigma_X^2 \\
&= \text{var}[X] \\
&= \mathbb{E}[X^2] - \mathbb{E}[X]^2,
\end{aligned}
$$

where $\sigma_X$ denotes the *standard deviation* of $X$; and var$[\cdot]$ denotes the *variance* operator that measures the dispersion of the probability density function with respect to its mean. The notion of variance is analogous to the concept of *moment of inertia* for a cross section. Together, $\mu_X$ and $\sigma_X$ are metrics describing the centroid and dispersion of a random variable. Another adimensional

$$
p_{X_1}(x_1) \begin{cases} p_{X_1}(1) = 0.1 \\ p_{X_1}(2) = 0.5 \\ p_{X_1}(3) = 0.4 \end{cases}
$$

$$
p_{X_2}(x_2) \begin{cases} p_{X_2}(1) = 0.8 \\ p_{X_2}(2) = 0.15 \\ p_{X_2}(3) = 0.05 \end{cases}
$$

$$p_{\mathbf{X}}(x_1, x_2) = p_{X_1}(x_1) \cdot p_{X_2}(x_2)$$

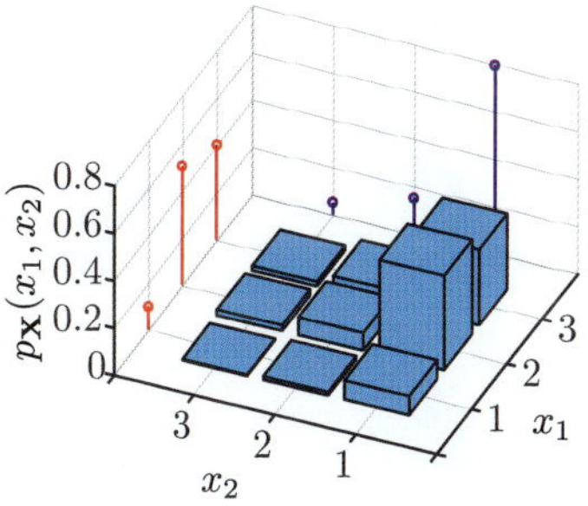

Figure 3.16: Examples of marginals and bivariate probability mass functions.

**Expected value**

$$
\begin{aligned}
\mathbb{E}[X] &= \int x \cdot f_X(x)dx \quad \text{(continuous)} \\
&= \sum_i x_i \cdot p_X(x_i) \quad \text{(discrete)}
\end{aligned}
$$

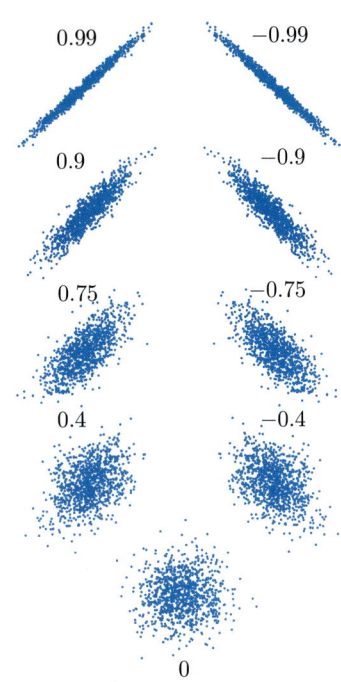

Figure 3.17: Examples of scatter plots between the realizations of two random variables for different correlation coefficients $\rho$.

Figure 3.18: Example of scatter plot where there is a quadratic dependence between the variables, yet the correlation coefficient $\rho \approx 0$.

[6] Lowen, A. C. and J. Steel (2014). Roles of humidity and temperature in shaping influenza seasonality. *Journal of Virology* 88(14), 7692–7695.

dispersion metric for describing a random variable is the coefficient of variation, $\delta_X = \frac{\sigma_X}{\mu_X}$. Note that $\delta_X$ only applies for $\mu_X \neq 0$.

Given two random variables $X, Y$, their *covariance*, $\text{cov}(X, Y)$, is defined by the expectation of the product of the mean-centered variables,

$$\begin{aligned} \mathbb{E}[(X - \mu_X)(Y - \mu_Y)] &= \text{cov}(X, Y) \\ &= \mathbb{E}[XY] - \mathbb{E}[X] \cdot \mathbb{E}[Y] \\ &= \rho_{XY} \cdot \sigma_X \cdot \sigma_Y. \end{aligned}$$

The *correlation coefficient* $\rho_{XY}$ can take a value between -1 and 1, which quantifies the *linear dependence* between $X$ and $Y$,

$$\rho_{XY} = \frac{\text{cov}(X, Y)}{\sigma_X \sigma_Y}, \quad -1 \leq \rho_{XY} \leq +1.$$

A positive (negative) correlation indicates that a large outcome for $X$ is associated with a high probability for a large (small) outcome for $Y$. Figure 3.17 presents examples of scatter plots generated for different correlation coefficients.

In the special case where $X$ and $Y$ are independent, the correlation is zero, $X \perp\!\!\!\perp Y \implies \rho_{XY} = 0$. Note that the inverse is not true; a correlation coefficient equal to zero does not guarantee the independence, $\rho_{ij} = 0 \not\Longrightarrow X_i \perp\!\!\!\perp X_j$. This happens because correlation only measures the linear dependence between a pair of random variables; two random variables can be nonlinearly dependent, yet have a correlation coefficient equal to zero. Figure 3.18 presents an example of a scatter plot with quadratic dependence yet no linear dependence, so $\rho \approx 0$.

Correlation also does not imply *causality*. For example, the number of flu cases is negatively correlated with the temperature; when the seasonal temperatures drop during winter, the number of flu cases increases. Nonetheless, the cold itself is not causing the flu; someone isolated in a cold climate is unlikely to contract the flu because the virus is itself unlikely to be present in the environment. Instead, studies have shown that the flu virus has a higher transmissibility in the cold and dry conditions that are prevalent during winter. See, for example, Lowen and Steel.[6]

For a set of $n$ random variables $X_1, X_2, \cdots, X_n$, the *covariance matrix* defines the dispersion of each variable through its variance located on the main diagonal, and the dependence between variables through the pairwise covariance located on the off-diagonal terms,

$$\boldsymbol{\Sigma} = \begin{bmatrix} \sigma_{X_1}^2 & \cdots & \rho_{1n} \sigma_{X_1} \sigma_{X_n} \\ & \vdots & \vdots \\ \text{sym.} & & \sigma_{X_n}^2 \end{bmatrix}.$$

A covariance matrix is symmetric (sym.), and each term is defined following $[\mathbf{\Sigma}]_{ij} = \text{cov}(X_i, X_j) = \rho_{ij}\sigma_{X_i}\sigma_{X_j}$. Because a variable is linearly correlated with itself ($\rho = 1$), the main diagonal terms reduce to $[\mathbf{\Sigma}]_{ii} = \sigma^2_{X_i}$. A covariance matrix has to be *positive semi-definite* (see §2.4.2) so the variances on the main diagonal must be $> 0$. In order to avoid singular cases, there should be no linearly dependent variables, that is, $-1 < \rho_{ij} < 1, \forall i \neq j$.

## 3.4   Functions of Random Variables

Let us consider a continuous random variable $X \sim f_X(x)$ and a *monotonic* deterministic function $y = g(x)$. The function's output $Y$ is a random variable because it takes as input the random variable $X$. The PDF $f_Y(y)$ is defined knowing that for each infinitesimal part of the domain $dx$, there is a corresponding $dy$, and the probability over both domains must be equal,

$$
\begin{aligned}
\Pr(y < Y \leq y + dy) &= \Pr(x < X \leq x + dx) \\
\underbrace{f_Y(y)\,dy}_{\geq 0} &= \underbrace{f_X(x)\,dx}_{\geq 0}.
\end{aligned}
$$

The *change-of-variable* rule for $f_Y(y)$ is defined by

$$
\begin{aligned}
f_Y(y) &= f_X(x)\left|\frac{dx}{dy}\right| \\
&= f_X(x)\left|\frac{dy}{dx}\right|^{-1} \\
&= f_X(g^{-1}(y))\left|\frac{dg(g^{-1}(y))}{dx}\right|^{-1},
\end{aligned}
$$

where multiplying by $\frac{dx}{dy}$ accounts for the change in the size of the neighborhood of $x$ with respect to $y$, and where the absolute value ensures that $f_Y(y) \geq 0$. For a function $y = g(x)$ and its inverse $x = g^{-1}(y)$, the *gradient* is obtained from

$$
\frac{dy}{dx} \equiv \frac{dg(x)}{dx} \equiv \frac{dg(\overbrace{g^{-1}(y)}^{=x})}{dx}.
$$

Figure 3.19 presents an example of nonlinear transformation $y = g(x)$. Notice how, because of the nonlinear transformation, the maximum for $f_X(x^*)$ and the maximum for $f_Y(y^*)$ do not occur for the same locations, that is, $y^* \neq g(x^*)$.

Given a set of $n$ random variables $\mathbf{x} \in \mathbb{R}^n : \mathbf{X} \sim f_{\mathbf{X}}(\mathbf{x})$, we can generalize the transformation rule for an $n$ to $n$ multivariate function $\mathbf{y} = g(\mathbf{x})$, as illustrated in figure 3.20a for a case where $n = 2$. As with the univariate case, we need to account for the change in

A *monotonic* function $g(x)$ takes one variable as input and returns one variable as output and is strictly either increasing or decreasing.

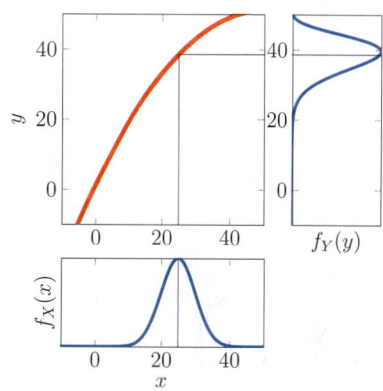

Figure 3.19: Example of 1-D nonlinear transformation $y = g(x)$. Notice how the nonlinear transformation causes the modes (i.e., the most likely values) to be different in the $x$ and $y$ spaces.

A transformation from a space $x$ to another space $y$ requires taking into account the change in the size of the neighborhood

$$
f_Y(y) = f_X(x)\left|\frac{dx}{dy}\right|.
$$

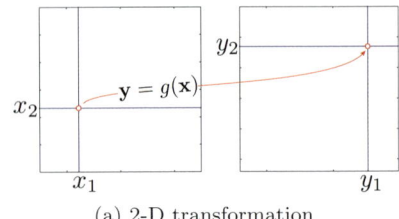

(a) 2-D transformation

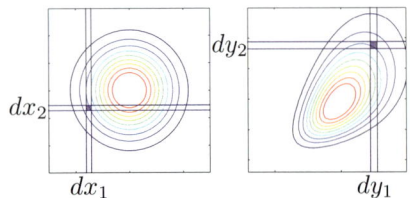

(b) Effect of a 2-D transformation on the neighborhood size

Figure 3.20: Illustration of a 2-D transformation.

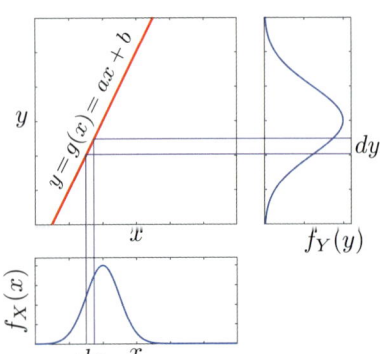

(a) Generic linear transformation

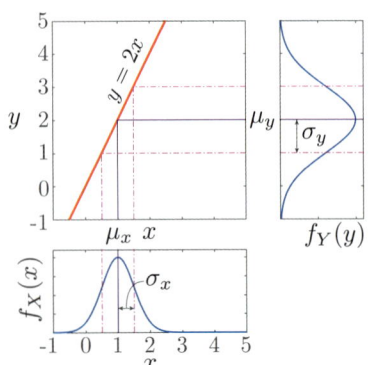

(b) Linear transformation $y = 2x$

Figure 3.21: Examples of transformations through a linear function.

the neighborhood size when going from the original to the transformed space, as illustrated in figure 3.20b. The transformation is then defined by

$$
\begin{aligned}
f_{\mathbf{Y}}(\mathbf{y})d\mathbf{y} &= f_{\mathbf{X}}(\mathbf{x})d\mathbf{x} \\
f_{\mathbf{Y}}(\mathbf{y}) &= f_{\mathbf{X}}(\mathbf{x})\left|\frac{d\mathbf{x}}{d\mathbf{y}}\right|,
\end{aligned}
$$

where $\left|\frac{d\mathbf{x}}{d\mathbf{y}}\right|$ is the inverse of the determinant of the *Jacobian* matrix,

$$
\begin{aligned}
\left|\frac{d\mathbf{x}}{d\mathbf{y}}\right| &= |\det \mathbf{J}_{\mathbf{y},\mathbf{x}}|^{-1} \\
\left|\frac{d\mathbf{y}}{d\mathbf{x}}\right| &= |\det \mathbf{J}_{\mathbf{y},\mathbf{x}}| .
\end{aligned}
$$

The Jacobian is an $n \times n$ matrix containing the partial derivatives of $y_k$ with respect to $x_l$, evaluated at $\mathbf{x}$ so that $[\mathbf{J}_{\mathbf{y},\mathbf{x}}]_{k,l} = \frac{\partial y_k}{\partial x_l}$,

$$
\mathbf{J}_{\mathbf{y},\mathbf{x}} = \begin{bmatrix} \frac{\partial y_1}{\partial x_1} & \cdots & \frac{\partial y_1}{\partial x_n} \\ \vdots & \ddots & \vdots \\ \frac{\partial y_n}{\partial x_1} & \cdots & \frac{\partial y_n}{\partial x_n} \end{bmatrix} = \begin{bmatrix} \nabla g_1(\mathbf{x}) \\ \vdots \\ \nabla g_n(\mathbf{x}) \end{bmatrix}.
$$

Note that each row of the Jacobian matrix corresponds to the *gradient vector* evaluated at $\mathbf{x}$,

$$
\nabla g(\mathbf{x}) = \begin{bmatrix} \frac{\partial g(\mathbf{x})}{\partial x_1} & \cdots & \frac{\partial g(\mathbf{x})}{\partial x_n} \end{bmatrix}.
$$

The determinant (see §2.4.1) of the Jacobian is a scalar quantifying the size of the neighborhood of $d\mathbf{y}$ with respect to $d\mathbf{x}$.

### 3.4.1  Linear Functions

Figure 3.21b illustrates how a function $y = 2x$ transforms a random variable $X$ with mean $\mu_X = 1$ and standard deviation $\sigma_X = 0.5$ into $Y$ with mean $\mu_Y = 2$ and standard deviation $\sigma_y = 1$. In the machine learning context, it is common to employ linear functions of random variables $y = g(x) = ax + b$, as illustrated in figure 3.21a. Given a random variable $X$ with mean $\mu_X$ and variance $\sigma_X^2$, the change in the neighborhood size simplifies to

$$
\left|\frac{dy}{dx}\right| = |a|.
$$

In such a case, because of the linear property of the expectation operation (see §3.3.5),

$$
\mu_Y = g(\mu_X) = a\mu_X + b, \quad \sigma_Y = |a|\sigma_X.
$$

Let us consider a set of $n$ random variables $\mathbf{X}$ defined by its mean vector and covariance matrix,

$$
\mathbf{X} = \begin{bmatrix} X_1 \\ \vdots \\ X_n \end{bmatrix}, \ \boldsymbol{\mu}_{\mathbf{X}} = \begin{bmatrix} \mu_{X_1} \\ \vdots \\ \mu_{X_n} \end{bmatrix}, \ \boldsymbol{\Sigma}_{\mathbf{X}} = \begin{bmatrix} \sigma^2_{X_1} & \cdots & \rho_{1n}\sigma_{X_1}\sigma_{X_n} \\ & \ddots & \vdots \\ \text{sym.} & & \sigma^2_{X_n} \end{bmatrix},
$$

and the variables $\mathbf{Y} = [Y_1\ Y_2\ \cdots\ Y_n]^{\mathsf{T}}$ obtained from a linear function $\mathbf{Y} = \mathbf{g}(\mathbf{X}) = \mathbf{AX} + \mathbf{b}$ so that

$$
\underbrace{\begin{bmatrix} \ \\ \ \end{bmatrix}_{n\times 1}}_{\mathbf{Y}} = \underbrace{\begin{bmatrix} \ \\ \ \end{bmatrix}_{n\times n}}_{\mathbf{A}=\mathbf{J}_{\mathbf{y},\mathbf{x}}} \times \underbrace{\begin{bmatrix} \ \\ \ \end{bmatrix}_{n\times 1}}_{\mathbf{X}} + \underbrace{\begin{bmatrix} \ \\ \ \end{bmatrix}_{n\times 1}}_{\mathbf{b}}.
$$

**Note:** For linear functions $\mathbf{Y} = \mathbf{AX} + \mathbf{b}$, the Jacobian $\mathbf{J}_{\mathbf{y},\mathbf{x}}$ is the matrix $\mathbf{A}$ itself.

The function outputs $\mathbf{Y}$ (i.e., the mean vector), covariance matrix, and the joint covariance are then described by

$$
\left. \begin{array}{l} \boldsymbol{\mu}_{\mathbf{Y}} = \mathbf{g}(\boldsymbol{\mu}_{\mathbf{X}}) = \mathbf{A}\boldsymbol{\mu}_{\mathbf{X}} + \mathbf{b} \\ \boldsymbol{\Sigma}_{\mathbf{Y}} = \mathbf{A}\boldsymbol{\Sigma}_{\mathbf{X}}\mathbf{A}^{\mathsf{T}} \\ \boldsymbol{\Sigma}_{\mathbf{XY}} = \boldsymbol{\Sigma}_{\mathbf{X}}\mathbf{A}^{\mathsf{T}} \end{array} \right\} \quad \begin{bmatrix} \mathbf{X} \\ \mathbf{Y} \end{bmatrix}, \ \begin{bmatrix} \boldsymbol{\mu}_{\mathbf{X}} \\ \boldsymbol{\mu}_{\mathbf{Y}} \end{bmatrix}, \ \begin{bmatrix} \boldsymbol{\Sigma}_{\mathbf{X}} & \boldsymbol{\Sigma}_{\mathbf{XY}} \\ \boldsymbol{\Sigma}^{\mathsf{T}}_{\mathbf{XY}} & \boldsymbol{\Sigma}_{\mathbf{Y}} \end{bmatrix}.
$$

If instead of having an $n \to n$ function, we have an $n \to 1$ function $y = g(\mathbf{X}) = \mathbf{a}^{\mathsf{T}}\mathbf{X} + b$, then the Jacobian simplifies to the gradient vector $\nabla g(\mathbf{x}) = \left[ \frac{\partial g(\mathbf{x})}{\partial x_1} \ \cdots \ \frac{\partial g(\mathbf{x})}{\partial x_n} \right]$, which is again equal to the vector $\mathbf{a}^{\mathsf{T}}$,

$$
\underbrace{\begin{bmatrix} \ \end{bmatrix}_{1\times 1}}_{Y} = \underbrace{\begin{bmatrix} \quad\quad \end{bmatrix}_{1\times n}}_{\mathbf{a}^{\mathsf{T}}=\nabla g(\mathbf{x})} \times \underbrace{\begin{bmatrix} \ \\ \ \end{bmatrix}_{n\times 1}}_{\mathbf{X}} + \underbrace{\begin{bmatrix} \ \end{bmatrix}_{1\times 1}}_{b}.
$$

The function output $Y$ is then described by

$$
\begin{aligned}
\mu_Y &= g(\boldsymbol{\mu}_{\mathbf{X}}) = \mathbf{a}^{\mathsf{T}}\boldsymbol{\mu}_{\mathbf{X}} + b \\
\sigma^2_Y &= \mathbf{a}^{\mathsf{T}}\boldsymbol{\Sigma}_{\mathbf{X}}\mathbf{a}.
\end{aligned}
$$

### 3.4.2  Linearization of Nonlinear Functions

Because of the analytic simplicity associated with linear functions of random variables, it is common to approximate nonlinear functions by linear ones using a Taylor series so that

The *Hessian* $\mathbf{H}(\boldsymbol{\mu}_{\mathbf{X}})$ is an $n \times n$ matrix containing the $2^{\text{nd}}$-order partial derivatives evaluated at $\boldsymbol{\mu}_{\mathbf{X}}$. See §5.2 for details.

$$
g(\mathbf{X}) \approx \underbrace{\overbrace{\underbrace{g(\boldsymbol{\mu}_{\mathbf{X}}) + \overbrace{\nabla g(\boldsymbol{\mu}_{\mathbf{X}})}^{\text{Gradient}}(\mathbf{X}-\boldsymbol{\mu}_{\mathbf{X}})}_{1^{\text{st}}\text{-order approximation}} + \frac{1}{2}(\mathbf{X}-\boldsymbol{\mu}_{\mathbf{X}})^{\mathsf{T}}\overbrace{\mathbf{H}(\boldsymbol{\mu}_{\mathbf{X}})}^{\text{Hessian}}(\mathbf{X}-\boldsymbol{\mu}_{\mathbf{X}})}^{2^{\text{nd}}\text{-order approximation}} + \cdots}_{m^{\text{th}}\text{-order approximation}}.
$$

In practice, the series are most often limited to the first-order approximation, so for a one-to-one function, it simplifies to

$$Y = \mathbf{g}(X) \approx aX + b.$$

Figure 3.22 presents an example of such a linear approximation for a one-to-one transformation. Linearizing at the expected value $\mu_x$ minimizes the approximation errors because the linearization is then centered in the region associated with a high probability content for $f_X(x)$. In that case, $a$ corresponds to the gradient of $g(x)$ evaluated at $\mu_X$,

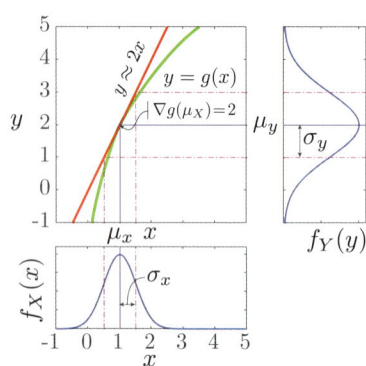

Figure 3.22: Example of a linearized nonlinear transformation.

$$a = \left[ \frac{dg(x)}{dx} \right]_{x=\mu_X}.$$

For the $n \to 1$ multivariate case, the linearized transformation leads to

$$\begin{aligned} Y = g(\mathbf{X}) &\approx \mathbf{a}^\mathsf{T}\mathbf{X} + b \\ &= \nabla g(\boldsymbol{\mu_X})(\mathbf{X} - \boldsymbol{\mu_X}) + g(\boldsymbol{\mu_X}), \end{aligned}$$

where $Y$ has a mean and variance equal to

$$\begin{aligned} \mu_Y &\approx g(\boldsymbol{\mu_X}) \\ \sigma_Y^2 &\approx \nabla g(\boldsymbol{\mu_X})\boldsymbol{\Sigma_X}\nabla g(\boldsymbol{\mu_X})^\mathsf{T}. \end{aligned}$$

For the $n \to n$ multivariate case, the linearized transformation leads to

$$\begin{aligned} \mathbf{Y} = \mathbf{g}(\mathbf{X}) &\approx \mathbf{A}\mathbf{X} + \mathbf{b} \\ &= \mathbf{J_{Y,X}}(\boldsymbol{\mu_X})(\mathbf{X} - \boldsymbol{\mu_X}) + \mathbf{g}(\boldsymbol{\mu_X}), \end{aligned}$$

where $Y$ is described by the mean vector and covariance matrix,

$$\begin{aligned} \boldsymbol{\mu_Y} &\cong \mathbf{g}(\boldsymbol{\mu_X}) \\ \boldsymbol{\Sigma_Y} &\cong \mathbf{J_{Y,X}}(\boldsymbol{\mu_X})\boldsymbol{\Sigma_X}\mathbf{J_{Y,X}^\mathsf{T}}(\boldsymbol{\mu_X}). \end{aligned}$$

For multivariate nonlinear functions, the gradient or Jacobian is evaluated at the expected value $\boldsymbol{\mu_X}$.

*Exercises*

P3.1 Draw on the Venn diagram: $\overline{E_1 \cup E_2} \cap E_3 \cup E_1 \cap E_2$.

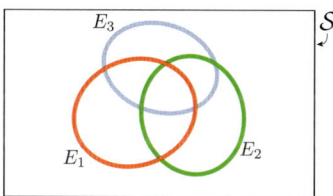

P3.2 Explain the difference between independent and mutually exclusive events.

P3.3 Compute $\Pr(E_1 \cup E_2)$ for two events $E_1 : \Pr(E_1) = 0.2$ and $E_2 : \Pr(E_2) = 0.5$, for which $\Pr(E_1 E_2) = 0.1$.

P3.4 Given two discrete random variables $X : x \in \{x_1, x_2, x_3\}$ and $Y : y \in \{y_1, y_2, y_3\}$ with the joint PMF defined below, compute the marginal probabilities $p(x)$ and $p(y)$ and the conditional probabilities $p(x|y)$.

|       | $x_1$ | $x_2$ | $x_3$ |
|-------|-------|-------|-------|
| $y_1$ | 0.1   | 0.05  | 0.2   |
| $y_2$ | 0.2   | 0.01  | 0.1   |
| $y_3$ | 0.2   | 0.04  | 0.1   |

P3.5 Explain the difference between $X$ and $x$ for a random variable.

P3.6 Draw the PMF and CMF of $X$ as defined in P3.4.

P3.7 Why can a PDF $f_X(x)$ be $> 1$ and a PMF $p_X(x)$ cannot?

P3.8 Why is a CDF $F_X(x)$ nondecreasing?

P3.9 Explain why two random variables can have a correlation coefficient $\rho = 0$ and still not be independent.

P3.10 In what case does $f_{X_1 X_2}(x_1, x_2) = f_{X_1}(x_1) \cdot f_{X_2}(x_2)$ hold?

P3.11 If $F_X(20) = 0.5$ and $F_X(10) = 0.1$, what is $\Pr(10 \le X \le 20)$?

P3.12 For $X : \mathbb{E}[X] = 20, \operatorname{var}[X] = 10^2$ and a deterministic function $Y = 5X + 30$, compute $\mathbb{E}[Y]$ and $\operatorname{var}[Y]$.

P3.13 Given a slender column with a resistance defined by $F$, $K$ is a random variable $K \sim \mathcal{N}(k; \mu_K, \sigma_K^2)$ with $\{\mu_K = 0.75, \sigma_K = 0.1\}$, $E = 200{,}000\,\text{MPa}$, $L = 10{,}000\,\text{mm}$, and $I = 10^7\,\text{mm}^4$. (a) Compute $\mathbb{E}[F]$ and $\operatorname{var}[F]$ by linearizing $F$. (b) Draw the PDF for $F$ using the change of variable rule.

$$F = \frac{\pi^2 E I}{(KL)^2}$$

# 4
# Probability Distributions

The definition of probability distributions $f_X(x)$ was left aside in chapter 3. This chapter presents the formulation and properties for the probability distributions employed in this book: the Normal distribution for $x \in \mathbb{R}$, the log-normal for $x \in \mathbb{R}^+$, and the Beta for $x \in (0, 1)$.

## 4.1 Normal Distribution

The most widely employed probability distribution is the *Normal*, also known as the *Gaussian*, distribution. In this book, the names *Gaussian* and *Normal* are employed interchangeably when describing a probability distribution. This section covers the mathematical foundation for the univariate and multivariate Normal and then details the properties explaining its widespread usage.

### 4.1.1 Univariate Normal

The probability density function (PDF) for a Normal random variable is defined over the real numbers $x \in \mathbb{R}$. $X \sim \mathcal{N}(x; \mu, \sigma^2)$ is parameterized by its *mean* $\mu$ and *variance* $\sigma^2$, so its PDF is

$$f_X(x) = \mathcal{N}(x; \mu, \sigma^2) = \frac{1}{\sqrt{2\pi}\sigma} \exp\left(-\frac{1}{2}\left(\frac{x-\mu}{\sigma}\right)^2\right).$$

Figure 4.1 presents an example of PDF and cumulative distribution function (CDF) with parameters $\mu = 0$ and $\sigma = 1$. The mode— that is, the most likely value—corresponds to the mean. Changing the mean $\mu$ causes a translation of the distribution. Increasing the standard deviation $\sigma$ causes a proportional increase in the PDF's dispersion. The Normal CDF is presented in figure 4.1b. Its formulation is obtained through integration, where the integral can

**Univariate Normal**
$x \in \mathbb{R} : X \sim \mathcal{N}(x; \mu, \sigma^2)$

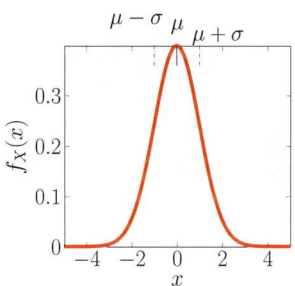

(a) Probability density function (PDF)

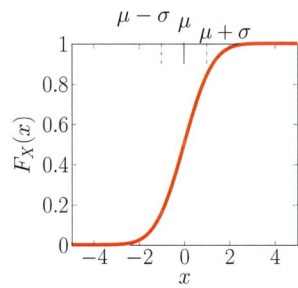

(b) Cumulative distribution function (CDF)

Figure 4.1: Representation of the univariate Normal for $\mu = 0$, $\sigma = 1$.

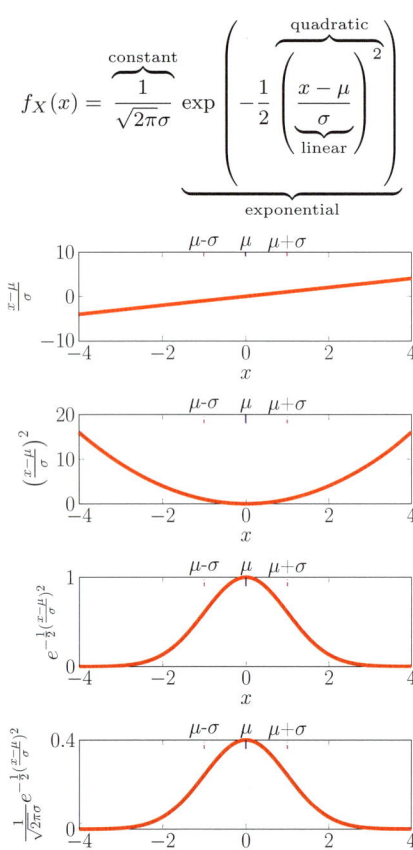

$$f_X(x) = \overbrace{\frac{1}{\sqrt{2\pi}\sigma}}^{\text{constant}} \exp\left( \underbrace{-\frac{1}{2}\left( \overbrace{\underbrace{\frac{x-\mu}{\sigma}}_{\text{linear}}}^{\text{quadratic}} \right)^2}_{\text{exponential}} \right)$$

Figure 4.2: Illustration of the univariate Normal probability density function formulation for $\mu = 0$, $\sigma = 1$.

be formulated using the *error function* erf($\cdot$),

$$
\begin{aligned}
F_X(x) &= \int_{-\infty}^{x} \frac{1}{\sqrt{2\pi}\sigma} \exp\left( -\frac{1}{2}\left( \frac{x'-\mu}{\sigma} \right)^2 \right) dx' \\
&= \frac{1}{2}\left( 1 + \text{erf}\left( \frac{x-\mu}{\sigma\sqrt{2}} \right) \right).
\end{aligned}
$$

Figure 4.2 illustrates the successive steps taken to construct the univariate Normal PDF. Within the innermost parenthesis of the PDF formulation is a linear function $\frac{x-\mu}{\sigma}$, which centers $x$ on the mean $\mu$ and normalizes it with the standard deviation $\sigma$. This first term is then squared, leading to a positive number over all its domain except at the mean, where it is equal to zero. Taking the negative exponential of this second term leads to a bell-shaped curve, where the value equals one ($\exp(0) = 1$) at the mean $x = \mu$ and where there are inflexion points at $\mu \pm \sigma$. At this step, the curve is proportional to the final Normal PDF. Only the normalization constant is missing to ensure that $\int_{-\infty}^{\infty} f(x)dx = 1$. The normalization constant is obtained by integrating the exponential term,

$$\int_{-\infty}^{+\infty} \exp\left( -\frac{1}{2}\left( \frac{x-\mu}{\sigma} \right)^2 \right) dx = \sqrt{2\pi}\sigma. \qquad (4.1)$$

Dividing the exponential term by the normalization constant in equation 4.1 results in the final formulation for the Normal PDF. Note that for $x = \mu$, $f(\mu) \neq 1$ because the PDF has been normalized so its integral is one.

### 4.1.2   Multivariate Normal

The joint probability density function (PDF) for two Normal random variables $\{X_1, X_2\}$ is given by

$$f_{X_1X_2}(x_1,x_2) = \frac{1}{2\pi\sigma_1\sigma_2\sqrt{1-\rho^2}} \exp\left( -\frac{1}{2(1-\rho^2)}\left( \left( \frac{x_1-\mu_1}{\sigma_1} \right)^2 + \left( \frac{x_2-\mu_2}{\sigma_2} \right)^2 - 2\rho\left( \frac{x_1-\mu_1}{\sigma_1} \right)\left( \frac{x_2-\mu_2}{\sigma_2} \right) \right) \right).$$

There are three terms within the parentheses inside the exponential. The first two are analogous to the quadratic terms for the univariate case. The third one includes a new parameter $\rho$ describing the correlation coefficient between $X_1$ and $X_2$. Together, these three terms describe the equation of a 2-D ellipse centered at $[\mu_1\ \mu_2]^\mathsf{T}$.

In a more general way, the probability density function for $n$ random variables $\mathbf{X} = [X_1\ X_2\ \cdots\ X_n]^\mathsf{T}$ is described by $\mathbf{x} \in \mathbb{R}^n$ : $\mathbf{X} \sim \mathcal{N}(\mathbf{x}; \boldsymbol{\mu_X}, \boldsymbol{\Sigma_X})$, where $\boldsymbol{\mu_X} = [\mu_1\ \mu_2\ \cdots\ \mu_n]^\mathsf{T}$ is a vector

**Multivariate Normal**
$\mathbf{x} \in \mathbb{R}^n : \mathbf{X} \sim \mathcal{N}(\mathbf{x}; \boldsymbol{\mu_X}, \boldsymbol{\Sigma_X})$

containing mean values and $\mathbf{\Sigma_X}$ is the covariance matrix,

$$\mathbf{\Sigma_X} = \mathbf{D_X R_X D_X} = \begin{bmatrix} \sigma_1^2 & \rho_{12}\sigma_1\sigma_2 & \cdots & \rho_{1n}\sigma_1\sigma_n \\ & \sigma_2^2 & \cdots & \rho_{2n}\sigma_2\sigma_n \\ & & \vdots & \vdots \\ \text{sym.} & & & \sigma_n^2 \end{bmatrix}_{n \times n}.$$

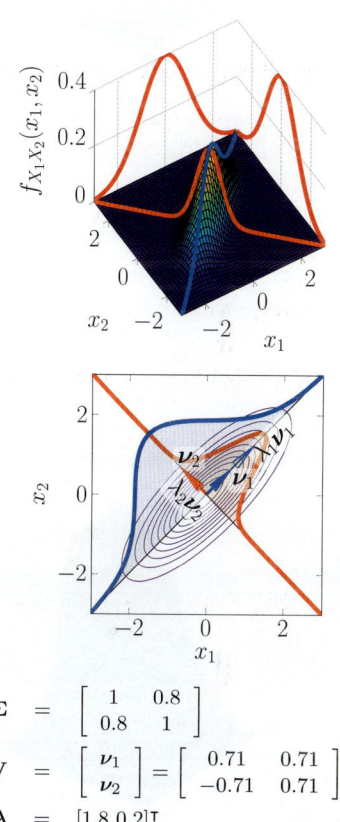

$\mathbf{D_X}$   : Standard deviation matrix
$\mathbf{R_X}$   : Correlation matrix
$\mathbf{\Sigma_X}$   : Covariance matrix

$\mathbf{D_X}$ is the *standard deviation matrix* containing the standard deviation of each random variable on its main diagonal, and $\mathbf{R_X}$ is the symmetric (sym.) *correlation matrix* containing the correlation coefficient for each pair of random variables,

$$\mathbf{D_X} = \begin{bmatrix} \sigma_1 & 0 & 0 & 0 \\ & \sigma_2 & 0 & 0 \\ & & \ddots & 0 \\ \text{sym.} & & & \sigma_n \end{bmatrix}, \mathbf{R_X} = \begin{bmatrix} 1 & \rho_{12} & \cdots & \rho_{1n} \\ & 1 & \cdots & \rho_{2n} \\ & & \vdots & \rho_{n-1n} \\ \text{sym.} & & & 1 \end{bmatrix}.$$

Note that a variable is linearly correlated with itself so the main diagonal terms for the correlation matrix are $[\mathbf{R_X}]_{ii} = 1, \forall i$. The multivariate Normal joint PDF is described by

$$f_{\mathbf{X}}(\mathbf{x}) = \frac{1}{(2\pi)^{n/2}(\det \mathbf{\Sigma_X})^{1/2}} \exp\left(-\frac{1}{2}(\mathbf{x} - \boldsymbol{\mu_X})^{\mathsf{T}}\mathbf{\Sigma_X}^{-1}(\mathbf{x} - \boldsymbol{\mu_X})\right),$$

where the terms inside the exponential describe an $n$-dimensional ellipsoid centered at $\boldsymbol{\mu_X}$. The directions of the principal axes of this ellipsoid are described by the eigenvector (see §2.4.2) of the covariance matrix $\mathbf{\Sigma_X}$, and their lengths by the eigenvalues. Figure 4.3 presents an example of a covariance matrix decomposed into its eigenvector and eigenvalues. The curves overlaid on the joint PDF describe the marginal PDFs in the eigen space.

For the multivariate Normal joint PDF formulation, the term on the left of the exponential is again the normalization constant, which now includes the determinant of the covariance matrix. As presented in §2.4.1, the determinant quantifies how much the covariance matrix $\mathbf{\Sigma_X}$ is scaling the space $\mathbf{x}$. Figure 4.4 presents examples of bivariate Normal PDF and CDF with parameters $\mu_1 = 0$, $\sigma_1 = 2$, $\mu_2 = 0$, $\sigma_2 = 1$, and $\rho = 0.6$. For the bivariate CDF, notice how evaluating the upper bound for one variable leads to the marginal CDF, represented by the bold red line, for the other variable.

$$\mathbf{\Sigma} = \begin{bmatrix} 1 & 0.8 \\ 0.8 & 1 \end{bmatrix}$$

$$\mathbf{V} = \begin{bmatrix} \boldsymbol{\nu}_1 \\ \boldsymbol{\nu}_2 \end{bmatrix} = \begin{bmatrix} 0.71 & 0.71 \\ -0.71 & 0.71 \end{bmatrix}$$

$$\boldsymbol{\lambda} = [1.8 \ 0.2]^{\mathsf{T}}$$

Figure 4.3: Example of bivariate PDF with $\mu_1 = \mu_2 = 0$, $\sigma_1 = \sigma_2 = 1$, and $\rho = 0.8$, for which the covariance matrix is decomposed into its eigenvector and eigenvalues.

### 4.1.3 Properties

A multivariate Normal random variable follow several properties. Here, we insist on six:

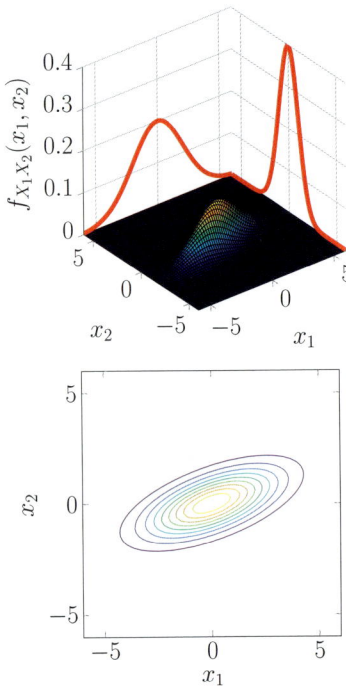

(a) Bivariate Normal PDF

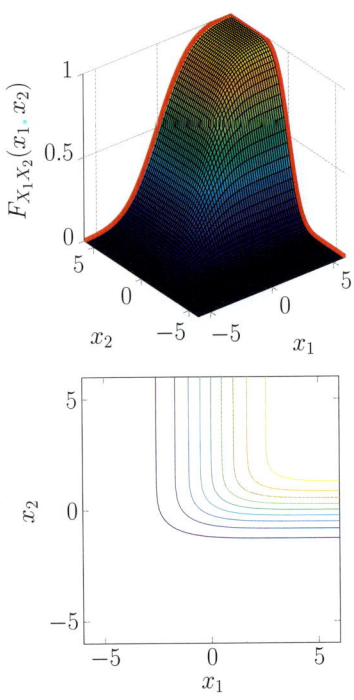

(b) Bivariate Normal CDF

Figure 4.4: Examples of bivariate PDF and CDF for $\mu_1 = \mu_2 = 0$, $\sigma_1 = 2$, $\sigma_2 = 1$, and $\rho = 0.6$.

1. It is *completely defined* by its *mean vector* $\boldsymbol{\mu_X}$ and *covariance matrix* $\boldsymbol{\Sigma_X}$.

2. Its *marginal distributions are also Normal*, and the PDF of any marginal is given by

$$x_i : X_i \sim \mathcal{N}(x_i; [\boldsymbol{\mu_X}]_i, [\boldsymbol{\Sigma_X}]_{ii}).$$

3. The *absence of correlation implies statistical independence*. Note that this is not generally true for other types of random variables (see §3.3.5),

$$\rho_{ij} = 0 \Leftrightarrow X_i \perp\!\!\!\perp X_j.$$

4. The *central limit theorem* (CLT) states that, under some conditions, the asymptotic distribution obtained from the normalized sum of *independent identically distributed* (iid) random variables (normally distributed or not) is Normal. Given $X_i$, $\forall i \in \{1, \cdots, n\}$, a set of iid random variables with expected value $\mathbb{E}[X_i] = \mu_X$ and finite variance $\sigma_X^2$, the PDF of $Y = \sum_{i=1}^{n} X_i$ approaches $\mathcal{N}(n\mu_X, n\sigma_X^2)$, for $n \to \infty$. More formally, the CLT states that

$$\sqrt{n}\left(\tfrac{Y}{n} - \mu_X\right) \xrightarrow{d} \mathcal{N}(0, \sigma_X^2),$$

where $\xrightarrow{d}$ means *converges in distribution*. In practice, when observing the outcomes of real-life phenomena, it is common to obtain empirical distributions that are similar to the Normal distribution. We can see the parallel where these phenomena are themselves issued from the superposition of several phenomena. This property is key in explaining the widespread usage of the Normal probability distribution.

5. The output from *linear functions of Normal random variables are also Normal*. Given $\mathbf{x} : \mathbf{X} \sim \mathcal{N}(\mathbf{x}; \boldsymbol{\mu_X}, \boldsymbol{\Sigma_X})$ and a linear function $\mathbf{y} = \mathbf{A}\mathbf{x} + \mathbf{b}$, the properties of linear transformations described in §3.4.1 allow obtaining

$$\mathbf{Y} \sim \mathcal{N}(\mathbf{y}; \mathbf{A}\boldsymbol{\mu_X} + \mathbf{b}, \mathbf{A}\boldsymbol{\Sigma_X}\mathbf{A}^\mathsf{T}).$$

Let us consider the simplified case of a linear function $z = x + y$ for two random variables $x : X \sim \mathcal{N}(x; \mu_X, \sigma_X^2)$, $y : Y \sim \mathcal{N}(y; \mu_Y, \sigma_Y^2)$. Their sum is described by

$$Z \sim \mathcal{N}(z; \underbrace{\mu_X + \mu_Y}_{\mu_Z}, \underbrace{\sigma_X^2 + \sigma_Y^2 + 2\rho_{XY}\sigma_X\sigma_Y}_{\sigma_Z^2}).$$

In the case where both variables are statistically independent $X \perp Y$, the variance of their sum is equal to the sum of their respective variance. For the general case describing the sum of a set of $n$ correlated normal random variables $X_i$ such that $\mathbf{X} \sim \mathcal{N}(\mathbf{x}; \boldsymbol{\mu}_{\mathbf{X}}, \boldsymbol{\Sigma}_{\mathbf{X}})$,

$$
\begin{aligned}
Z &= \sum_{i=1}^{n} X_i \\
&\sim \mathcal{N}\left(z; \sum_{i=1}^{n}[\boldsymbol{\mu}_{\mathbf{X}}]_i, \sum_{i=1}^{n}\sum_{j=1}^{n}[\boldsymbol{\Sigma}]_{ij}\right).
\end{aligned} \tag{4.2}
$$

As we will see in the next chapters, the usage of linear models is widespread in machine learning because of the analytical tractability of linear functions of Normal random variables.

6. *Conditional distributions are Normal.* For instance, we can partition an ensemble of $n$ random variables $\mathbf{X}$ in two subsets so that

$$
\mathbf{X} = \begin{bmatrix} \mathbf{X}_i \\ \mathbf{X}_j \end{bmatrix}, \ \boldsymbol{\mu} = \begin{bmatrix} \boldsymbol{\mu}_i \\ \boldsymbol{\mu}_j \end{bmatrix}, \ \boldsymbol{\Sigma} = \begin{bmatrix} \boldsymbol{\Sigma}_i & \boldsymbol{\Sigma}_{ij} \\ \boldsymbol{\Sigma}_{ji} & \boldsymbol{\Sigma}_j \end{bmatrix},
$$

where $\boldsymbol{\Sigma}_i$ describes the covariance matrix for the $i^{\text{th}}$ subset of random variables and $\boldsymbol{\Sigma}_{ij} = \boldsymbol{\Sigma}_{ji}^{\mathsf{T}}$ describes the covariance between the random variables belonging to subsets $i$ and $j$. It is mentioned in §3.3.4 that a conditional probability density function is obtained from the division of a joint PDF by a marginal PDF. The same concept applies to define the conditional PDF of $\mathbf{X}_i$ given a vector of observations $\mathbf{X}_j = \mathbf{x}_j$,

$$
f_{\mathbf{X}_i|\mathbf{x}_j}(\mathbf{x}_i | \underbrace{\mathbf{X}_j = \mathbf{x}_j}_{\text{observations}}) = \frac{f_{\mathbf{X}_i\mathbf{X}_j}(\mathbf{x}_i, \mathbf{x}_j)}{f_{\mathbf{X}_j}(\mathbf{x}_j)} = \mathcal{N}(\mathbf{x}_i; \boldsymbol{\mu}_{i|j}, \boldsymbol{\Sigma}_{i|j}),
$$

where the conditional mean and covariance are

$$
\begin{aligned}
\boldsymbol{\mu}_{i|j} &= \boldsymbol{\mu}_i + \boldsymbol{\Sigma}_{ij}\boldsymbol{\Sigma}_j^{-1}(\mathbf{x}_j - \boldsymbol{\mu}_j) \\
\boldsymbol{\Sigma}_{i|j} &= \boldsymbol{\Sigma}_i - \boldsymbol{\Sigma}_{ij}\boldsymbol{\Sigma}_j^{-1}\boldsymbol{\Sigma}_{ij}^{\mathsf{T}}.
\end{aligned} \tag{4.3}
$$

If we simplify this setup for only two random variables $X_1$ and $X_2$, where we want the conditional PDF of $X_1$ given an observation $X_2 = x_2$, then equation 4.3 simplifies to

$$
\begin{aligned}
\mu_{1|2} &= \mu_1 + \rho\sigma_1 \frac{x_2 - \mu_2}{\sigma_2} \\
\sigma_{1|2}^2 &= \sigma_1^2(1 - \rho^2).
\end{aligned}
$$

In the special case where the prior mean $\mu_1 = \mu_2 = 0$ and the prior standard deviations $\sigma_1 = \sigma_2 > 0$, then the conditional mean simplifies to the observation $x_2$ times the correlation coefficient $\rho$. Note that the conditional variance $\sigma_{1|2}^2$ is independent of the observed value $x_2$; it only depends on the prior variance and the correlation coefficient. For the special case where $\rho = 1$, then $\sigma_{1|2}^2 = 0$.

(a) Multi-beam bridge span. (Photo: Archives Radio-Canada)

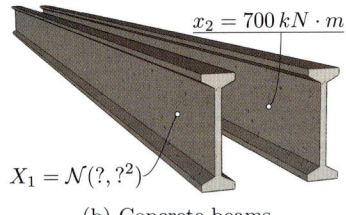

(b) Concrete beams

Figure 4.5: Example of dependence between the resistance of beams.

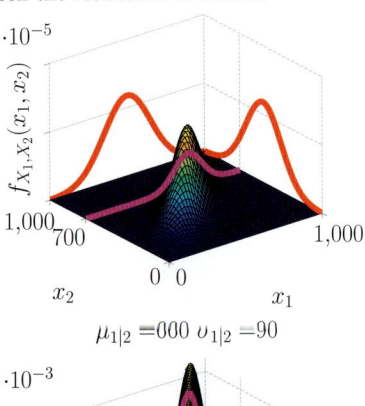

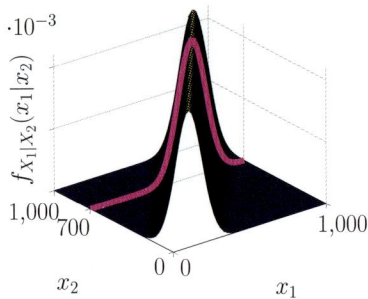

Figure 4.6: Joint prior PDF $f_{X_1 X_2}(x_1, x_2)$ and conditional PDF $f_{X_1|x_2}(x_1|x_2)$ describing the resistance of beams.

Figure 4.7: Steel cables each made from multiple wires. (This example is adapted from Armen Der Kiureghian's CE229 course at UC Berkeley.)

## 4.1.4   Example: Conditional Distributions

For the beam example illustrated in figure 4.5, our *prior knowledge for the resistance* $\{X_1, X_2\}$ of two adjacent beams is

$$
\begin{aligned}
X_1 &\sim \mathcal{N}(x_1; 500, 150^2) \quad [\text{kN·m}] \\
X_2 &\sim \mathcal{N}(x_2; 500, 150^2) \quad [\text{kN·m}],
\end{aligned}
$$

and we know that the beam resistances are correlated with $\rho_{12} = 0.8$. Such a correlation could arise because both beams were fabricated with the same process, in the same factory. This prior knowledge is described by the joint bivariate Normal PDF,

$$
f_{X_1 X_2}(x_1, x_2) = \mathcal{N}(\mathbf{x}; \boldsymbol{\mu}_{\mathbf{X}}, \boldsymbol{\Sigma}_{\mathbf{X}}) \begin{cases} \boldsymbol{\mu}_{\mathbf{X}} = \begin{bmatrix} 500 \\ 500 \end{bmatrix} \\ \boldsymbol{\Sigma}_{\mathbf{X}} = \begin{bmatrix} 150^2 & 0.8 \cdot 150^2 \\ 0.8 \cdot 150^2 & 150^2 \end{bmatrix}. \end{cases}
$$

If we observe that the resistance of the second beam $x_2 = 700$ kN·m, we can employ conditional probabilities to estimate the PDF of the strength $X_1$, given the observation $x_2$,

$$
f_{X_1|x_2}(x_1|x_2) = \mathcal{N}(x_1; \mu_{1|2}, \sigma^2_{1|2}),
$$

where

$$
\mu_{1|2} = 500 + 0.8 \times 150 \frac{\overbrace{700}^{\text{observation}} - 500}{150} = 660 \text{ kN·m}
$$

$$
\sigma_{1|2} = 150\sqrt{1 - 0.8^2} = 90 \text{ kN·m}.
$$

Figure 4.6 presents the joint and conditional PDFs corresponding to this example. For the joint PDF, the highlighted pink slice corresponding to $x_2 = 700$ is proportional to the conditional probability $f_{X_1|x_2}(x_1|x_2 = 700)$. If we want to obtain the conditional distribution from the joint PDF, we have to divide it by the marginal PDF $f_{X_2}(x_2 = 700)$. This ensures that the conditional PDF for $x_1$ integrates to 1. This example is trivial, yet it sets the foundations for the more advanced models that will be presented in the following chapters.

## 4.1.5   Example: Sum of Normal Random Variables

Figure 4.7 presents steel cables where each one is made from dozens of individual wires. Let us consider a cable made of 50 steel wires, each having a resistance $x_i : X_i \sim \mathcal{N}(x_i; 10, 3^2)$ kN. We use equation 4.2 to compare the cable resistance $X_{\text{cable}} = \sum_{i=1}^{50} X_i$ depending on the correlation coefficient $\rho_{ij}$. With the hypothesis that $X_i \perp\!\!\!\perp X_j \Leftrightarrow \rho_{ij} = 0$, all nondiagonal terms of the covariance matrix $[\boldsymbol{\Sigma}_{\mathbf{X}}]_{ij} = 0, \forall i \neq j$, which leads to

$$X_{\text{cable}} \sim \mathcal{N}(x; 50 \times 10\,\text{kN}, \underbrace{50 \times (3\,\text{kN})^2}_{\sigma_{X_{\text{cable}}} = 3\sqrt{50} \approx 21\,\text{kN}}).$$

With the hypothesis $\rho_{ij} = 1$, all terms in $[\boldsymbol{\Sigma}_{\mathbf{X}}]_{ij} = (3\,\text{kN})^2, \forall i, j$, so that

$$X_{\text{cable}} \sim \mathcal{N}(x; 50 \times 10\,\text{kN}, \underbrace{50^2 \times (3\,\text{kN})^2}_{\sigma_{X_{\text{cable}}} = 3\,\text{kN} \times 50 = 150\,\text{kN}}).$$

Figure 4.8 presents the resulting PDFs for the cable resistance, given each hypothesis. These results show that if the uncertainty in the resistance for each wire is independent, there will be some cancellation; some wires will have a resistance above the mean, and some will have a resistance below. The resulting coefficient of variation for $\rho = 0$ is $\delta_{\text{cable}} = \frac{21}{500} = 0.11$, which is approximately three times smaller than $\delta_{\text{wire}} = \frac{3}{10} = 0.3$, the variability associated with each wire. In the opposite case, if the resistance is linearly correlated ($\rho = 1$), the uncertainty adds up as you increase the number of wires, so $\delta_{\text{cable}} = \frac{150}{500} = \delta_{\text{wire}}$.

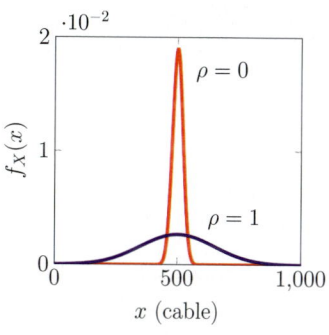

Figure 4.8: Probability density function of the cable resistance depending on the correlation between the strength of each wire.

## 4.2   Log-Normal Distribution

The *log-normal* distribution is obtained by transforming the Normal distribution through the function $\ln x$. Because the logarithm function is only defined for positive values, the domain of the log-normal distribution is $x \in \mathbb{R}^+$.

### 4.2.1   Univariate Log-Normal

The random variable $X \sim \ln\mathcal{N}(x; \lambda, \zeta)$ is log-normal if $\ln X \sim \mathcal{N}(\ln x; \lambda, \zeta^2)$ is Normal. Given the transformation function $x' = \ln x$, the *change of variable* rule presented in §3.4 requires that

$$\begin{aligned}
\overbrace{f_{X'}(x')}^{\mathcal{N}(x'; \lambda, \zeta^2)} dx' &= f_X(x)\, dx \\
f_{X'}(x') \left| \tfrac{dx'}{dx} \right| &= \underbrace{f_X(x)}_{\ln\mathcal{N}(x; \lambda, \zeta)},
\end{aligned}$$

where the derivative of $\ln x$ with respect to x is

$$\frac{dx'}{dx} = \frac{d\ln x}{dx} = \frac{1}{x}.$$

Therefore, the analytic formulation for the log-normal PDF is given by the product of the transformation's derivative and the Normal

**Univariate log-normal**

$$x \in \mathbb{R}^+ : X \sim \ln\mathcal{N}(x; \lambda, \zeta)$$
$$x' \in \mathbb{R} : X' = \ln X \sim \mathcal{N}(\ln x; \lambda, \zeta^2)$$

| | |
|---|---|
| $\mu_X$ | : mean of $X$ |
| $\lambda$ | : mean of $\ln X$ ($= \mu_{\ln X}$) |
| $\sigma_X^2$ | : variance of $X$ |
| $\zeta^2$ | : variance of $\ln X$ ($= \sigma_{\ln X}^2$) |

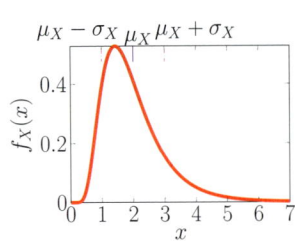

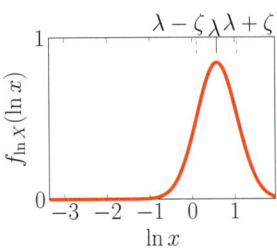

Figure 4.9: Univariate log-normal probability density function for $\{\mu_X = 2, \sigma_X = 1\}$ and $\{\lambda = 0.58, \zeta = 0.47\}$.

**Multivariate log-normal**
$\mathbf{x} \in (\mathbb{R}^+)^n : \mathbf{X} \sim \ln \mathcal{N}(\mathbf{x}; \boldsymbol{\mu}_{\ln \mathbf{X}}, \boldsymbol{\Sigma}_{\ln \mathbf{X}})$

PDF evaluated for $x' = \ln x$,

$$
\begin{aligned}
f_X(x) &= \frac{1}{x} \cdot \mathcal{N}(\ln x; \lambda, \zeta^2) \\
&= \frac{1}{x} \cdot \frac{1}{\sqrt{2\pi}\zeta} \exp\left( -\frac{1}{2}\left( \frac{\ln x - \lambda}{\zeta} \right)^2 \right), \quad x > 0.
\end{aligned}
$$

The univariate log-normal PDF is parameterized by the mean $(\mu_{\ln X} = \lambda)$ and variance $(\sigma^2_{\ln X} = \zeta^2)$ defined in the log-transformed space $(\ln x)$. The mean $\mu_X$ and variance $\sigma^2_X$ of the log-normal random variable can be transformed in the log-space using the relations

$$
\begin{aligned}
\lambda &= \mu_{\ln X} = \ln \mu_X - \frac{\zeta^2}{2} \\
\zeta &= \sigma_{\ln X} = \sqrt{\ln\left(1 + \left(\frac{\sigma_X}{\mu_X}\right)^2\right)} = \sqrt{\ln(1 + \delta_X^2)}.
\end{aligned}
\tag{4.4}
$$

Note that for $\delta_X < 0.3$, the standard deviation in the log-space is approximately equal to the coefficient of variation in the original space, $\zeta \approx \delta_X$. Figure 4.9 presents an example of log-normal PDF plotted (a) in the original space and (b) in the log-transformed space. The mean and standard deviation are $\{\mu_X = 2, \sigma_X = 1\}$ in the original space and $\{\lambda = 0.58, \zeta = 0.47\}$ in the log-transformed space.

### 4.2.2  Multivariate Log-Normal

$X_1, X_2, \cdots, X_n$ are jointly log-normal if $\ln X_1, \ln X_2, \cdots, \ln X_n$ are jointly Normal. The multivariate log-normal PDF is parameterized by the mean values $(\mu_{\ln X_i} = \lambda)$, variances $(\sigma^2_{\ln X_i} = \zeta^2)$, and correlation coefficients $(\rho_{\ln X_i \ln X_j})$ defined in the log-transformed space. Correlation coefficients in the log-space $\rho_{\ln X_i \ln X_j}$ are related to the correlation coefficients in the original space $\rho_{X_i X_j}$ using the relation

$$
\rho_{\ln X_i \ln X_j} = \frac{1}{\zeta_i \zeta_j} \ln(1 + \rho_{X_i X_j} \delta_{X_i} \delta_{X_j}),
$$

where $\rho_{\ln X_i \ln X_j} \approx \rho_{X_i X_j}$ for $\delta_{X_i}, \delta_{X_j} \ll 0.3$. The PDF for two random variables $\{X_1, X_2\}$ such that $\{x_1, x_2\} > 0$ is

$$
f_{X_1 X_2}(x_1, x_2) = \frac{1}{x_1 x_2 \sqrt{2\pi}\zeta_1 \zeta_2 \sqrt{1 - \rho_{\ln}^2}} \exp\left( -\frac{1}{2(1 - \rho_{\ln}^2)}\left( \left(\frac{\ln x_1 - \lambda_1}{\zeta_1}\right)^2 + \left(\frac{\ln x_2 - \lambda_2}{\zeta_2}\right)^2 - 2\rho_{\ln}\left(\frac{\ln x_1 - \lambda_1}{\zeta_1}\right)\left(\frac{\ln x_2 - \lambda_2}{\zeta_2}\right)^2 \right) \right).
$$

Figure 4.10 presents an example of bivariate log-normal PDF with parameters $\mu_1 = \mu_2 = 1.5$, $\sigma_1 = \sigma_2 = 0.5$, and $\rho = 0.9$. The general formulation for the multivariate log-normal PDF is

$$
\begin{aligned}
f_{\mathbf{X}}(\mathbf{x}) &= \ln \mathcal{N}(\mathbf{x}; \boldsymbol{\mu}_{\ln \mathbf{X}}, \boldsymbol{\Sigma}_{\ln \mathbf{X}}) \\
&= \frac{1}{(\prod_{i=1}^{n} x_i)(2\pi)^{n/2}(\det \boldsymbol{\Sigma}_{\ln \mathbf{X}})^{1/2}} \exp\left(-\tfrac{1}{2}(\ln \mathbf{x} - \boldsymbol{\mu}_{\ln \mathbf{X}})^{\mathsf{T}} \boldsymbol{\Sigma}_{\ln \mathbf{X}}^{-1} (\ln \mathbf{x} - \boldsymbol{\mu}_{\ln \mathbf{X}})\right),
\end{aligned}
$$

where $\boldsymbol{\mu}_{\ln \mathbf{X}}$ and $\boldsymbol{\Sigma}_{\ln \mathbf{X}}$ are respectively the mean vector and covariance matrix defined in the log-space.

### 4.2.3   Properties

Because the log-normal distribution is obtained through a transformation of the Normal distribution, it inherits several of its properties.

1. It is *completely defined by* its *mean vector* $\boldsymbol{\mu}_{\ln \mathbf{X}}$ and *covariance matrix* $\boldsymbol{\Sigma}_{\ln \mathbf{X}}$.

2. Its *marginal distributions are also log-normal*, and the PDF of any marginal is given by

$$
x_i : X_i \sim \ln \mathcal{N}(x_i; [\boldsymbol{\mu}_{\ln \mathbf{X}}]_i, [\boldsymbol{\Sigma}_{\ln \mathbf{X}}]_{ii}).
$$

3. The *absence of correlation implies statistical independence.* Remember that this is not generally true for other types of random variables (see §3.3.5),

$$
\rho_{ij} = 0 \Leftrightarrow X_i \perp\!\!\!\perp X_j.
$$

4. *Conditional distributions are log-normal*, so the PDF of $\mathbf{X}_i$ given an observation $\mathbf{X}_j = \mathbf{x}_j$ is given by

$$
f_{\mathbf{X}_i | \mathbf{x}_j}(\mathbf{x}_i | \underbrace{\mathbf{X}_j = \mathbf{x}_j}_{\text{observations}}) = \ln \mathcal{N}(\mathbf{x}_i; \boldsymbol{\mu}_{\ln i|j}, \boldsymbol{\Sigma}_{\ln i|j}),
$$

where the conditional mean vector and covariance are

$$
\begin{aligned}
\boldsymbol{\mu}_{\ln i|j} &= \boldsymbol{\mu}_{\ln i} + \boldsymbol{\Sigma}_{\ln ij} \boldsymbol{\Sigma}_j^{-1} (\ln \mathbf{x}_j - \boldsymbol{\mu}_{\ln j}) \\
\boldsymbol{\Sigma}_{\ln i|j} &= \boldsymbol{\Sigma}_{\ln i} - \boldsymbol{\Sigma}_{\ln ij} \boldsymbol{\Sigma}_j^{-1} \boldsymbol{\Sigma}_{\ln ij}^{\mathsf{T}}.
\end{aligned}
\tag{4.5}
$$

5. The *multiplication of jointly log-normal random variables is jointly log-normal* so that for $X \sim \ln \mathcal{N}(x; \lambda_X, \zeta_X)$ and $Y \sim \ln \mathcal{N}(y; \lambda_Y, \zeta_Y)$, where $X \perp\!\!\!\perp Y$,

$$
\left.\begin{aligned}
Z &= X \cdot Y \\
&\sim \ln \mathcal{N}(z; \lambda_Z, \zeta_Z)
\end{aligned}\right\}
\begin{aligned}
\lambda_Z &= \lambda_X + \lambda_Y \\
\zeta_Z^2 &= \zeta_X^2 + \zeta_Y^2.
\end{aligned}
$$

Because the product of log-normal random variables can be transformed in the sum of Normal random variables, the properties of the *central limit theorem* presented in §4.1.3 still hold.

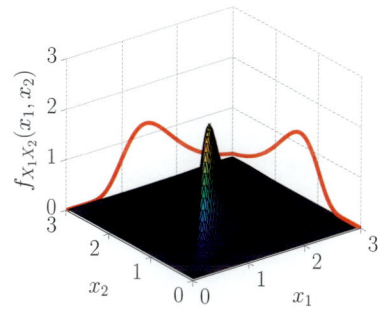

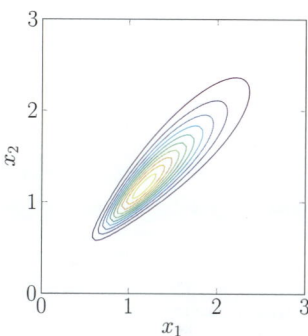

Figure 4.10: Example of bivariate log-normal probability density function for $\mu_1 = \mu_2 = 1.5$, $\sigma_1 = \sigma_2 = 0.5$, and $\rho = 0.9$.

## 4.3  Beta Distribution

The Beta distribution is defined over the interval $(0, 1)$. It can be scaled by the transformation $x' = x \cdot (b - a) + a$ to model bounded quantities within any range $(a, b)$. The Beta probability density function (PDF) is defined by

$$f_X(x) = \mathcal{B}(x; \alpha, \beta) = \frac{x^{\alpha-1}(1-x)^{\beta-1}}{\mathrm{B}(\alpha, \beta)} \begin{cases} \alpha > 0 \\ \beta > 0 \\ \mathrm{B}(\alpha, \beta): \text{ Beta function,} \end{cases}$$

where $\alpha$ and $\beta$ are the two distribution parameters, and the Beta function $\mathrm{B}(\alpha, \beta)$ is the normalization constant so that

$$\mathrm{B}(\alpha, \beta) = \int_0^1 x^{\alpha-1}(1-x)^{\beta-1} dx.$$

A common application of the Beta PDF is to employ the interval $(0, 1)$ to model the probability density of a probability itself. Let us consider two mutually exclusive and collectively exhaustive events, for example, any event $A$ and its complement $\overline{A}$, $\mathcal{S} = \{A, \overline{A}\}$. If the probability that the event $A$ occurs is uncertain, it can be described by a random variable so that

$$\begin{cases} \Pr(A) = X \\ \Pr(\overline{A}) = 1 - X, \end{cases}$$

where $x \in (0, 1) : X \sim \mathcal{B}(x; \alpha, \beta)$. The parameter $\alpha$ can be interpreted as *pseudo-counts* representing the number of observations of the event $A$, and $\beta$ is the number of observations of the complementary event $\overline{A}$. This relation between *pseudo-counts* and the Beta distribution, as well as practical applications, are further detailed in chapter 6. Figure 4.11 presents examples of Beta PDFs for three sets of parameters. Note how for $\alpha = \beta = 1$, the Beta distribution is analogous to the Uniform distribution $\mathcal{U}(x; 0, 1)$.

**Beta**

$x \in (0, 1) : X \sim \mathcal{B}(x; \alpha, \beta)$

$\mathbb{E}[X] = \dfrac{\alpha}{\alpha + \beta}$

$\mathrm{var}[X] = \dfrac{\alpha\beta}{(\alpha + \beta)^2(\alpha + \beta + 1)}$

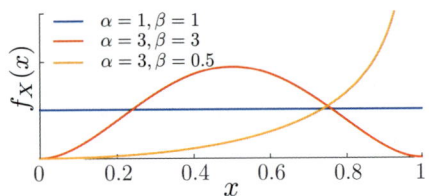

Figure 4.11: Three examples of the Beta probability density function evaluated for different sets of parameters $\{\alpha, \beta\}$.

*Exercises*

P4.1  For the joint PDFs $\mathcal{N}([x_1 \ x_2]^\mathsf{T}; \boldsymbol{\mu_X}, \boldsymbol{\Sigma_X})$,
 a) Compute $\sigma_{X_1}, \sigma_{X_2}, \rho$; and
 b) Draw the joint PDF contours if $\boldsymbol{\mu_X} = [0 \ 0]^\mathsf{T}$.

$$\boldsymbol{\Sigma_X} = \begin{bmatrix} 1 & 0 \\ 0 & 4 \end{bmatrix}$$

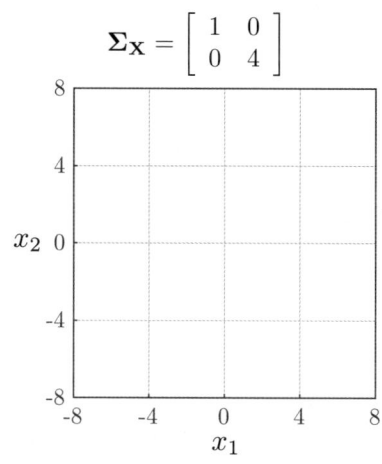

$$\boldsymbol{\Sigma_X} = \begin{bmatrix} 4 & -3.6 \\ -3.6 & 4 \end{bmatrix}$$

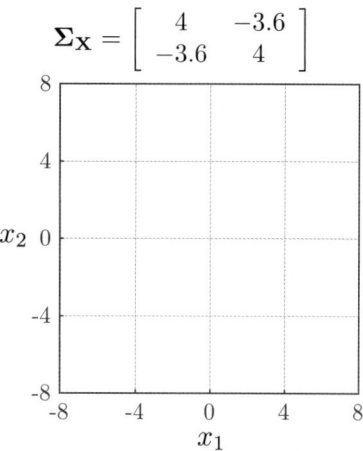

P4.2  Explain the role of the five parameters $\mu_{X_1}, \sigma_{X_1}, \mu_{X_2}, \sigma_{X_2}$, and $\rho$ describing the bivariate Normal PDF.

P4.3  Explain what happens to a bivariate Normal PDF if its covariance matrix is singular.

P4.4  Given $Z = X + Y$, so that $X \sim \mathcal{N}(x; 0, 1^2)$ and $Y \sim \mathcal{N}(y; 2, 5^2)$, compute $\mu_Z$ and $\sigma_Z^2$ if
 a) $\rho = 0$, b) $\rho = 1$.

P4.5  For a log-normal random variable $X \sim \ln\mathcal{N}(x; \lambda, \zeta)$, explain the difference between $\{\mu_X, \sigma_X\}$ and $\{\lambda, \zeta\}$ (reminder: $\lambda \equiv \mu_{\ln X}$ and $\zeta \equiv \sigma_{\ln X}$).

P4.6  Plot the PDF and CDF for $X \sim \mathcal{N}(x; 0, 1^2)$.

P4.7  Plot the PDF and CDF for $Y \sim \ln\mathcal{N}(y; \lambda, \zeta)$, if $\mathbb{E}[Y] = 1$ and $\text{Var}[Y] = 0.25$.

P4.8  Plot the PDF and CDF for $Z \sim \mathcal{B}(z; \alpha, \beta)$, if $\mathbb{E}[Z] = 0.66$ and $\text{Var}[Z] = 0.03$.

P4.9  From the marginal PDFs in P4.6 and P4.8, plot the contours of the joint PDF $f(y, z)$ if $Y \perp\!\!\!\perp Z$.

P4.10  The prior knowledge of the resistance $(X_1, X_2, X_3, X_4, X_5)$ for five adjacent beams is

$$X_i \sim \mathcal{N}(x_i; 500, 150^2) \ [\text{kN} \cdot \text{m}].$$

Knowing that the resistance of beams is correlated $\rho_{ij} = 0.8, \ \forall i \neq j$, what is the mean and covariance for strengths $X_1, X_2$, and $X_3$, given that we have observed that $x_4 = 695 \, \text{kN} \cdot \text{m}$ and $x_5 = 675 \, \text{kN} \cdot \text{m}$?

# 5
# Convex Optimization

When building a model in the context of machine learning, we often seek optimal model parameters $\boldsymbol{\theta}$, in the sense where they *maximize* the prior probability (or probability density) of predicting observed data. Here, we denote by $\tilde{f}(\boldsymbol{\theta})$ the target function we want to maximize. Optimal parameter values $\boldsymbol{\theta}^*$ are those that maximize the function $\tilde{f}(\boldsymbol{\theta})$,

$$\boldsymbol{\theta}^* = \arg\max_{\boldsymbol{\theta}} \tilde{f}(\boldsymbol{\theta}). \tag{5.1}$$

With a small caveat that will be covered below, *convex optimization* methods can be employed for the maximization task in equation 5.1. The key aspect of convex optimization methods is that, under certain conditions, they are guaranteed to reach optimal values for convex functions. Figure 5.1 presents examples of convex and non-convex sets. For a set to be *convex*, you must be able to link any two points belonging to it without being outside of this set. Figure 5.1b presents a case where this property is not satisfied. For a *convex function*, the segment linking any pair of its points lies above or is equal to the function. Conversely, for a *concave function*, the opposite holds: the segment linking any pair of points lies below or is equal to the function. A concave function can be transformed into a convex one by taking the negative of it. Therefore, a maximization problem formulated as a *concave optimization* can be formulated in terms of a *convex optimization* following

$$\boldsymbol{\theta}^* = \underbrace{\arg\max_{\boldsymbol{\theta}} \tilde{f}(\boldsymbol{\theta})}_{\text{Concave optimization}} \equiv \underbrace{\arg\min_{\boldsymbol{\theta}} -\tilde{f}(\boldsymbol{\theta})}_{\text{Convex optimization}}.$$

In this chapter, we refer to *convex optimization* even if we are interested in maximizing a concave function, rather than minimizing a convex one. This choice is justified by the prevalence of *convex optimization* in the literature. Moreover, note that for several machine learning methods, we seek $\boldsymbol{\theta}^*$ based on a minimization problem

**Note:** In the context of this book, we are only concerned by target functions $\tilde{f}(\boldsymbol{\theta}) \in \mathbb{R}$ that are continuous and twice differentiable.

**Note:** In order to be mathematically rigorous, equation 5.1 should employ the operator $\in$ rather than $=$ to recognize that $\boldsymbol{\theta}^*$ belongs to a set of possible solutions maximizing $\tilde{f}(\boldsymbol{\theta})$.

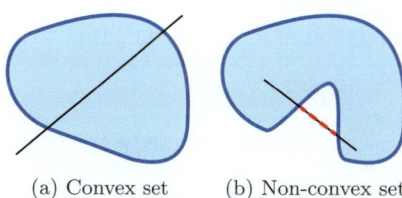

(a) Convex set　　　(b) Non-convex set

Figure 5.1: Examples of a convex and a non-convex set.

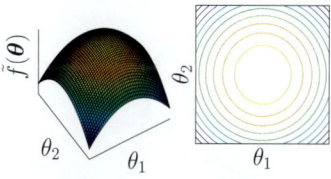

(a) $\tilde{f}(\boldsymbol{\theta})$: Concave, $-\tilde{f}(\boldsymbol{\theta})$: Convex

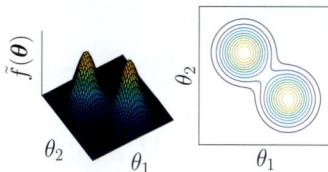

(b) $\tilde{f}(\boldsymbol{\theta})$: Non-concave, $-\tilde{f}(\boldsymbol{\theta})$: Non-convex

Figure 5.2: Representations of convex/concave and non-convex/non-concave functions.

[1] Bertsekas, D. P., A. Nedi, and A. E. Ozdaglar (2003). *Convex analysis and optimization.* Athena Scientific; Chong, E. K. P. and S. H. Zak (2013). *An introduction to optimization* (4th ed.). Wiley; and Nocedal, J. and S. Wright (2006). *Numerical optimization.* Springer Science & Business Media.

**Derivative**
$$\tilde{f}'(\theta) \equiv \frac{d\tilde{f}(\theta)}{d\theta}$$

**Gradient**
$$\nabla \tilde{f}(\boldsymbol{\theta}) \equiv \nabla_{\boldsymbol{\theta}} \tilde{f}(\boldsymbol{\theta})$$
$$= \left[ \frac{\partial \tilde{f}(\boldsymbol{\theta})}{\partial \theta_1} \; \frac{\partial \tilde{f}(\boldsymbol{\theta})}{\partial \theta_2} \; \cdots \; \frac{\partial \tilde{f}(\boldsymbol{\theta})}{\partial \theta_n} \right]^{\mathsf{T}}$$

**Maximum of a concave function**
$$\theta^* = \arg \max_{\theta} \tilde{f}(\theta) : \frac{d\tilde{f}(\theta^*)}{d\theta} = 0$$

$\lambda$ is also known as the *learning rate* or *step length*.

[2] Armijo, L. (1966). Minimization of functions having Lipschitz continuous first partial derivatives. *Pacific Journal of Mathematics 16*(1), 1–3.

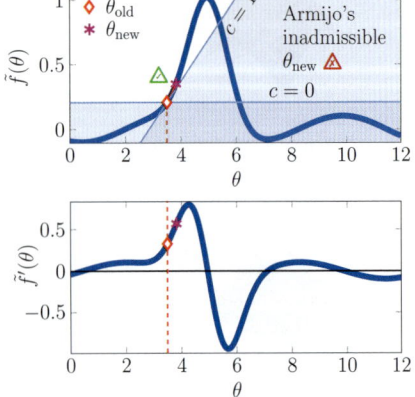

Figure 5.3: Example of application of the Armijo rule to test if $\theta_{\text{new}}$ has sufficiently increased the objective function in comparison with $\theta_{\text{old}}$.

where $-\tilde{f}(\boldsymbol{\theta})$ is a function of the difference between observed values and those predicted by a model. Figure 5.2 presents examples of convex/concave and non-convex/non-concave functions. Non-convex/non-concave functions such as the one in figure 5.2b may have several *local optima*. Many functions of practical interest are non-convex/non-concave. As we will see in this chapter, convex optimization methods can also be employed for non-convex/non-concave functions given that we choose a proper starting location. This chapter presents the gradient ascent and Newton-Raphson methods, as well as practical tools to be employed with them. For full-depth details regarding optimization methods, the reader should refer to dedicated textbooks.[1]

## 5.1   Gradient Ascent

A *gradient* is a vector containing the partial derivatives of a function with respect to its variables. For a continuous function, the maximum is located at the point where its gradient equals zero. *Gradient ascent* is based on the principle that as long as we move in the direction of the gradient, we are moving toward a maximum. For the unidimensional case, we choose to move to a new position $\theta_{\text{new}}$ defined as the old value $\theta_{\text{old}}$ plus a search direction $d$ defined by a scaling factor $\lambda$ times the derivative estimated at $\theta_{\text{old}}$,

$$\theta_{\text{new}} = \theta_{\text{old}} + \underbrace{\lambda \cdot \tilde{f}'(\theta_{\text{old}})}_{d}.$$

A common practice for setting $\lambda$ is to employ *backtracking line search* where a new position is accepted if the *Armijo rule*[2] is satisfied so that

$$\tilde{f}(\theta_{\text{new}}) \geq \tilde{f}(\theta_{\text{old}}) + c \cdot d\tilde{f}'(\theta_{\text{old}}), \text{ with } c \in (0, 1). \quad (5.2)$$

Figure 5.3 presents a comparison of the application of equation 5.2 with the two extreme cases, $c = 0$ and $c = 1$. For $c = 1$, $\theta_{\text{new}}$ is only accepted if $\tilde{f}(\theta_{\text{new}})$ lies above the plane defined by the tangent at $\theta_{\text{old}}$. For $c = 0$, $\theta_{\text{new}}$ is only accepted if $\tilde{f}(\theta_{\text{new}}) > \tilde{f}(\theta_{\text{old}})$. The larger $c$ is, the stricter is the Armijo rule for ensuring that sufficient progress is made by the current step. With backtracking line search, we start from an initial value of $\lambda_0$ and reduce it until equation 5.2 is satisfied. Algorithm 1 presents a minimal version of the gradient ascent with backtracking line search.

---

**Algorithm 1:** Gradient ascent with backtracking line search

---

1  initialize $\lambda = \lambda_0$, $\theta_{\text{old}} = \theta_0$, define $\epsilon$, $c$, $\tilde{f}(\theta)$

2  **while** $|\tilde{f}'(\theta_{old})| > \epsilon$ **do**

3      compute $\begin{cases} \tilde{f}(\theta_{\text{old}}) & \text{(Function value)} \\ \tilde{f}'(\theta_{\text{old}}) & \text{(1}^{\text{st}}\text{ derivative)} \end{cases}$

4      compute $\theta_{\text{new}} = \theta_{\text{old}} + \underbrace{\lambda \tilde{f}'(\theta_{\text{old}})}_{d}$

5      **if** $\tilde{f}(\theta_{new}) < \tilde{f}(\theta_{old}) + c \cdot d\,\tilde{f}'(\theta_{old})$ **then**

6         assign $\lambda = \lambda/2$        (Backtracking)

7         Goto 4

8      assign $\lambda = \lambda_0$, $\theta_{\text{old}} = \theta_{\text{new}}$

9  $\theta^* = \theta_{\text{old}}$

---

Figure 5.4 presents the first two steps of the application of algorithm 1 to a non-convex/non-concave function with an initial value $\theta_0 = 3.5$ and a scaling factor $\lambda_0 = 3$. For the second step, the scaling factor $\lambda$ has to be reduced twice in order to satisfy the Armijo rule. One of the difficulties with gradient ascent is that the convergence speed depends on the choice of $\lambda_0$. If $\lambda_0$ is too small, several steps will be wasted and convergence will be slow. If $\lambda_0$ is too large, the algorithm may not converge.

Figure 5.5 presents a limitation common to all convex optimization methods when applied to functions involving local maxima; if the starting location $\theta_0$ is not located on the slope segment leading to the global maximum, the algorithm will most likely miss it and converge to a local maximum. The task of selecting a proper value $\boldsymbol{\theta}_0$ is nontrivial because in most cases, it is not possible to visualize $\tilde{f}(\boldsymbol{\theta})$. This issue can be tackled by attempting multiple starting locations $\boldsymbol{\theta}_0$ and by using domain knowledge to identify proper starting locations.

Gradient ascent can be applied to search for the maximum of a multivariate function by replacing the univariate derivative by the gradient so that

$$\boldsymbol{\theta}_{\text{new}} = \boldsymbol{\theta}_{\text{old}} + \lambda \cdot \nabla_{\boldsymbol{\theta}} \tilde{f}(\boldsymbol{\theta}_{\text{old}}).$$

As illustrated in figure 5.6, because gradient ascent follows the direction where the gradient is maximal, it often displays an oscillatory pattern. This issue can be mitigated by introducing a *momentum* term in the calculation of $\boldsymbol{\theta}_{\text{new}}$,[3]

$$\begin{aligned} \mathbf{v}_{\text{new}} &= \gamma \cdot \mathbf{v}_{\text{old}} + \lambda \cdot \nabla_{\boldsymbol{\theta}} \tilde{f}(\boldsymbol{\theta}_{\text{old}}), \\ \boldsymbol{\theta}_{\text{new}} &= \boldsymbol{\theta}_{\text{old}} + \mathbf{v}_{\text{new}} \end{aligned}$$

where $\mathbf{v}$ can be interpreted as a velocity that carries the momentum from the previous iterations.

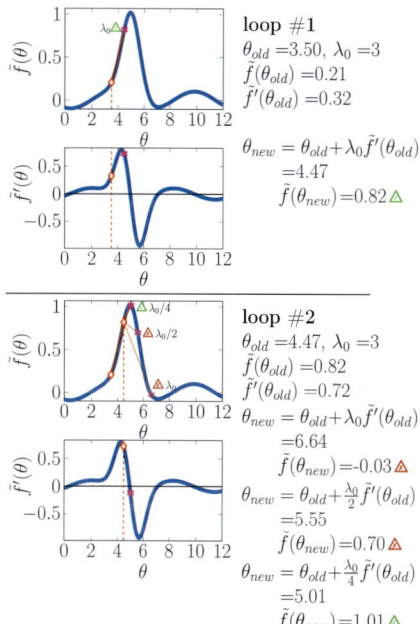

loop #1
$\theta_{old} = 3.50$, $\lambda_0 = 3$
$\tilde{f}(\theta_{old}) = 0.21$
$\tilde{f}'(\theta_{old}) = 0.32$

$\theta_{new} = \theta_{old} + \lambda_0 \tilde{f}'(\theta_{old})$
    $= 4.47$
$\tilde{f}(\theta_{new}) = 0.82$ △

loop #2
$\theta_{old} = 4.47$, $\lambda_0 = 3$
$\tilde{f}(\theta_{old}) = 0.82$
$\tilde{f}'(\theta_{old}) = 0.72$
$\theta_{new} = \theta_{old} + \lambda_0 \tilde{f}'(\theta_{old})$
    $= 6.64$
$\tilde{f}(\theta_{new}) = -0.03$ △
$\theta_{new} = \theta_{old} + \frac{\lambda_0}{2} \tilde{f}'(\theta_{old})$
    $= 5.55$
$\tilde{f}(\theta_{new}) = 0.70$ △
$\theta_{new} = \theta_{old} + \frac{\lambda_0}{4} \tilde{f}'(\theta_{old})$
    $= 5.01$
$\tilde{f}(\theta_{new}) = 1.01$ △

Figure 5.4: Example of application of gradient ascent with backtracking for finding the maximum of a function.

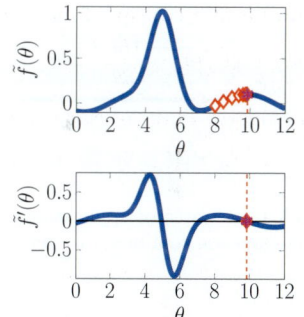

Figure 5.5: Example of application of gradient ascent converging to a local maximum for a function.

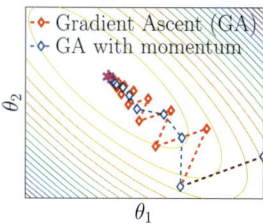

Figure 5.6: Comparison of gradient ascent with and without momentum.

[3] Rumelhart, D. E., G. E. Hinton, and R. J. Williams (1986). Learning representations by back-propagating errors. *Nature 323*, 533–536.

**Second-order derivatives**

$$\tilde{f}''(\theta) \equiv \frac{d^2\tilde{f}(\theta)}{d\theta^2}$$

$$\tilde{f}_i''(\boldsymbol{\theta}) \equiv \frac{\partial^2\tilde{f}(\boldsymbol{\theta})}{\partial\theta_i^2}$$

**Hessian**

$$\left[\mathbf{H}[\tilde{f}(\boldsymbol{\theta})]\right]_{ij} = \frac{\partial^2\tilde{f}(\boldsymbol{\theta})}{\partial\theta_i\partial\theta_j}$$

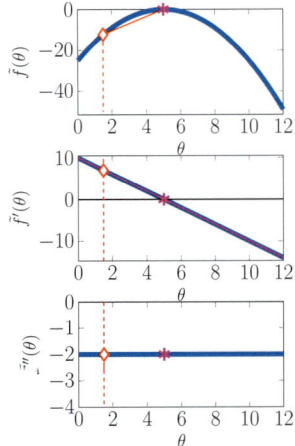

Figure 5.7: Example of application of Newton-Raphson to a quadratic function.

## 5.2   Newton-Raphson

The *Newton-Raphson* method allows us to adaptively scale the search direction vector using the second-order derivative $\tilde{f}''(\theta)$. Knowing that the maximum of a function corresponds to the point where the gradient is zero, $\tilde{f}'(\theta) = 0$, we can find this maximum by formulating a linearized gradient equation using the second-order derivative of $\tilde{f}(\theta)$ and then set it equal to zero. The analytic formulation for the linearized gradient function (see §3.4.2) approximated at the current location $\theta_{\text{old}}$ is

$$\tilde{f}'(\theta) \approx \tilde{f}''(\theta_{\text{old}}) \cdot (\theta - \theta_{\text{old}}) + \tilde{f}'(\theta_{\text{old}}). \tag{5.3}$$

We can estimate $\theta_{\text{new}}$ by setting equation 5.3 equal to zero, and then by solving for $\theta$, we obtain

$$\theta_{\text{new}} = \theta_{\text{old}} - \frac{\tilde{f}'(\theta_{\text{old}})}{\tilde{f}''(\theta_{\text{old}})}. \tag{5.4}$$

Let us consider the case where we want to find the maximum of a *quadratic function* (i.e., $\propto x^2$), as illustrated in figure 5.7. In the case of a quadratic function, the algorithm converges to the exact solution in one iteration, no matter the starting point, because the gradient of a quadratic function is exactly described by the linear function in equation 5.3.

Algorithm 2 presents a minimal version of the Newton-Raphson method with backtracking line search. Note that at line 6, there is again a scaling factor $\lambda$, which is employed because the Newton-Raphson method is exact only for quadratic functions. For more general non-convex/non-concave functions, the linearized gradient is an approximation such that a value of $\lambda = 1$ will not always lead to a $\theta_{\text{new}}$ satisfying the Armijo rule in equation 5.2.

Figure 5.8 presents the application of algorithm 2 to a non-convex/non-concave function with an initial value $\theta_0 = 3.5$ and a scaling factor $\lambda_0 = 1$. For each loop, the pink solid line represents the linearized gradient function formulated in equation 5.3. Notice how, for the first two iterations, the second derivative $\tilde{f}''(\theta) > 0$. Having a positive second derivative indicates that the linearization of $\tilde{f}'(\theta)$ equals zero for a minimum rather than for a maximum. One simple option in this situation is to define $\lambda = -\lambda$ in order to ensure that the next step moves in the same direction as the gradient. The convergence with Newton-Raphson is typically faster than with gradient ascent.

---

**Algorithm 2:** Newton-Raphson with backtracking line search

1   initialize $\lambda = \lambda_0 = 1$, $\theta_{old} = \theta_0$, define $\epsilon$, $c$, $\tilde{f}(\theta)$

2   **while** $|\tilde{f}'(\theta_{old})| > \epsilon$ **do**

3     compute:   $\theta_{old} \rightarrow$

$$\begin{cases} \tilde{f}(\theta_{old}) & \text{(Function evaluation)} \\ \tilde{f}'(\theta_{old}) & \text{(First derivative)} \\ \tilde{f}''(\theta_{old}) & \text{(Second derivative)} \end{cases}$$

4     **if** $\tilde{f}''(\theta_{old}) > 0$ **then**

5      $\lambda = -\lambda$

6     compute $\theta_{new} = \theta_{old} - \lambda \cdot \underbrace{\dfrac{\tilde{f}'(\theta_{old})}{\tilde{f}''(\theta_{old})}}_{d}$

7     **if** $\tilde{f}(\theta_{new}) < \tilde{f}(\theta_{old}) + c \cdot d\tilde{f}'(\theta_{old})$ **then**

8      assign $\lambda = \lambda/2$             (Backtracking)

9      Goto 6

10     assign $\lambda = \lambda_0$, $\theta_{old} = \theta_{new}$

11   $\theta^* = \theta_{old}$

---

The Newton-Raphson algorithm can be employed for identifying the optimal values $\boldsymbol{\theta}^* = [\theta_1^* \; \theta_2^* \; \cdots \; \theta_n^*]^{\mathsf{T}}$ in domains having multiple dimensions. The equation 5.4 developed for univariate cases can be extended for $n$-dimensional domains by following

$$\boldsymbol{\theta}_{new} = \boldsymbol{\theta}_{old} - \mathbf{H}[\tilde{f}(\boldsymbol{\theta}_{old})]^{-1} \cdot \nabla \tilde{f}(\boldsymbol{\theta}_{old}). \tag{5.5}$$

$\mathbf{H}[\tilde{f}(\boldsymbol{\theta})]$ denotes the $n \times n$ *Hessian* matrix containing the second-order partial derivatives for the function $\tilde{f}(\boldsymbol{\theta})$ evaluated at $\boldsymbol{\theta}$. The Hessian is a symmetric matrix where each term is defined by

$$\begin{aligned} \left[\mathbf{H}[\tilde{f}(\boldsymbol{\theta})]\right]_{ij} &= \frac{\partial^2 \tilde{f}(\boldsymbol{\theta})}{\partial \theta_i \partial \theta_j} \\ &= \frac{\partial}{\partial \theta_i}\left(\frac{\partial \tilde{f}(\boldsymbol{\theta})}{\partial \theta_j}\right) \\ &= \frac{\partial}{\partial \theta_j}\left(\frac{\partial \tilde{f}(\boldsymbol{\theta})}{\partial \theta_i}\right). \end{aligned}$$

When the terms on the main diagonal of the Hessian matrix are positive, it is an indication that our linearized gradient points toward a minimum rather than a maximum. As we did in algorithm 2, it is then possible to move toward the maximum by reversing the search direction. One issue with gradient-based multi-dimensional optimization is *saddle points*. Figure 5.9 presents an example of a saddle point in a function for which one second-order partial derivative is negative $\left(\frac{\partial^2 \tilde{f}(\boldsymbol{\theta})}{\partial \theta_1^2} < 0\right)$ and the other is positive $\left(\frac{\partial^2 \tilde{f}(\boldsymbol{\theta})}{\partial \theta_2^2} > 0\right)$.

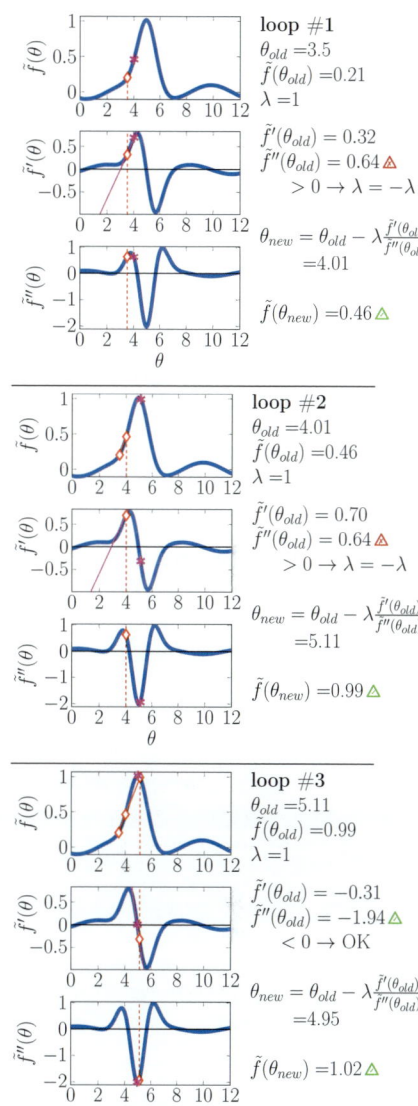

loop #1
$\theta_{old} = 3.5$
$\tilde{f}(\theta_{old}) = 0.21$
$\lambda = 1$

$\tilde{f}'(\theta_{old}) = 0.32$
$\tilde{f}''(\theta_{old}) = 0.64 \triangle$
$> 0 \rightarrow \lambda = -\lambda$

$\theta_{new} = \theta_{old} - \lambda \frac{\tilde{f}'(\theta_{old})}{\tilde{f}''(\theta_{old})}$
$= 4.01$

$\tilde{f}(\theta_{new}) = 0.46 \triangle$

loop #2
$\theta_{old} = 4.01$
$\tilde{f}(\theta_{old}) = 0.46$
$\lambda = 1$

$\tilde{f}'(\theta_{old}) = 0.70$
$\tilde{f}''(\theta_{old}) = 0.64 \triangle$
$> 0 \rightarrow \lambda = -\lambda$

$\theta_{new} = \theta_{old} - \lambda \frac{\tilde{f}'(\theta_{old})}{\tilde{f}''(\theta_{old})}$
$= 5.11$

$\tilde{f}(\theta_{new}) = 0.99 \triangle$

loop #3
$\theta_{old} = 5.11$
$\tilde{f}(\theta_{old}) = 0.99$
$\lambda = 1$

$\tilde{f}'(\theta_{old}) = -0.31$
$\tilde{f}''(\theta_{old}) = -1.94 \triangle$
$< 0 \rightarrow$ OK

$\theta_{new} = \theta_{old} - \lambda \frac{\tilde{f}'(\theta_{old})}{\tilde{f}''(\theta_{old})}$
$= 4.95$

$\tilde{f}(\theta_{new}) = 1.02 \triangle$

Figure 5.8: Example of application of the Newton-Raphson algorithm with backtracking line search for finding the maximum of a function.

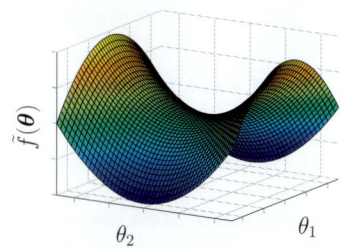

Figure 5.9: Example of saddle point $\boldsymbol{\theta}$ where $\frac{\partial^2 \tilde{f}(\boldsymbol{\theta})}{\partial \theta_1^2} < 0$ and $\frac{\partial^2 \tilde{f}(\boldsymbol{\theta})}{\partial \theta_2^2} > 0$.

In such a case, one option is to regularize the Hessian matrix by substracting a constant $\alpha$ on its main diagonal in order to ensure that all terms are negative:

$$\tilde{\mathbf{H}}[\tilde{f}(\boldsymbol{\theta})] = \mathbf{H}[\tilde{f}(\boldsymbol{\theta})] - \alpha\mathbf{I}.$$

Avoiding being trapped in saddle points is a key challenge in optimization. The reader interested in more advanced strategies for that purpose should consult specialized literature.[4]

In practice, second-order methods such as Newton-Raphson can be employed when we have access to an analytical formulation for evaluating the Hessian, $\mathbf{H}[\tilde{f}(\boldsymbol{\theta})]$. Otherwise, if we have to employ numerical derivatives (see §5.4) to estimate the Hessian, the computational demand quickly becomes prohibitive as the number of variables increases. For example, if the number of variables is $n = 100$, by considering the symmetry in the Hessian matrix, there are $100 + \frac{1}{2}(100 \times 100 - 100) = 5050$ second-order partial derivatives to estimate for each Newton-Raphson iteration. Even for computationally efficient functions $\tilde{f}(\boldsymbol{\theta})$, the necessary number of evaluations becomes a challenge. In addition, for a large $n$, there is a substantial effort required to invert the Hessian in equation 5.5. Therefore, even if Newton-Raphson is more efficient in terms of the number of iterations required to reach convergence, this does not take into account the time necessary to obtain the second-order derivatives. When dealing with a large number of variables or for functions $\tilde{f}(\boldsymbol{\theta})$ that are computationally expensive, we may revert to using momentum-based gradient-ascent methods.

## 5.3   Coordinate Ascent

One alternative to the methods we have seen for multivariate optimization is to perform the search for each variable separately, which corresponds to solving a succession of 1D optimization problems. This approach is known as *coordinate optimization*.[5] This approach is known to perform well when there is no strong coupling between parameters with respect to the values of the objective function. Algorithm 3 presents a minimal version of the coordinate ascent Newton-Raphson using backtracking line search. Figure 5.10 presents two intermediate steps illustrating the application of algorithm 3 to a bidimensional non-convex/non-concave function.

[4] Nocedal, J. and S. Wright (2006). *Numerical optimization*. Springer Science & Business Media; Dauphin, Y. N., R. Pascanu, C. Gulcehre, K. Cho, S. Ganguli, and Y. Bengio (2014). Identifying and attacking the saddle point problem in high-dimensional non-convex optimization. In *Advances in Neural Information Processing Systems*, 27, 2933–2941; and Goodfellow, I., Y. Bengio, and A. Courville (2016). *Deep learning*. MIT Press.

[5] Wright, S. J. (2015). Coordinate descent algorithms. *Mathematical Programming 151*(1), 3–34; and Friedman, J., T. Hastie, H. Höfling, and R. Tibshirani (2007). Pathwise coordinate optimization. *The Annals of Applied Statistics 1*(2), 302–332.

---

**Algorithm 3:** Coordinate ascent using Newton-Raphson

1  initialize $\lambda = \lambda_0 = 1$, $\boldsymbol{\theta}_{\text{old}} = \boldsymbol{\theta}_0 = [\theta_1\ \theta_2\ \cdots\ \theta_n]^\mathsf{T}$
2  define $\epsilon$, $c$, $\tilde{f}(\boldsymbol{\theta})$
3  **while** $\|\nabla_{\boldsymbol{\theta}} \tilde{f}(\boldsymbol{\theta}_{old})\| > \epsilon$ **do**
4  $\quad$ **for** $i \in \{1 : n\}$ **do**
5  $\quad\quad$ compute $\begin{cases} \tilde{f}(\boldsymbol{\theta}_{\text{old}}) & \text{(Function evaluation)} \\ \tilde{f}'_i(\boldsymbol{\theta}_{\text{old}}) & (i^{\text{th}}\ 1^{\text{st}}\ \text{partial derivative}) \\ \tilde{f}''_i(\boldsymbol{\theta}_{\text{old}}) & (i^{\text{th}}\ 2^{\text{nd}}\ \text{partial derivative}) \end{cases}$
6  $\quad\quad$ **if** $\tilde{f}''_i(\boldsymbol{\theta}_{old}) > 0$ **then**
7  $\quad\quad\quad$ $\lambda = -\lambda$
8  $\quad\quad$ compute $[\boldsymbol{\theta}_{\text{new}}]_i = [\boldsymbol{\theta}_{\text{old}}]_i - \lambda \underbrace{\dfrac{\tilde{f}'_i(\boldsymbol{\theta}_{\text{old}})}{\tilde{f}''_i(\boldsymbol{\theta}_{\text{old}})}}_{d}$
9  $\quad\quad$ **if** $\tilde{f}(\boldsymbol{\theta}_{new}) < \tilde{f}(\boldsymbol{\theta}_{old}) + c \cdot d\,\tilde{f}'_i(\boldsymbol{\theta}_{old})$ **then**
10 $\quad\quad\quad$ assign $\lambda = \lambda/2$ $\qquad\qquad$ (Backtracking)
11 $\quad\quad\quad$ Goto 8
12 $\quad\quad$ assign $\lambda = \lambda_0$, $\boldsymbol{\theta}_{\text{old}} = \boldsymbol{\theta}_{\text{new}}$
13 $\boldsymbol{\theta}^* = \boldsymbol{\theta}_{\text{new}}$

---

## 5.4 Numerical Derivatives

One aspect that was omitted in the previous sections is *How do we obtain the first- and second-order derivatives?* When a twice differentiable formulation for $\tilde{f}(\boldsymbol{\theta})$ exists, $\frac{\partial \tilde{f}(\boldsymbol{\theta})}{\partial \theta_i}$ and $\frac{\partial^2 \tilde{f}(\boldsymbol{\theta})}{\partial \theta_i^2}$ can be expressed analytically. When analytic formulations are not available, derivatives can be estimated numerically using either a *forward*, *backward*, or *central differentiation* scheme.[6] Here, we only focus on the *central differentiation method*. Note that forward and backward differentiations are not as accurate as central, yet they are computationally cheaper. As illustrated in figure 5.11, first- and second-order partial derivatives of $\tilde{f}(\boldsymbol{\theta})$ with respect to the $i^{\text{th}}$ element of a vector $\boldsymbol{\theta} = [\theta_1\ \theta_2\ \cdots\ \theta_n]^\mathsf{T}$ are given by

$$\tilde{f}'_i(\boldsymbol{\theta}) = \frac{\partial \tilde{f}(\boldsymbol{\theta})}{\partial \theta_i} \approx \frac{\tilde{f}(\boldsymbol{\theta} + \mathbb{I}(i)\Delta\theta) - \tilde{f}(\boldsymbol{\theta} - \mathbb{I}(i)\Delta\theta)}{2\Delta\theta}$$

$$\tilde{f}''_i(\boldsymbol{\theta}) = \frac{\partial \tilde{f}(\boldsymbol{\theta})}{\partial \theta_i^2} \approx \frac{\dfrac{\tilde{f}(\boldsymbol{\theta} + \mathbb{I}(i)\Delta\theta) - \tilde{f}(\boldsymbol{\theta})}{\Delta\theta} - \dfrac{\tilde{f}(\boldsymbol{\theta}) - \tilde{f}(\boldsymbol{\theta} - \mathbb{I}(i)\Delta\theta)}{\Delta\theta}}{\Delta\theta}$$

$$= \frac{\tilde{f}(\boldsymbol{\theta} + \mathbb{I}(i)\Delta\theta) - 2\tilde{f}(\boldsymbol{\theta}) + \tilde{f}(\boldsymbol{\theta} - \mathbb{I}(i)\Delta\theta)}{(\Delta\theta)^2},$$

where $\Delta\theta \ll \theta$ is a small perturbation to the value $\theta_i$ and $\mathbb{I}(i)$ is a

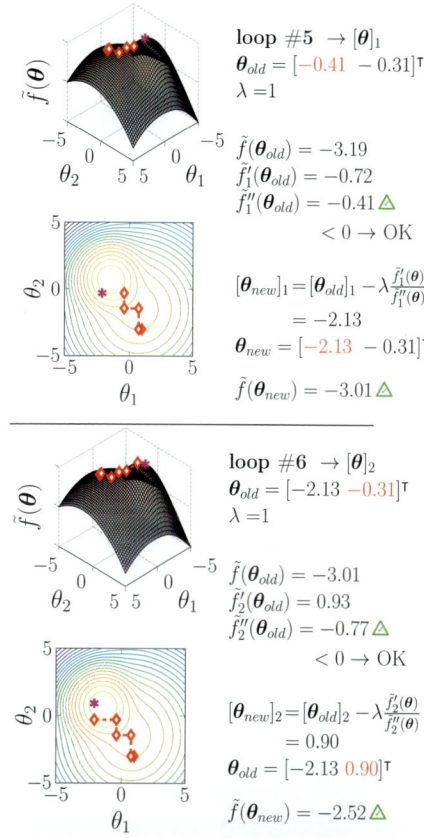

loop #5 $\to [\boldsymbol{\theta}]_1$
$\boldsymbol{\theta}_{old} = [-0.41\ -0.31]^\mathsf{T}$
$\lambda = 1$

$\tilde{f}(\boldsymbol{\theta}_{old}) = -3.19$
$\tilde{f}'_1(\boldsymbol{\theta}_{old}) = -0.72$
$\tilde{f}''_1(\boldsymbol{\theta}_{old}) = -0.41\ \triangle$
$\qquad\qquad < 0 \to \text{OK}$

$[\boldsymbol{\theta}_{new}]_1 = [\boldsymbol{\theta}_{old}]_1 - \lambda \frac{\tilde{f}'_1(\boldsymbol{\theta})}{\tilde{f}''_1(\boldsymbol{\theta})}$
$\qquad = -2.13$
$\boldsymbol{\theta}_{new} = [-2.13\ -0.31]^\mathsf{T}$

$\tilde{f}(\boldsymbol{\theta}_{new}) = -3.01\ \triangle$

loop #6 $\to [\boldsymbol{\theta}]_2$
$\boldsymbol{\theta}_{old} = [-2.13\ -0.31]^\mathsf{T}$
$\lambda = 1$

$\tilde{f}(\boldsymbol{\theta}_{old}) = -3.01$
$\tilde{f}'_2(\boldsymbol{\theta}_{old}) = 0.93$
$\tilde{f}''_2(\boldsymbol{\theta}_{old}) = -0.77\ \triangle$
$\qquad\qquad < 0 \to \text{OK}$

$[\boldsymbol{\theta}_{new}]_2 = [\boldsymbol{\theta}_{old}]_2 - \lambda \frac{\tilde{f}'_2(\boldsymbol{\theta})}{\tilde{f}''_2(\boldsymbol{\theta})}$
$\qquad = 0.90$
$\boldsymbol{\theta}_{old} = [-2.13\ 0.90]^\mathsf{T}$

$\tilde{f}(\boldsymbol{\theta}_{new}) = -2.52\ \triangle$

Figure 5.10: Example of application of the coordinate ascent Newton-Raphson algorithm with backtracking for finding the maximum of a 2-D function.

[6] Nocedal, J. and S. Wright (2006). *Numerical optimization.* Springer Science & Business Media; and Abramowitz, M. and I. A. Stegun (1972). *Handbook of mathematical functions with formulas, graphs, and mathematical table.* National Bureau of Standards, Applied Mathematics.

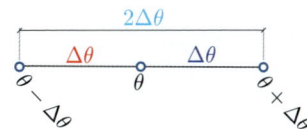

Figure 5.11: Illustration of 1-D numerical derivatives.

**Examples**

$\mathbb{I}(3) = [0\ 0\ 1\ 0\ \cdots\ 0]^\mathsf{T}$
$\mathbb{I}(1) = [1\ 0\ 0\ 0\ \cdots\ 0]^\mathsf{T}$

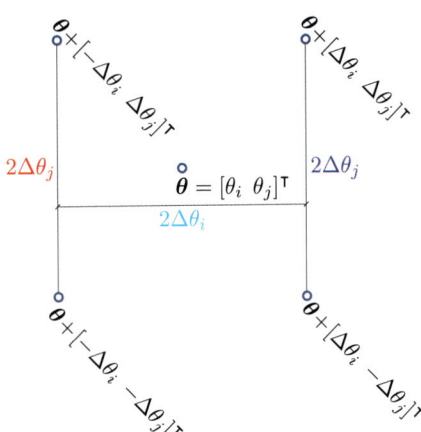

Figure 5.12: Illustration of 2-D partial numerical derivatives.

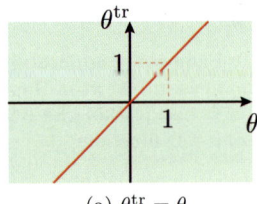

(a) $\theta^{\mathrm{tr}} = \theta$

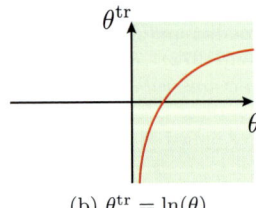

(b) $\theta^{\mathrm{tr}} = \ln(\theta)$

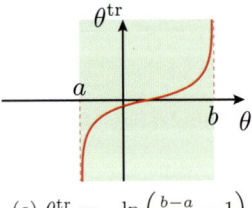

(c) $\theta^{\mathrm{tr}} = -\ln\left(\frac{b-a}{\theta-a} - 1\right)$

Figure 5.13: Examples of transformation functions.

$n \times 1$ indicator vector, for which all values are equal to 0, except the $i^{\mathrm{th}}$, which is equal to one.

As illustrated in figure 5.12, numerical derivatives can also be employed to estimate each term of the Hessian matrix,

$$\left[\mathbf{H}[\tilde{f}(\boldsymbol{\theta})]\right]_{ij} \approx \frac{\frac{\partial \tilde{f}(\boldsymbol{\theta}+\Delta\theta)}{\partial \theta_j} - \frac{\partial \tilde{f}(\boldsymbol{\theta}-\Delta\theta)}{\partial \theta_j}}{2\Delta\theta_i},$$

where terms on the numerator are defined as

$$\frac{\partial \tilde{f}(\boldsymbol{\theta}+\Delta\theta)}{\partial \theta_j} = \frac{\tilde{f}(\boldsymbol{\theta}+\mathbb{I}(i)\Delta\theta_i+\mathbb{I}(j)\Delta\theta_j) - \tilde{f}(\boldsymbol{\theta}+\mathbb{I}(i)\Delta\theta_i-\mathbb{I}(j)\Delta\theta_j)}{2\Delta\theta_j}$$

$$\frac{\partial \tilde{f}(\boldsymbol{\theta}-\Delta\theta)}{\partial \theta_j} = \frac{\tilde{f}(\boldsymbol{\theta}-\mathbb{I}(i)\Delta\theta_i+\mathbb{I}(j)\Delta\theta_j) - \tilde{f}(\boldsymbol{\theta}-\mathbb{I}(i)\Delta\theta_i-\mathbb{I}(j)\Delta\theta_j)}{2\Delta\theta_j}.$$

In practice, the full Hessian matrix can only be estimated numerically when the number of variables $n$ is small or when evaluating $\tilde{f}(\boldsymbol{\theta})$ is computationally cheap.

## 5.5   Parameter-Space Transformation

When optimizing using either the gradient ascent or the Newton-Raphson method, we are likely to run into difficulties for parameters $\theta$ that are not defined in an unbounded space. In such a case, the efficiency is hindered because the algorithms may propose new positions $\theta_{\mathrm{new}}$ that lie outside the valid domain. When trying to identify optimal parameters for probability distributions such as those described in chapter 4, common domains for parameters are as follows:

- Mean parameters: $\mu \in \mathbb{R}$

- Standard deviations: $\sigma \in \mathbb{R}^+$

- Correlation coefficients: $\rho \in (-1, 1)$

- Probability: $\Pr(X = x) \in (0, 1)$

One solution to this problem is to perform the optimization in a transformed space $\theta^{\mathrm{tr}}$ such that

$$\theta^{\mathrm{tr}} = g(\theta) \in \mathbb{R}.$$

For each $\theta$, the choice of transformation function $g(\theta)$ depends on its domain. Figure 5.13 presents examples of transformations for $\theta \in \mathbb{R}$, $\theta \in \mathbb{R}^+$, and $\theta \in (a, b)$. Note that in the simplest case, where $\theta \in \mathbb{R}$, no transformation is required, so $\theta^{\mathrm{tr}} = \theta$.

For $\theta \in \mathbb{R}^+$, a common transformation is to take the logarithm $\theta^{\mathrm{tr}} = \ln(\theta)$, and its inverse transformation is $\theta = e^{\theta^{\mathrm{tr}}}$. The analytical derivatives for the transformation and its inverse are

$$\frac{d\theta^{\mathrm{tr}}}{d\theta} = \frac{1}{\theta}, \ \frac{d\theta}{d\theta^{\mathrm{tr}}} = e^{\theta^{\mathrm{tr}}}.$$

For parameters bounded in an interval $\theta \in (a, b)$, a possible transformation is the scaled *logistic sigmoid* function

$$\theta^{\mathrm{tr}} = -\ln\left(\frac{b-a}{\theta-a} - 1\right),$$

and its inverse is given by

$$\theta = \frac{b-a}{1 + e^{-\theta^{\mathrm{tr}}}} + a.$$

The derivative of the transformation and its inverse are

$$\frac{d\theta}{d\theta^{\mathrm{tr}}} = \frac{a-b}{(\theta-a)(\theta-b)}$$

$$\frac{d\theta^{\mathrm{tr}}}{d\theta} = \frac{b-a}{(1 + e^{-\theta^{\mathrm{tr}}})^2} e^{-\theta^{\mathrm{tr}}}.$$

Note that the derivative of these transformations will be employed in chapter 7 when performing parameter-space transformations using the change-of-variable rule we have seen in §3.4. The transformations presented here are not unique, as many other functions can be employed. For further details about parameter space transformations, the reader is invited to refer to Gelman et al.[7]

[7] Gelman, A., J. B. Carlin, H. S. Stern, and D. B. Rubin (2014). *Bayesian data analysis* (3rd ed.). CRC Press.

## Exercises

P5.1  Explain the differences between gradient ascent and Newton-Raphson and how each of them works.

P5.2  In what case is gradient ascent better suited than Newton-Raphson?

P5.3  With convex optimization algorithms, what is the purpose of performing the optimization in a transformed space?

P5.4  Implement a gradient ascent algorithm to find the value $x$ that maximizes the function $\tilde{f}(\theta) = \exp(-(\theta - 5)^2) + 0.1\sin(\theta - 2)$ in the interval $(0, 12)$. Use as a starting point $\theta_0 = -3$. Check your results with figure 5.4.

P5.5  Implement a Newton-Raphson algorithm to find the value $\theta$ that maximizes the function and setup from P5.4. Check your results with figure 5.8.

P5.6  Repeat P5.5 to confirm that you reach the same result while using numerical derivatives rather than analytical ones.

P5.7  Implement a 2-D coordinate-ascent Newton-Raphson algorithm to find the value $\boldsymbol{\theta}$ that maximizes the function $\tilde{f}(\boldsymbol{\theta}) = 0.6 \cdot \mathcal{N}\left(\boldsymbol{\theta}, [-1.5\ 0.75]^{\mathsf{T}}, \mathrm{diag}([1\ 1]^2)\right) + 0.4 \cdot \mathcal{N}\left(\boldsymbol{\theta}, [0.75\ -1.5]^{\mathsf{T}}, \mathrm{diag}([2\ 2]^2)\right)$. Use as a starting point $\boldsymbol{\theta}_0 = [1\ -3]^{\mathsf{T}}$. Check your results with figure 5.10.

# Part II

# Bayesian Estimation

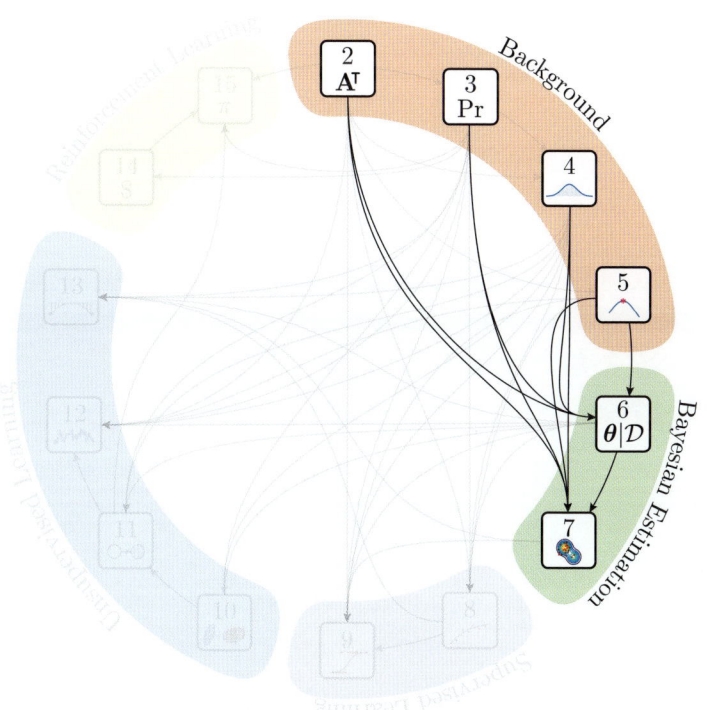

# 6

# *Learning from Data*

When scalars $x$ are employed to model physical phenomena, we can use empirical observations $\mathbf{y}$ to learn about the probability (or probability density) for each value that $x$ can take. When $X$ is a random variable, we can use observations to learn about the parameters $\boldsymbol{\theta}$ describing its probability density (or mass) function $f(x; \boldsymbol{\theta})$. This chapter presents the general Bayesian formulation, as well as approximate methods derived from it, for learning about unknown state variables and parameters, given known data. The applications presented in this chapter focus on simple models in order to keep the attention on the Bayesian methods themselves. We will see in chapters 8–13 how this theory can be generalized for applications to machine learning models having complex architectures. For an in-depth review of Bayesian methods applied to data analysis, the reader may refer to specialized textbooks such as those by Gelman et al.[1] or Box and Tiao.[2]

## 6.1 *Bayes*

Learning from data can be seen as an indirect problem; we observe some quantities $\mathbf{y}$, and we want to employ this information to infer knowledge about hidden-state variables $\mathbf{x}$ and model parameters $\boldsymbol{\theta}$, where $\mathbf{y}$ depends on $\mathbf{x}$ and $\boldsymbol{\theta}$ through $f(\mathbf{y}|\mathbf{x}; \boldsymbol{\theta})$. In chapter 3, we saw that such a dependence can be treated using *conditional probabilities*.

In the first part of this chapter, we solely focus on the estimation of hidden-state variables $x$. Given two random variables $X \sim f(x)$ and $Y \sim f(y)$, their joint probability density function (PDF) is obtained by the product of a conditional and a marginal PDF,

$$f(x,y) = \begin{cases} f(x|y) \cdot f(y) \\ f(y|x) \cdot f(x). \end{cases} \tag{6.1}$$

**Learning context**

$$p(\text{unknown}|\text{known}) \begin{cases} \nearrow y\text{: Observation} \\ \\ \searrow x\text{: Constant} \\ \searrow X\text{: Random variable} \\ \searrow \boldsymbol{\theta}\text{: } X \sim f(x; \boldsymbol{\theta}) \end{cases}$$

[1] Gelman, A., J. B. Carlin, H. S. Stern, and D. B. Rubin (2014). *Bayesian data analysis* (3rd ed.). CRC Press.

[2] Box, G. E. P. and G. C. Tiao (1992). *Bayesian inference in statistical analysis*. Wiley.

**Note:** The term *hidden* describes a state variable that is not directly observed.

**Notation:** ; and | are both employed to describe a conditional probability or probability density. The distinction is that ; is employed when $\boldsymbol{\theta}$ are parameters of a PDF or PMF, for example, $f(x; \boldsymbol{\theta}) = \mathcal{N}(x; \mu, \sigma^2)$, and | denotes a conditional dependence between variables, for example, $f(y|x)$. Besides these semantic distinctions, both are employed in the same way when it comes to performing calculations.

In the case where $y$ is known and $X$ is not, we can reorganize the terms of equation 6.1 in order to obtain *Bayes rule*,

$$f(x|y) = \frac{f(y|x) \cdot f(x)}{f(y)},$$

which describes the posterior PDF (i.e., our posterior knowledge) of $X$ given that we have observed $y$. Let us consider $X$, a random variable so that $X \sim f(x)$, and given a set of observations $\mathcal{D} = \{y_1, y_2, \cdots, y_\mathtt{D}\}$ that are realizations of the random variables $\mathbf{Y} = [Y_1 \; Y_2 \; \cdots \; Y_\mathtt{D}]^\intercal \sim f(\mathbf{y})$. Our posterior knowledge for $X$ given the observations $\mathcal{D}$ is described by the conditional PDF

$$\underbrace{f(x|\mathcal{D})}_{\text{posterior}} = \frac{\overbrace{f(\mathcal{D}|x)}^{\text{likelihood}} \cdot \overbrace{f(x)}^{\text{prior}}}{\underbrace{f(\mathcal{D})}_{\text{evidence}}}.$$

**Note:** From now on, the number of elements in a set or the number of variables in a vector is defined by a `typewriter`-font upper-case letter; for example, $\mathtt{D}$ is the number of observations in a set $\mathcal{D}$, and $\mathtt{X}$ is the number of variables in the vector $\mathbf{x} = [x_1 \; x_2 \; \cdots \; x_\mathtt{X}]^\intercal$.

**A set of observations**
$\mathcal{D} = \{y_1, y_2, \cdots, y_\mathtt{D}\}$
e.g., $\mathcal{D} = \{1.2, 3.8, \cdots, 0.22\}, (y \in \mathbb{R})$
$\quad\quad \mathcal{D} = \{3, 1, \cdots, 6\}, (y \in \mathbb{Z}^+)$
$\quad\quad \mathcal{D} = \{\text{blue, blue}, \cdots, \text{red}\},$
$\quad\quad\quad\quad (y \in \{\text{blue, red, green}\})$

*Prior*  $f(x)$ describes our *prior knowledge* for the values $x$ that a random variable $X$ can take. The prior knowledge can be expressed in multiple ways. For instance, it can be based on heuristics such as expert opinions. In the case where data is obtained sequentially, the posterior knowledge at time $t - 1$ becomes the prior at time $t$. In some cases, it also happens that no prior knowledge is available; then, we should employ a *non-informative* prior, that is, a prior that reflects an absence of knowledge (see §6.4.1 for further details on non-informative priors).

**Note:** A *uniform* prior and a *non-informative* prior are not the same thing. Some non-informative priors are uniform, but not all uniform priors are non-informative.

*Likelihood*  $f(\mathbf{Y} = \mathbf{y}|x) \equiv f(\mathcal{D}|x)$ describes the *likelihood*, or the conditional probability density, of observing the event $\{\mathbf{Y} = \mathcal{D}\}$, given the values that $x$ can take. Note that in the special case of an exact observation where $\mathcal{D} = y = x$, then

$$f(\mathcal{D}|x) = \begin{cases} 1, & y = x \\ 0, & \text{otherwise.} \end{cases}$$

In such a case, the observations are exact so the prior does not play a role; by observing $y$ you have all the information about $x$ because $y = x$. In the general case where $y : Y = \text{fct}(x)$, that is, $y$ is a realization from a stochastic process, which is a function of $x$—$f(\mathcal{D}|x)$ describes the *prior* probability density of observing $\mathcal{D}$ given a specific set of values for $x$.

Here, the term *prior* refers to our state of knowledge that has not yet been influenced by the observations in $\mathcal{D}$.

*Evidence*  $f(\mathbf{Y} = \mathbf{y}) \equiv f(\mathcal{D})$ is called the *evidence* and consists in a *normalization constant* ensuring that the posterior PDF integrates to 1. In §3.3 we saw that for both the discrete and the continuous cases,

$$\underbrace{\sum_x p(x|\mathcal{D}) = 1}_{\text{discrete case}}, \quad \underbrace{\int_{-\infty}^{\infty} f(x|\mathcal{D})dx = 1}_{\text{continuous case}}.$$

For that property to be true, the normalization constant is defined as the sum/integral over the entire domain of $x$, of the likelihood times the prior,

$$\underbrace{p(\mathcal{D}) = \sum_x p(\mathcal{D}|x) \cdot p(x)}_{\text{discrete case}}, \quad \underbrace{f(\mathcal{D}) = \int f(\mathcal{D}|x) \cdot f(x)dx}_{\text{continuous case}}.$$

So far, the description of learning as an indirect problem is rather abstract. The next sections will apply this concept to simple problem configurations involving discrete and continuous random variables.

## 6.2  Discrete State Variables

This section presents how the probability of occurrence for discrete state variables are estimated from observations. In all the examples presented in this section, the uncertainty associated with our knowledge of a *hidden* random variable $X$ is epistemic because a *true value* $\check{x}$ exists, yet it is unknown. Moreover, $x$ is not directly observable; we can only observe $y$, where $p(y|x)$ describes the conditional probability of an observation $y$ given the value of the discrete hidden-state variable $x$.

### 6.2.1  Example: Disease Screening

Figure 6.1 illustrates the context of a rare disease for which the probability of occurrence in the general population is 1 percent, and a screening test such that if one has the disease, the probability of testing positive is $\Pr(\text{test}+|\text{disease}) = 0.8$; and if one does not have the disease, the probability of testing positive is $\Pr(\text{test}+|\neg\text{disease}) = 0.1$. The sample space for the variable describing the hidden state we are interested in is $x \in \{\text{disease}, \neg\text{disease}\}$, for which the prior probability is $\Pr(X = \text{disease}) = 0.01$ and $\Pr(X = \neg\text{disease}) = 1 - \Pr(X = \text{disease}) = 0.99$.

For a single observation, $\mathcal{D} = \{\text{test}+\}$, the posterior probability

**Discrete state variable**: A discrete random variable $X$ employed to represent the uncertainty regarding the state $x$ of a system.

*Reminder:* The term *hidden* refers to a variable that is not observed.

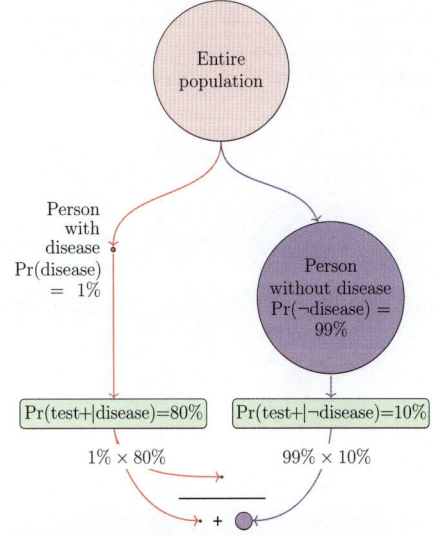

Figure 6.1: Visual interpretation of Bayes rule in the context of disease screening. (Adapted from John Henderson (CC BY).)

of having the disease is

$$\begin{aligned}
\Pr(\text{disease} \mid \text{test}+) &= \frac{\Pr(\text{test}+|\text{disease})\cdot\Pr(\text{disease})}{\Pr(\text{test}+|\text{disease})\cdot\Pr(\text{disease})+\Pr(\text{test}+|\neg\text{disease})\cdot\Pr(\neg\text{disease})} \\
&= \frac{1\%\times80\%}{(1\%\times80\%)+(99\%\times10\%)} \\
&= 7.5\%,
\end{aligned}$$

and the posterior probability of not having the disease is

$$\begin{aligned}
\Pr(\neg\text{disease} \mid \text{test}+) &= 1 - \Pr(\text{disease} \mid \text{test}+) \\
&= \frac{\Pr(\text{test}+|\neg\text{disease})\cdot\Pr(\neg\text{disease})}{\Pr(\text{test}+|\text{disease})\cdot\Pr(\text{disease})+\Pr(\text{test}+|\neg\text{disease})\cdot\Pr(\neg\text{disease})} \\
&= \frac{99\%\times10\%}{(1\%\times80\%)+(99\%\times10\%)} \\
&= 92.5\%.
\end{aligned}$$

If we push this example to the extreme case where the disease is so rare that only one human on Earth has it (see figure 6.2) and where we have a highly accurate screening test so that

$$\text{test}+ \rightarrow \begin{cases} \Pr(\text{test} + |\text{disease}) = 0.999 \\ \Pr(\text{test} + |\neg\text{disease}) = 0.001, \end{cases}$$

the question is *If you test positive, should you be worried?* The answer is no; the disease is so rare that even with the high test accuracy, if we tested every human on Earth, we would expect $\approx 0.001 \times (8 \times 10^9) = 8 \times 10^6$ false positive diagnoses. This example illustrates how in extreme cases, prior information may outweigh the knowledge you obtain from observations.

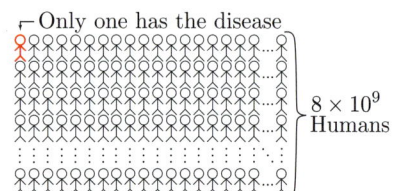

Figure 6.2: The example of a disease so rare that only one human on Earth has it.

**Note:** $\sum_x p(y = \text{🚨}|x)$ does not have to equal one for the likelihood; however, it does for the posterior, $\sum_x p(x|y) = 1$.

### 6.2.2   Example: Fire Alarm

It is common to observe that when a fire alarm is triggered in a public building, people tend to remain calm and ignore the alarm. Let us explore this situation using conditional probabilities. The sample space for the hidden state we are interested in is $x \in \{\text{fire}, \neg\text{fire}\} \equiv \{\text{🔥}, \triangle\}$ and the sample space for the observation is $y \in \{\text{alarm}, \neg\text{alarm}\} \equiv \{\text{🚨}, \text{🚨}\}$. If we assume that

$$p(\text{🚨}|x) = \begin{cases} \Pr(\text{🚨}|\text{🔥}) = 0.95 \\ \Pr(\text{🚨}|\triangle) = 0.05, \end{cases}$$

then the question is *Why does no one react when a fire alarm is triggered?* The issue here is that $p(\text{🚨}|x)$ describes the probability of an alarm given the state; in order to answer the question, we need to calculate $\Pr(\text{🔥}_t|\text{🚨})$, which depends on the prior probability that there was a fire at time $t-1$ as well as at the transition probabilities $p(x_t|x_{t-1})$, as illustrated in figure 6.3.

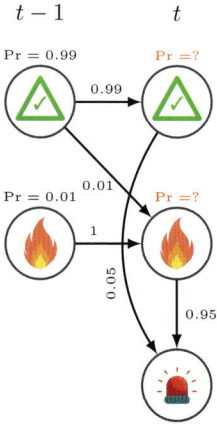

Figure 6.3: Fire alarm example.

We assume that at $t-1$, before the alarm is triggered, our prior knowledge is

$$p(x_{t-1}) = \begin{cases} \Pr(\text{🔥}_{t-1}) = 0.01 \\ \Pr(\triangle_{t-1}) = 0.99. \end{cases}$$

The transition probability between $t-1$ and $t$, at the moment when the alarm is triggered, is described by the conditional probabilities

$$p(x_t|x_{t-1}) = \begin{cases} \Pr(\triangle_t|\triangle_{t-1}) = 0.99 \\ \Pr(\text{🔥}_t|\triangle_{t-1}) = 0.01 \\ \Pr(\text{🔥}_t|\text{🔥}_{t-1}) = 1 \\ \Pr(\triangle_t|\text{🔥}_{t-1}) = 0. \end{cases}$$

By combining the prior knowledge at $t-1$ with the transition probabilities, we obtain the joint prior probability of hidden states at both time steps,

$$p(x_t, x_{t-1}) = \begin{cases} \begin{aligned} \Pr(\text{🔥}_t, \text{🔥}_{t-1}) &= \Pr(\text{🔥}_t|\text{🔥}_{t-1}) \cdot \Pr(\text{🔥}_{t-1}) \\ &= 1 \cdot 0.01 = 0.01 \\ \Pr(\text{🔥}_t, \triangle_{t-1}) &= \Pr(\text{🔥}_t|\triangle_{t-1}) \cdot \Pr(\triangle_{t-1}) \\ &= 0.01 \cdot 0.99 = 0.0099 \\ \Pr(\triangle_t, \triangle_{t-1}) &= \Pr(\triangle_t|\triangle_{t-1}) \cdot \Pr(\triangle_{t-1}) \\ &= 0.99 \cdot 0.99 = 0.9801 \\ \Pr(\triangle_t, \text{🔥}_{t-1}) &= \Pr(\triangle_t|\text{🔥}_{t-1}) \cdot \Pr(\text{🔥}_{t-1}) \\ &= 0 \cdot 0.01 = 0. \end{aligned} \end{cases}$$

We obtain the marginal prior probability of each state $x$ at a time $t$ by marginalizing this joint probability, that is, summing over states at $t-1$,

$$p(x_t) = \begin{cases} \begin{aligned} \Pr(\text{🔥}_t) &= \Pr(\text{🔥}_t, \text{🔥}_{t-1}) + \Pr(\text{🔥}_t, \triangle_{t-1}) \\ &= 0.01 + 0.0099 \\ &= 0.0199 \approx 0.02 \\ \Pr(\triangle_t) &= \Pr(\triangle_t, \triangle_{t-1}) + \Pr(\triangle_t, \text{🔥}_{t-1}) \\ &= 0.9801 \approx 0.98. \end{aligned} \end{cases}$$

If we combine the information from the prior knowledge at $t$ and the observation, we can compute the posterior probability of the state $x_t$, given that the alarm has been triggered,

$$\Pr(\text{🔥}_t|\text{🚨}) = \frac{\overbrace{\Pr(\text{🚨}|\text{🔥}_t) \cdot \Pr(\text{🔥}_t)}^{\Pr(\text{🚨},\text{🔥}_t)}}{\underbrace{(\Pr(\text{🚨}|\text{🔥}_t) \cdot \Pr(\text{🔥}_t)) + (\Pr(\text{🚨}|\triangle_t) \cdot \Pr(\triangle_t))}_{\Pr(\text{🚨})}}$$

$$= \frac{0.95 \cdot 0.02}{(0.95 \cdot 0.02) + (0.05 \cdot 0.98)}$$

$$= 0.28$$

$$\Pr(\triangle_t|\text{🚨}) = 1 - \Pr(\text{🔥}_t|\text{🚨}) = 0.72.$$

Overall, despite an alarm being triggered, the probability that there is a fire remains less than 30 percent.

We can explore what would happen at time $t + 1$ if, after the alarm has been triggered, a safety official informs us that there is a fire and that we need to evacuate. We assume here that the conditional probability of receiving such information given the state is

$$p(\text{🧑‍🚒}|x) = \left\{ \begin{array}{l} \Pr(\text{🧑‍🚒}|\text{🔥}) = 0.95 \\ \Pr(\text{🧑‍🚒}|\triangle) = 0.05, \end{array} \right.$$

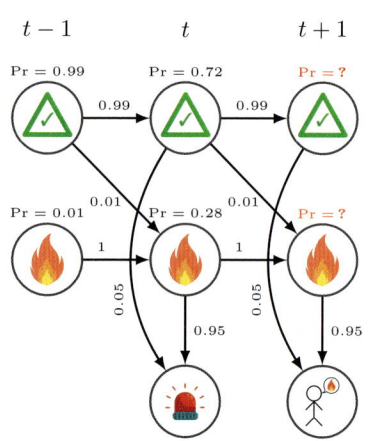

$t-1$ $\quad$ $t$ $\quad$ $t+1$

Pr = 0.99 $\quad$ Pr = 0.72 $\quad$ Pr = ?

Pr = 0.01 $\quad$ Pr = 0.28 $\quad$ Pr = ?

Figure 6.4: Example illustrating the role of the prior in sequential data acquisition and interpretation.

as illustrated in figure 6.4. Here we assume that given the state, the probability of receiving the information from the safety official is conditionally independent from the alarm, that is, $\Pr(\text{🧑‍🚒}|\text{🚨}, \text{🔥}) = \Pr(\text{🧑‍🚒}|\text{🔥})$. If, as we did for the estimation at time $t$, we combine the information from the prior knowledge at time $t$, the transition probabilities, and the observation at time $t+1$, we can compute the posterior probability of the hidden state $x_{t+1}$, given that the alarm has been triggered and that we received the information from a safety official. First, the marginal prior probabilities of each state at time $t+1$ are estimated from the posterior probability at $t$ combined with the transition probabilities so that

$$
\begin{aligned}
\Pr(\text{🔥}_{t+1}|\text{🚨}) &= \overbrace{\Pr(\text{🔥}_{t+1}|\triangle_t) \cdot \Pr(\triangle_t|\text{🚨})}^{\Pr(\text{🔥}_{t+1}, \triangle_t|\text{🚨})} + \overbrace{\Pr(\text{🔥}_{t+1}|\text{🔥}_t) \cdot \Pr(\text{🔥}_t|\text{🚨})}^{\Pr(\text{🔥}_{t+1}, \text{🔥}_t|\text{🚨})} \\
&= 0.01 \cdot 0.72 + 1 \cdot 0.28 \\
&= 0.29 \\
\Pr(\triangle_{t+1}|\text{🚨}) &= 1 - \Pr(\text{🔥}_{t+1}|\text{🚨}) = 0.71.
\end{aligned}
$$

Then, the posterior probability at $t+1$ is estimated following

$$
\begin{aligned}
\Pr(\text{🔥}_{t+1}|\text{🚨}, \text{🧑‍🚒}) &= \frac{\overbrace{\Pr(\text{🧑‍🚒}|\text{🔥}_{t+1}) \cdot \Pr(\text{🔥}_{t+1}|\text{🚨})}^{\Pr(\text{🧑‍🚒}, \text{🔥}_{t+1}|\text{🚨})}}{\underbrace{(\Pr(\text{🧑‍🚒}|\text{🔥}_{t+1}) \cdot \Pr(\text{🔥}_{t+1}|\text{🚨})) + (\Pr(\text{🧑‍🚒}|\triangle_{t+1}) \cdot \Pr(\triangle_{t+1}|\text{🚨}))}_{\Pr(\text{🧑‍🚒}|\text{🚨})}} \\
&= \frac{0.95 \cdot 0.29}{(0.95 \cdot 0.29) + (0.05 \cdot 0.71)} \\
&= 0.89
\end{aligned}
$$

$$\Pr(\triangle_t|\text{🚨}) = 1 - \Pr(\text{🔥}_t|\text{🚨}) = 0.11.$$

This example illustrates how when knowledge about a hidden state evolves over time as new information becomes available, the posterior at $t$ is employed to estimate the prior at $t+1$.

### 6.2.3  Example: Post-Earthquake Damage Assessment

Given an earthquake of intensity[3]

$$X : x \in \{\text{light (L), moderate (M), important (I)}\}$$

and a structure that can be in a state (e.g., see figure 6.5)

$$Y : y \in \{\text{damaged (D), undamaged (\overline{D})}\}.$$

These two random variables are illustrated using a Venn diagram in figure 6.6. The likelihood of damage given each earthquake intensity is

$$p(\mathsf{D}|x) = \begin{cases} \Pr(Y = \mathsf{D}|x = \mathsf{L}) &= 0.01 \\ \Pr(Y = \mathsf{D}|x = \mathsf{M}) &= 0.10 \\ \Pr(Y = \mathsf{D}|x = \mathsf{I}) &= 0.60, \end{cases}$$

and the prior probability of each intensity is given by

$$p(x) = \begin{cases} \Pr(X = \mathsf{L}) &= 0.90 \\ \Pr(X = \mathsf{M}) &= 0.08 \\ \Pr(X = \mathsf{I}) &= 0.02. \end{cases}$$

We are interested in inferring the intensity of an earthquake given that we have observed damaged buildings $y = \mathsf{D}$ following an event, so that

$$p(x|\mathsf{D}) = \frac{p(\mathsf{D}|x) \cdot p(x)}{p(\mathsf{D})}.$$

The first step to solve this problem is to compute the normalization constant,

$$p(y = \mathsf{D}) = \overbrace{p(\mathsf{D}|\mathsf{L}) \cdot p(\mathsf{L})}^{\Pr(Y=\mathsf{D}, X=\mathsf{L})} + \overbrace{p(\mathsf{D}|\mathsf{M}) \cdot p(\mathsf{M})}^{\Pr(Y=\mathsf{D}, X=\mathsf{M})} + \overbrace{p(\mathsf{D}|\mathsf{I}) \cdot p(\mathsf{I})}^{\Pr(Y=\mathsf{D}, X=\mathsf{I})}$$
$$= 0.01 \cdot 0.90 + 0.10 \cdot 0.08 + 0.6 \cdot 0.02$$
$$= 0.029.$$

Then, the posterior probability of each state is calculated following

$$\Pr(x = \mathsf{I}|y = \mathsf{D}) = \frac{p(\mathsf{D}|\mathsf{I}) \cdot p(\mathsf{I})}{p(\mathsf{D})} = \frac{0.60 \cdot 0.02}{0.029} = 0.41$$
$$\Pr(x = \mathsf{M}|y = \mathsf{D}) = \frac{p(\mathsf{D}|\mathsf{M}) \cdot p(\mathsf{M})}{p(\mathsf{D})} = \frac{0.10 \cdot 0.08}{0.029} = 0.28$$
$$\Pr(x = \mathsf{L}|y = \mathsf{D}) = \frac{p(\mathsf{D}|\mathsf{L}) \cdot p(\mathsf{L})}{p(\mathsf{D})} = \frac{0.01 \cdot 0.90}{0.029} = 0.31.$$

We see that despite the prior probability of a high-intensity earthquake being small (i.e., $\Pr(X = \mathsf{I}) = 0.02$), the posterior is significantly higher because of the observation indicating the presence of structural damage.

[3] This example is adapted from Armen Der Kiureghian's CE229 course at UC Berkeley.

Figure 6.5: Extensive damage after San Francisco's 1906 earthquake.

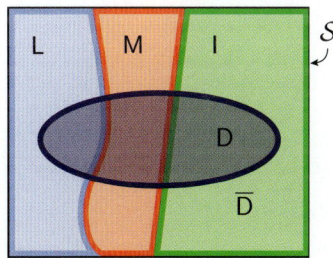

Figure 6.6: Venn diagram illustrating the sample space where the earthquake intensities light (L), moderate (M), and important (I) are mutually exclusive and collectively exhaustive events. The post-earthquake structural states of damaged (D) and undamaged ($\overline{\mathsf{D}}$) intersect the earthquake intensity events.

**Note:** For the likelihood $\sum_x p(\mathsf{D}|x) \neq 1$; however, for the prior $\sum_x p(x) = 1$, and for the posterior $\sum_x p(x|\mathsf{D}) = 1$.

## 6.3   Continuous State Variables

This section presents examples of applications of hidden-state estimation for continuous cases. Compared with discrete cases, continuous ones involve more abstract concepts in the definition of the likelihood $f(\mathcal{D}|\mathbf{x})$ and evidence $f(\mathcal{D})$. Like for the discrete case, if our model is $y = x$ and we observe $y$, then we know all there is to know about $x$ so the conditional probability of $y$ given $x$ is nonzero only for the true value $\check{x}$. In this section, we will cover cases involving imperfect observations rather than perfect ones.

### 6.3.1   Likelihood: $f(\mathcal{D}|\mathbf{x})$

In practice, we are interested in the case where $\check{x}$ is a true yet unknown hidden-state variable and where we have access to imprecise observations $y_i$, where the *observation model* is either

$$
\begin{aligned}
y_i &= \check{x} + v_i \quad \text{(direct observation)} \\
y_i &> \check{x} + v_i \quad \text{(upper-bound censored observation)} \\
y_i &< \check{x} + v_i \quad \text{(lower-bound censored observation),}
\end{aligned}
$$

where in all three cases the observation error, $v_i : V_i \sim \mathcal{N}(v; 0, \sigma_V^2)$, $V_i \perp\!\!\!\perp V_j, \forall i \neq j$. A direct observation corresponds to the standard case where the observation is the true state $\check{x}$ on which an observation error $v$ is added. $V$ is a Normal random variable employed to describe observation errors. $V$ is often chosen to be Normal in order to simplify the problems by maintaining analytical tractability for linear models (see §4.1.3). Lower- and upper-bound observations are called *censored* observations where we only know that the hidden-state variable is either respectively greater or smaller than the observed value $y$.

**Note:** $V_i \perp\!\!\!\perp V_j, \forall i \neq j$ indicates that the random variable $V_i$ is independent of $V_j$, for all $i \neq j$ (i.e., that all observation errors are independent of each other).

*Direct observations*   For a *direct observation* $y$, the likelihood is formulated from the additive observation model so that the PDF of $Y$ is described by the Normal PDF $Y \sim \mathcal{N}(y; x, \sigma_V^2)$. Because $\check{x}$ is unknown, the likelihood denoted by $f(y|x)$ or $\mathcal{L}(x|y)$ describes the prior probability density of observing $\{Y = y\}$ given $\{\check{x} = x\}$,

Hidden state: constant
$Y = x + V$
Measurement error: $\mathcal{N}(v; 0, \sigma_V^2)$
Observation: $\mathcal{N}(y; x, \sigma_V^2)$

$$
f(y|x) = \mathcal{L}(x|y) = \frac{1}{\sqrt{2\pi} \cdot \sigma_V} \exp\left(-\frac{1}{2}\left(\frac{y-x}{\sigma_V}\right)^2\right). \tag{6.2}
$$

The leftmost surface in figure 6.7 corresponds to equation 6.2 plotted with respect to both $x$ and $y$, for $\sigma_V = 1$. The pink dashed curve represents the explicit evaluation of $f(y|x = 0)$, where the evaluation at $x = 0$ is arbitrarily chosen for illustration purpose, and the red solid curve reverses the relationship to represent the likelihood of the observation $y = -1$ for any value $x$, where $f(y = -1|x) = \mathcal{L}(x|y = -1)$.

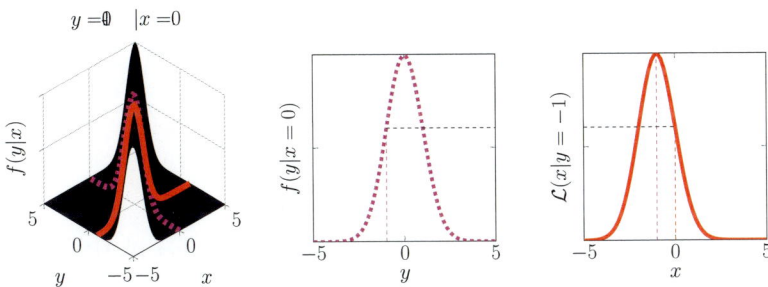

Figure 6.7: Representation of the likelihood function for a direct observation. The leftmost surface describes $f(y|x)$, the rightmost graph describes $f(y = -1|x) = \mathcal{L}(x|y = -1)$, and the center graph describes $f(y|x = 0)$.

Keep in mind that the likelihood is not a PDF with respect to $x$ so it does not integrate to 1. Accordingly, the rightmost graph in figure 6.7 is a function quantifying for each value $x$ how likely it is to observe the value $y = -1$.

*Upper-bound observations*   For an *upper-bound observation* $y > \check{x} + v$, the likelihood is described by the cumulative distribution function (CDF) of $Y \sim \mathcal{N}(y; x, \sigma_V^2)$,

$$
f(y|x) = \overbrace{\int_{-\infty}^{y} \frac{1}{\sqrt{2\pi}\sigma_V} \exp\left(-\frac{1}{2}\left(\frac{y'-x}{\sigma_V}\right)^2\right) dy'}^{\text{CDF}\mathcal{N}(y;x,\sigma_V^2)} \tag{6.3}
$$

$$
= \frac{1}{2} + \frac{1}{2}\mathrm{erf}\left(\frac{y-x}{\sqrt{2}\sigma_V}\right).
$$

The surface in figure 6.8 corresponds to equation 6.3 plotted with respect to $x$ and $y$, for $\sigma_V = 1$. The pink dashed curve represents the explicit evaluation of $f(y|x = 0)$, where the evaluation at $x = 0$ is arbitrarily chosen for illustration purpose, and the red solid curve reverses the relationship to represent the likelihood of the observation $y = -1$, given any value $x$, $\mathcal{L}(x|y = -1)$.

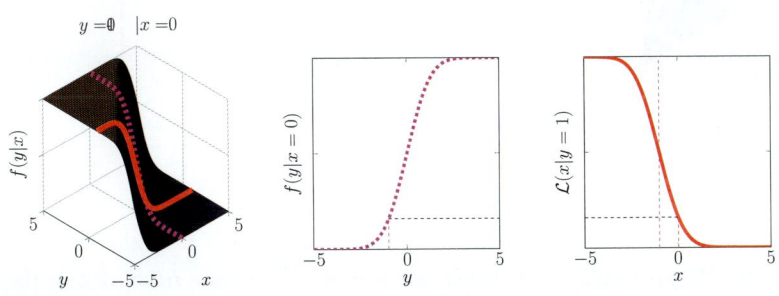

Figure 6.8: Representation of the likelihood function for an upper-bound censored observation. The leftmost surface describes $f(y|x)$, the rightmost graph describes $f(y = -1|x) = \mathcal{L}(x|y = -1)$, and the center graph describes $f(y|x = 0)$.

*Lower-bound observations*   For a *lower-bound observation* $y < \check{x} + v$, the likelihood is described by the complement of the cumulative distribution function of $Y \sim \mathcal{N}(y; x, \sigma_V^2)$,

$$
f(y|x) \;=\; 1 - \overbrace{\int_{-\infty}^{y} \frac{1}{\sqrt{2\pi}\sigma_V} \exp\left(-\frac{1}{2}\left(\frac{y'-x}{\sigma_V}\right)^2\right) dy'}^{\text{CDF}\mathcal{N}(y;x,\sigma_V^2)}
$$

$$
=\; 1 - \left(\frac{1}{2} + \frac{1}{2}\mathrm{erf}\left(\frac{y-x}{\sqrt{2}\sigma_V}\right)\right).
$$

(6.4)

The surface in figure 6.9a corresponds to equation 6.4 plotted with respect to $x$ and $y$, for $\sigma_V = 1$. The pink dashed curve represents the explicit evaluation of $f(y|x=0)$, where the evaluation at $x=0$ is arbitrarily chosen for illustration purpose, and again the red curve reverses the relationship to represent the likelihood of the observation $y = -1$, given any value $x$, $\mathcal{L}(x|y=-1)$. Note that in figure 6.9b, if instead of having $\sigma_V = 1$, $\sigma_V$ tends to zero, the smooth transition from low likelihood values to high ones would become a sharp jump from zero to one.

Figure 6.9: Representation of the likelihood function for a lower-bound censored observation. The leftmost surface describes $f(y|x)$, the rightmost graph describes $f(y=-1|x) = \mathcal{L}(x|y=-1)$, and the center graph describes $f(y|x=0)$.

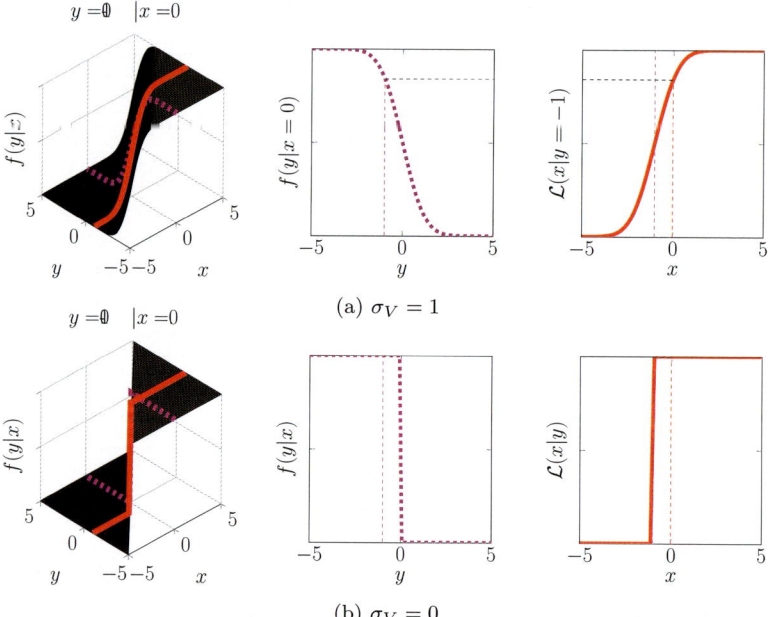

**Note:** The notion of conditional independence implies that given $x$, $Y_i$ is independent of $Y_j$, that is,

$$
Y_i \perp\!\!\!\perp Y_j | x \Leftrightarrow f(y_i|y_j, x) = f(y_i|x),
$$

so that $f(y_i, y_j|x) = f(y_i|x) \cdot f(y_j|x)$. This assumption is often satisfied because the observation model follows the form

$$
y = x + v,
$$

where $v$ is the realization of an observation error so that $V_i \perp\!\!\!\perp V_j, \forall i \neq j$.

*Multiple observations*   All cases above dealt with a single observation. When multiple observations are available, one must define the joint likelihood for all observations given $x$. With the assumption that observations are *conditionally independent* given $x$, the joint

PDF is obtained from the product of marginals,

$$
\begin{aligned}
f(\mathcal{D}|x) &= f(Y_1 = y_1, \cdots, Y_\mathtt{D} = y_\mathtt{D}|x) \\
&= \overbrace{\prod_{i=1}^{\mathtt{D}} f(Y_i = y_i|x)}^{\mathtt{D} \gg 1 \triangle} \\
&= \exp\left(\sum_{i=1}^{\mathtt{D}} \ln\left(f(Y_i = y_i|x)\right)\right).
\end{aligned} \tag{6.5}
$$

Note that practical difficulties related to numerical *underflow* or *overflow* arise when trying to evaluate the product of several marginal likelihoods $f(y_i|x)$. This is why in equation 6.5, the product of the marginals is replaced with the equivalent yet numerically more robust formulation using the exponential of the sum of the log marginals.

### 6.3.2   Evidence: $f(\mathcal{D})$

The evidence acts as a normalization constant obtained by integrating the product of the likelihood and the prior over all possible values of $x$,

$$
f(\mathcal{D}) = \int f(\mathcal{D}|x) \cdot f(x) dx.
$$

This integral is typically difficult to evaluate because it is not analytically tractable. Figure 6.10 presents a naive way of approaching this integral using the *rectangle rule*, where the domain containing a significant probability content is subdivided in $\mathtt{S}$ bins of width $\Delta x_s$ in order to approximate

$$
\hat{f}(\mathcal{D}) = \sum_{s=1}^{\mathtt{S}} f(\mathcal{D}|x_s) \cdot f(x_s) \Delta x_s.
$$

This approach is reasonable for trivial 1-D problems; however, it becomes computationally intractable for problems in larger dimensions (i.e., for $\mathbf{x} \in \mathbb{R}^{\mathtt{X}}$) because the number of grid combinations to evaluate increases exponentially with the number of dimensions $\mathtt{X}$ to integrate on. Section 6.5 presents an introduction to sampling methods that allow overcoming this challenge. Chapter 7 will introduce more advanced sampling methods allowing the estimation of the normalization constants for high-dimensional domains.

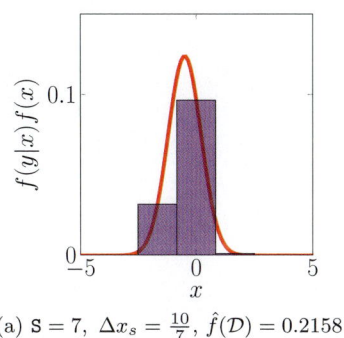

(a) $\mathtt{S} = 7$, $\Delta x_s = \frac{10}{7}$, $\hat{f}(\mathcal{D}) = 0.2158$

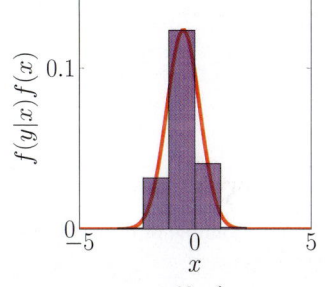

(b) $\mathtt{S} = 10$, $\Delta x_s = \frac{10}{10}$, $\hat{f}(\mathcal{D}) = 0.2198$

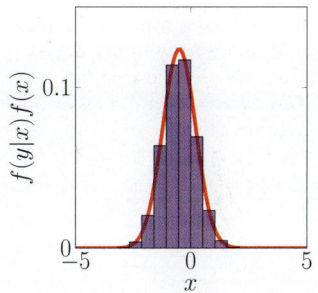

(c) $\mathtt{S} = 20$, $\Delta x_s = \frac{10}{20}$, $\hat{f}(\mathcal{D}) = 0.2197$

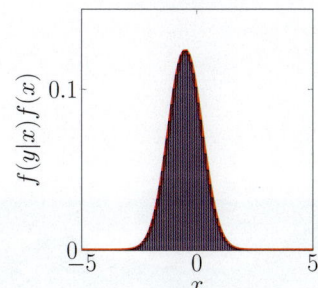

(d) $\mathtt{S} = 100$, $\Delta x_s = \frac{10}{100}$, $\hat{f}(\mathcal{D}) = 0.2197$

Figure 6.10: Examples of application of the rectangle integration rule.

### 6.3.3   Posterior: $f(\mathbf{x}|\mathcal{D})$

The posterior is the product of the likelihood function times the prior PDF divided by the normalization constant,

$$f(x|\mathcal{D}) = \frac{f(\mathcal{D}|x) \cdot f(x)}{f(\mathcal{D})}.$$

Let us consider two direct observations $\mathcal{D} = \{y_1 = -1, y_2 = 0\}$ obtained from the observation model

$$y_i = x + v_i, \ v_i : V_i \sim \mathcal{N}(v; 0, 1^2), \ V_i \perp\!\!\!\perp V_j, \forall i \neq j.$$

The likelihood is thus given by $f(\mathcal{D}|x) = \mathcal{N}(y_1; x, 1^2) \cdot \mathcal{N}(y_2; x, 1^2)$. We assume the prior PDF follows a uniform PDF within the range from -5 to 5, $f(x) = \mathcal{U}(x; -5, 5)$. As presented in figure 6.10, the normalization constant is estimated using the rectangle integration rule to $\hat{f}(\mathcal{D}) = 0.2197$. Figure 6.11 presents the prior, the likelihood, and the posterior for this example.

Figure 6.11: Example of prior, likelihood, and posterior.

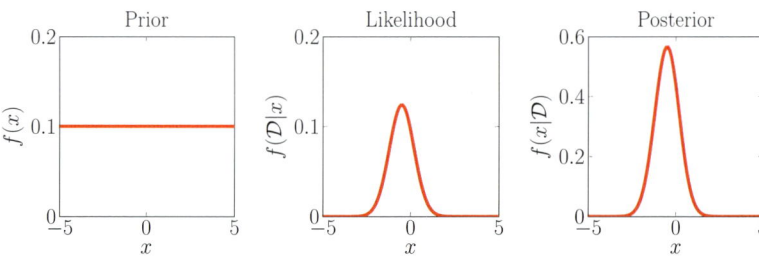

### 6.3.4   Number of Observations and Identifiability

In the extreme case where no observation is available (i.e., $\mathcal{D} = \emptyset$), the posterior is equal to the prior $f(\mathbf{x}|\emptyset) = f(\mathbf{x})$. In order to explore what happens at the other end of the spectrum where an infinite amount of data is available, it is required to distinguish between *identifiable* and *non-identifiable* problems. A problem is *identifiable* when, given an infinite number of direct observations, we are theoretically able to retrieve the true hidden state of a deterministic system $\check{x}$. For *non-identifiable* problems, even an infinite number of direct observations does not allow retrieving the true hidden state of the system $\check{x}$, because multiple equally valid solutions exist.

*Example: Non-identifiable problem*   Take the example illustrated in figure 6.12, where we want to estimate the contaminant concentration in two streams $x_1$ and $x_2$ and where we only have access to contaminant concentration observations

$$y_i = x_1 + x_2 + v_i, \ v_i : V_i \sim \mathcal{N}(v; 0, 2^2), \ V_i \perp\!\!\!\perp V_j, \forall i \neq j,$$

Figure 6.12: Example of two streams merging in a single river.
(Photo: Ashwini Chaudhary)

which are collected at the output of a river after both streams have merged. We assume that our prior knowledge is uniform over positive concentration values, that is, $f(x_1, x_2) \propto 1$. The resulting posteriors are presented in figure 6.13 for $\mathtt{D} = 1$ and $\mathtt{D} = 100$ observations, along with the true values marked by a red cross. This problem is intrinsically non-identifiable because, even when the epistemic uncertainty has been removed by a large number of observations, an infinite number of combinations of $x_1$ and $x_2$ remain possible.

Non-identifiable problems can be regularized using prior knowledge. If, for example, we know that the prior probability for the contaminant concentration in the first stream follows $p(x_1) = \mathcal{N}(x_1; 10, 5^2)$, the posterior will now have a single mode, as presented in figure 6.14.

*Well-constrained identifiable problems*   For *well-constrained* problems where a true value $\check{x}$ exists, given an infinite number of direct observations ($\mathtt{D} \to \infty$) that are conditionally independent given $x$, the posterior tends to a Dirac delta function, which is nonzero only for $x = \check{x}$,

$$f(x|\mathcal{D}) = \frac{f(\mathcal{D}|x) \cdot f(x)}{f(\mathcal{D})} \to \delta(x - \overbrace{\check{x}}^{\text{true value}}).$$
$$\underbrace{\phantom{\delta(x - \check{x})}}_{\text{Dirac delta function}}$$

Figure 6.15 presents an example of such a case where the posterior has collapsed over a single value. The posterior expected value then corresponds to the true value

$$\mathbb{E}[X|\mathcal{D}] \to \underbrace{\check{x}}_{\text{true value}},$$

and the posterior variance tends to zero because no epistemic uncertainty remains about the value of $x$,

$$\mathrm{var}[X|\mathcal{D}] \to \underbrace{0}_{\check{x}:\text{ constant}}.$$

Note that in most practical cases we do not have access to an infinite number of observations. Moreover, keep in mind that the less data we have access to, the more important it is to consider the uncertainty in the posterior $f(x|\mathcal{D})$.

## 6.4  Parameter Estimation

In this section, we will explore how to extend Bayesian estimation for the parameters $\boldsymbol{\theta}$ defining the probability density function of

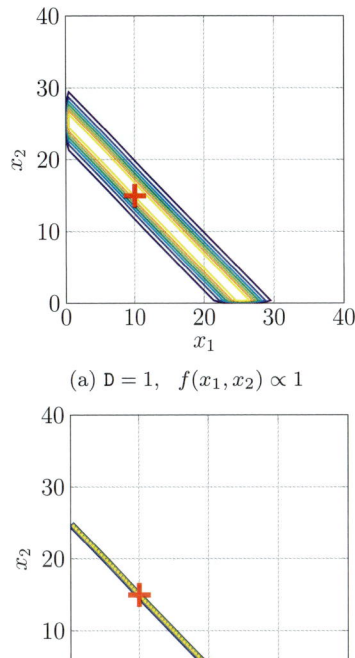

(a) $\mathtt{D} = 1$,   $f(x_1, x_2) \propto 1$

(b) $\mathtt{D} = 100$,   $f(x_1, x_2) \propto 1$

Figure 6.13: Posterior contours for an example of non-identifiable problem.

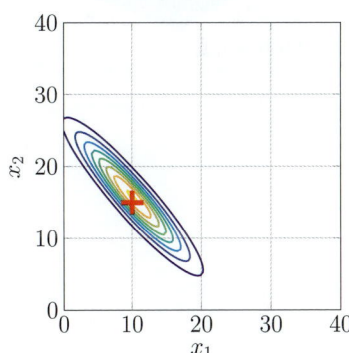

Figure 6.14: Posterior contours for an example of non-identifiable problem for $\mathtt{D} = 1$ and which is constrained by the prior knowledge $f(x_1, x_2) \propto \mathcal{N}(x_1; 10, 5^2)$.

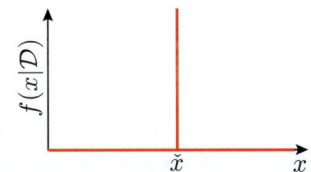

Figure 6.15: Example of posterior PDF tending toward the Dirac delta function.

**Notation**

$x_i : X \sim f_X(x; \boldsymbol{\theta})$

$\boldsymbol{\theta}$: PDF's parameters

$\{x_1, x_2, \cdots, x_\mathrm{D}\}$: realizations of the hidden variable $X$

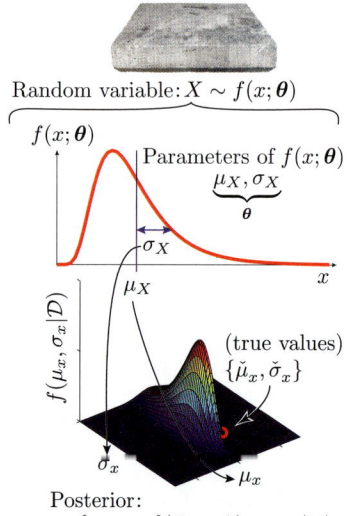

Figure 6.16: Example of Bayesian parameter estimation.

**Non-informative prior**

$\boldsymbol{\theta} = \{\mu_X, \sigma_X\}$, $\mathcal{N}(x; \mu_X, \sigma_X^2)$

$f(\mu_X) \propto 1$

$f(\sigma_X) \propto \frac{1}{\sigma_X}$

$f(\mu_X, \sigma_X) \propto \frac{1}{\sigma_X}$, if $\mu_X \perp\!\!\!\perp \sigma_X$

a hidden random variable $x : X \sim f(x) \equiv f(x; \boldsymbol{\theta})$. This setup is relevant when the system responses are stochastic. For example, even in the absence of observation errors, different samples from a concrete batch made from the same mix of components would have different resistances due to the uncertainty associated with the intrinsic heterogeneity of the material. The posterior probability density function sought is now

$$f(\boldsymbol{\theta}|\mathcal{D}) = \frac{f(\mathcal{D}|\boldsymbol{\theta}) \cdot f(\boldsymbol{\theta})}{f(\mathcal{D})},$$

where $f(\boldsymbol{\theta})$ is the prior PDF of unknown parameters, $f(\mathcal{D}|\boldsymbol{\theta})$ is the likelihood of observations $\mathcal{D} = \{y_1, \cdots, y_\mathrm{D}\}$, and $f(\boldsymbol{\theta}|\mathcal{D})$ is the posterior PDF for unknown parameters $\boldsymbol{\theta}$. After having estimated the posterior PDF for parameters $f(\boldsymbol{\theta}|\mathcal{D})$, we can *marginalize* this uncertainty in $f(x; \boldsymbol{\theta})$ in order to obtain the *posterior predictive* PDF for the hidden variable $X$,

$$f(x|\mathcal{D}) = \int f(x; \boldsymbol{\theta}) \cdot f(\boldsymbol{\theta}|\mathcal{D}) d\boldsymbol{\theta}. \qquad (6.6)$$

Figure 6.16 presents an example where we want to estimate the joint posterior probability of the mean $\mu_X$ and standard deviation $\sigma_X$ defining the PDF of $X$, which describes the variability in the resistance of a given concrete mix. Here, the concrete resistance $X$ is assumed to be a random variable. It means that unlike in previous sections, no true value exists for the resistance because it is now a stochastic quantity. Despite this, true values can still exist for the parameters of $f(x; \boldsymbol{\theta})$, that is, $\{\check{\mu}_x, \check{\sigma}_X\}$. This example will be further explored in §6.5.3.

### 6.4.1   Prior: $f(\boldsymbol{\theta})$

$f(\boldsymbol{\theta})$ describes our prior knowledge for the values that $\boldsymbol{\theta}$ can take. Prior knowledge can be based on heuristics (expert knowledge), on the posterior PDF obtained from previous data, or on a non-informative prior (i.e., absence of prior knowledge). A *non-informative* prior for parameters such as the mean of a Normal PDF is described by a uniform density $f(\mu) \propto 1$. For standard deviations, the non-informative prior typically assumes that all orders of magnitudes for $\sigma$ have equal probabilities. This hypothesis implies that $f(\ln \sigma) \propto 1$. By using the change of variable rule presented in §3.4, it leads to $f(\sigma) \propto \frac{1}{\sigma}$, because the derivative of this transformation is $\frac{d \ln \sigma}{d\sigma} = \frac{1}{\sigma}$. Note that the two non-informative priors described for $\mu$ and $\sigma$ are *improper* because $\int f(\theta) d\theta = \infty$, which does not respect the fundamental requirement that a PDF integrates to 1, as presented in §3.3. For this reason we cannot draw samples from an improper prior. Nevertheless, when used as a prior, an improper PDF can yield to a proper posterior (see §7.5).

### 6.4.2   Likelihood: $f(\mathcal{D}|\theta)$

$f(\mathcal{D}|\boldsymbol{\theta}) \equiv f(\mathbf{Y} = \mathbf{y}|\boldsymbol{\theta}) \equiv f(\mathbf{y}|\boldsymbol{\theta}) \equiv \mathcal{L}(\boldsymbol{\theta}|\mathbf{y})$ are all equivalent ways of describing the likelihood of observing specific values $\mathbf{y}$, given $\boldsymbol{\theta}$. In the special case where $y = x$ so that realizations of $x : X \sim f(x; \boldsymbol{\theta})$ are directly observed, the likelihood is

$$f(y|\boldsymbol{\theta}) = f(y; \boldsymbol{\theta}) = f(x; \boldsymbol{\theta}).$$

In a more general case where observations $\mathcal{D} = \{y_i, \forall i \in \{1 : \mathtt{D}\}\}$ are realizations of an additive observation model $y_i = x_i + v_i$, where $v_i : V \sim \mathcal{N}(v; 0, \sigma_V^2)$ and $x_i : X \sim \mathcal{N}(x; \underbrace{\mu_X, \sigma_X^2}_{\boldsymbol{\theta}})$, the likelihood is

$$f(y|\boldsymbol{\theta}) = \mathcal{N}(y; \mu_X, \sigma_X^2 + \sigma_V^2).$$

This choice of an *additive observation model* using normal random variables allows having an analytically tractable formulation for the likelihood. If we want to employ a multiplicative observation model, $y = x \cdot v$, then choosing $v : V \sim \ln\mathcal{N}(v; 0, \sigma_{\ln V}^2)$ and $x : X \sim \ln\mathcal{N}(x; \underbrace{\mu_{\ln X}, \sigma_{\ln X}^2}_{\boldsymbol{\theta}})$ as lognormal random variables also preserves the analytical tractability for the likelihood that follows

$$f(y|\boldsymbol{\theta}) = \ln\mathcal{N}(y; \mu_{\ln X}, \sigma_{\ln X}^2 + \sigma_{\ln V}^2).$$

*Hidden-state R.V.*: $\mathcal{N}(x; \mu_X, \sigma_X^2)$

$Y = X + V$

*Measurement error*: $\mathcal{N}(v; 0, \sigma_V^2)$

*Observation*: $\mathcal{N}(y; \mu_X, \sigma_X^2 + \sigma_V^2)$

### 6.4.3   Posterior PDF: $f(\theta|\mathcal{D})$

In §6.3.4, when the number of *independent* observations $\mathtt{D} \to \infty$, we saw that the posterior for $X$ tends to a Dirac delta function $f(\mathbf{x}|\mathcal{D}) \to \delta(\mathbf{x} - \check{\mathbf{x}})$. The situation is different for the estimation of parameters describing the PDF of a hidden random variable (R.V.); when $\mathtt{D} \to \infty$, it is now the posterior PDF for $\boldsymbol{\theta}$ that tends to a Dirac delta function as presented in figure 6.17,

**Note:** The term *independent* observations is employed as a shortcut for the more rigorous term that should be in this case *conditionally independent* given $\mathbf{x}$.

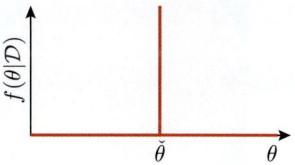

$$f(\boldsymbol{\theta}|\mathcal{D}) = \frac{f(\mathcal{D}|\boldsymbol{\theta}) \cdot f(\boldsymbol{\theta})}{f(\mathcal{D})} \to \underbrace{\delta(\boldsymbol{\theta} - \overbrace{\check{\boldsymbol{\theta}}}^{\text{true value}})}_{\text{Dirac delta function}},$$

where the posterior expected value tends toward the true parameter value and the variance of the posterior tends to zero,

Figure 6.17: The posterior PDF for parameters reduces to a Dirac delta function when the number of independent observations tends to infinity.

$$\mathbb{E}[\boldsymbol{\theta}|\mathcal{D}] \to \underbrace{\check{\boldsymbol{\theta}}}_{\text{true value}}$$

$$\text{var}[\boldsymbol{\theta}|\mathcal{D}] \to \underbrace{0}_{\check{\boldsymbol{\theta}}:\,\text{constant}}.$$

In such a situation, the epistemic uncertainty associated with parameter values $\boldsymbol{\theta}$ vanishes and the uncertainty about $X$ reaches its irreducible level, $f(x|\mathcal{D}) \to f(x; \check{\boldsymbol{\theta}})$.

In practical cases where the number of observations $\mathtt{D} < \infty$, uncertainty remains about the parameters $f(\boldsymbol{\theta}|\mathcal{D})$, so $f(x|\mathcal{D})$ is typically more diffused than $f(x; \check{\boldsymbol{\theta}})$. The uncertainty associated with parameters $f(\boldsymbol{\theta}|\mathcal{D})$ can be *marginalized* as described in equation 6.6 in order to obtain the *posterior predictive* PDF for the hidden random variable. This integral is typically not analytically tractable, so it is common to resort to the *Monte Carlo sampling methods* presented in the next section in order to compute the posterior predictive expected value and variance for $f(x|\mathcal{D})$.

## 6.5 Monte Carlo

*Monte Carlo* sampling is a numerical integration method that allows performing Bayesian estimation for analytically intractable cases. In this section, we will first see how Monte Carlo can be employed as an integration method. Then, we will explore how this method fits in the context of Bayesian estimation.

### 6.5.1 Monte Carlo Integration

We employ the example of a unit diameter circle with area $a = xr^2 = 0.785$ to demonstrate how we can approximate this area with Monte Carlo numerical integration. A circle of diameter $d = 1$ ($r = 0.5$) is described by the equation

$$(x - 0.5)^2 + (y - 0.5)^2 = r^2.$$

This equation is employed in the definition of the *indicator function*

$$I(x,y) = \begin{cases} 1 & \text{if } (x - 0.5)^2 + (y - 0.5)^2 \le r^2 \\ 0 & \text{otherwise.} \end{cases}$$

We then define a bivariate uniform probability density in the range $\{x,y\} \in (0,1)$, where $f_{XY}(x,y) = 1$. The circle area $a$ can be formulated as the integral of the product of the indicator function times $f_{XY}(x,y)$,

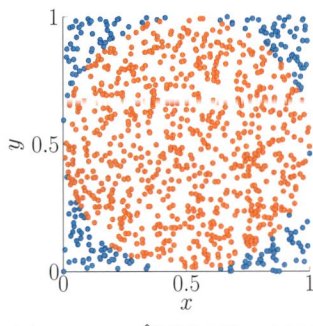

(a) $\mathtt{S} = 100, \hat{\mathbb{E}}[I(X,Y)] = 0.800$

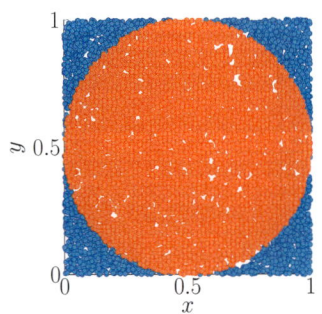

(b) $\mathtt{S} = 1000, \hat{\mathbb{E}}[I(X,Y)] = 0.759$

(c) $\mathtt{S} = 10000, \hat{\mathbb{E}}[I(X,Y)] = 0.785$

Figure 6.18: Examples of Monte Carlo numerical integration for which the theoretical answer is $a = \pi r^2 = 0.785$.

$$\underbrace{a}_{\text{area}} = \int_y \int_x I(x,y) \cdot f_{XY}(x,y) dx dy \tag{6.7}$$
$$= \mathbb{E}[I(X,Y)].$$

We saw in §3.3.5 that the integral in equation 6.7 corresponds to the *expectation* of the indicator function $I(X, Y)$. Given $\mathsf{S}$ random samples $\{x_s, y_s\}$, $s \in \{1, 2, \cdots, \mathsf{S}\}$ generated from a uniform probability density $f_{XY}(x, y)$, the expectation operation is defined as

$$a = \mathbb{E}[I(X, Y)] = \lim_{\mathsf{S} \to \infty} \frac{1}{\mathsf{S}} \sum_{s=1}^{\mathsf{S}} I(x_s, y_s).$$

In practice, because the number of samples $\mathsf{S}$ is finite, our estimation will remain an approximation of the expectation,

$$a \approx \hat{\mathbb{E}}[I(X, Y)] = \frac{1}{\mathsf{S}} \sum_{s=1}^{\mathsf{S}} I(x_s, y_s).$$

**Note:** The hat in $\hat{\mathbb{E}}[I(X, Y)]$ denotes that the quantity is an approximation of $\mathbb{E}[I(X, Y)]$.

Figure 6.18 presents realizations for $\mathsf{S} = \{10^2, 10^3, 10^4\}$ samples as well as their approximation of the area $a$. As the number of samples increases, the approximation error decreases and $\hat{\mathbb{E}}[I(X, Y)]$ tends to the true value of $a = \pi r^2 = 0.785$.

The interesting property of Monte Carlo integration is that the estimation quality, measured by the variance $\text{var}\left[\hat{\mathbb{E}}[I(X, Y)]\right]$, depends on the *number of samples* and *not on the number of dimensions*.[4] This property is key for the application of the Monte Carlo method in high-dimensional domains.

[4] MacKay, D. J. C. (1998). Introduction to Monte Carlo methods. In *Learning in graphical models*, pp. 175–204. Sprigner.

### 6.5.2 Monte Carlo Sampling: Continuous State Variables

Monte Carlo methods are now applied to the Bayesian estimation context presented in §6.3 where we are interested in estimating continuous state variables. The Monte Carlo integration method is first employed for approximating the evidence

$$\begin{aligned} f(\mathcal{D}) &= \int f(\mathcal{D}|x) \cdot f(x) dx \\ &= \mathbb{E}[f(\mathcal{D}|X)]. \end{aligned}$$

**Monte Carlo sampling**
$h(x)$: Sampling PDF
$x_s : X \sim h(x)$
$s \in \{1, 2, \cdots, \mathsf{S}\}$

A Monte Carlo approximation of the evidence can be computed using *importance sampling* where

$$\hat{f}(\mathcal{D}) = \frac{1}{\mathsf{S}} \sum_{s=1}^{\mathsf{S}} \frac{f(\mathcal{D}|x_s) \cdot f(x_s)}{h(x_s)}, \ x_s : X \sim h(x) : \text{Sampling PDF},$$

where samples $x_s : X \sim h(x)$ are drawn from any sampling distribution such that $h(x) > 0$ for any $x$ for which the PDF on the numerator $f(\mathcal{D}|x) \cdot f(x) > 0$. If we choose the prior PDF as sampling distribution, $h(x) = f(x)$, the formulation simplifies to

$$\hat{f}(\mathcal{D}) = \frac{1}{\mathsf{S}} \sum_{s=1}^{\mathsf{S}} f(\mathcal{D}|x_s), \ x_s : X \sim h(x) = f(x)$$

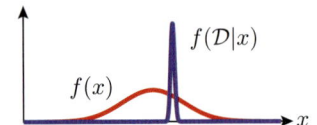

Figure 6.19: Example of difficult situation encountered when trying to sample from the prior.

because the prior and the sampling distribution cancel out.

The performance of the Monte Carlo integration depends on how close our sampling distribution $h(x)$ is to the posterior $f(x|\mathcal{D})$. Note that sampling from the prior, $h(x) = f(x)$, is often not a viable option when the number of observations D is large. This is because in such a case, the likelihood is the product of several terms so the probability content typically ends up being concentrated in a small region that is not well described by the prior, as illustrated in figure 6.19.

The posterior mean and variance can be approximated with Monte Carlo integration using

$$
\begin{aligned}
\mathbb{E}[X|\mathcal{D}] &= \int x \cdot f(x|\mathcal{D})dx \\
&\approx \frac{1}{\mathsf{S}} \sum_{s=1}^{\mathsf{S}} x_s \cdot \frac{f(\mathcal{D}|x_s) \cdot f(x_s)}{f(\mathcal{D})} \cdot \frac{1}{h(x_s)}, \quad x_s : X \sim h(x) \\
&\approx \frac{1}{\mathsf{S}} \sum_{s=1}^{\mathsf{S}} x_s \cdot \frac{f(\mathcal{D}|x_s)}{f(\mathcal{D})}, \quad\quad\quad\quad x_s : X \sim h(x) = f(x)
\end{aligned}
$$

and

$$
\begin{aligned}
\mathrm{var}[X|\mathcal{D}] &= \int (x - \mathbb{E}[X|\mathcal{D}])^2 \cdot f(x|\mathcal{D})dx \\
&\approx \frac{1}{\mathsf{S}-1} \sum_{s=1}^{\mathsf{S}} (x_s - \mathbb{E}[X|\mathcal{D}])^2 \cdot \frac{f(\mathcal{D}|x_s) \cdot f(x_s)}{f(\mathcal{D})} \cdot \frac{1}{h(x_s)}, \quad x_s : X \sim h(x) \\
&\approx \frac{1}{\mathsf{S}-1} \sum_{s=1}^{\mathsf{S}} (x_s - \mathbb{E}[X|\mathcal{D}])^2 \cdot \frac{f(\mathcal{D}|x_s)}{f(\mathcal{D})}, \quad\quad x_s : X \sim h(x) = f(x).
\end{aligned}
$$

For both the posterior mean and variance, when choosing the prior as the sampling distribution, $h(x) = f(x)$, it cancels out from the formulation.

Figure 6.20: Example of Bayesian estimation with censored observations. (Photo: Elevate)

*Example: Concentration estimation*   Figure 6.20 depicts a laboratory test for measuring a deterministic mercury concentration $\check{x}$. Observations follow the observation model

$$
\begin{cases}
y_i = \check{x} + v_i, & \text{if } \check{x} + v_i \geq 10\mu g/L \\
y_i = \texttt{error}, & \text{if } \check{x} + v_i < 10\mu g/L,
\end{cases}
$$

where the measuring device has a minimum threshold and the mutually independent observation errors are described by

$$
v_i : V_i \sim \mathcal{N}(v; 0, 2^2), V_i \perp\!\!\!\perp V_j, \forall i \neq j.
$$

We want to estimate the posterior PDF describing the concentration if the observation obtained is $\mathcal{D} = \{\texttt{error}\}$ for two hypotheses regarding the prior knowledge: (1) uniform in the interval $(0, 50)$, $f(x) = \mathcal{U}(x; 0, 50)$, and (2) a normal PDF with mean 25 and standard deviation 10, $f(x) = \mathcal{N}(x; 25, 10^2)$. Figure 6.21 presents the results for the (a) uniform prior and (b) normal prior. Note that only 100, out of the $10^5$ Monte Carlo samples that were employed to estimate the evidence $f(\mathcal{D})$, are displayed. This example illustrates how information can be extracted from censored observations. Note also how the choice of prior influences the posterior knowledge. The choice of prior is particularly important here because limited information is carried by the censored observations.

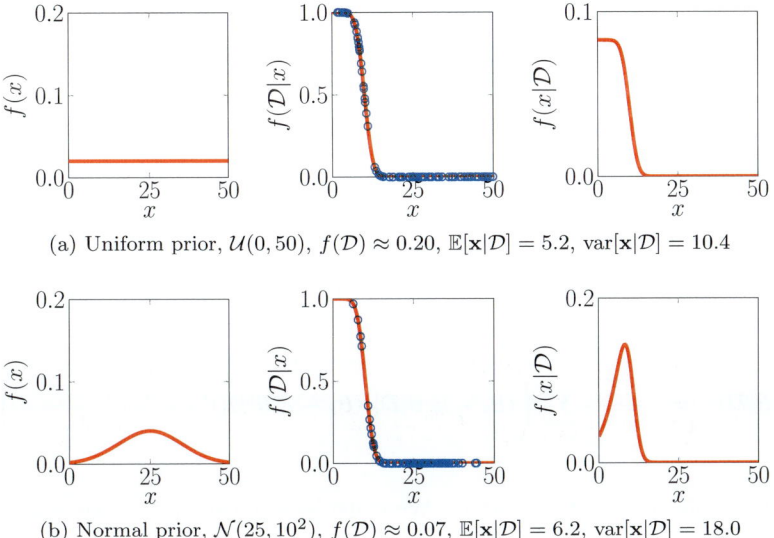

(a) Uniform prior, $\mathcal{U}(0, 50)$, $f(\mathcal{D}) \approx 0.20$, $\mathbb{E}[\mathbf{x}|\mathcal{D}] = 5.2$, var$[\mathbf{x}|\mathcal{D}] = 10.4$

(b) Normal prior, $\mathcal{N}(25, 10^2)$, $f(\mathcal{D}) \approx 0.07$, $\mathbb{E}[\mathbf{x}|\mathcal{D}] = 6.2$, var$[\mathbf{x}|\mathcal{D}] = 18.0$

Figure 6.21: Comparative example of a uniform and normal prior PDF for performing Bayesian estimation with censored observations.

We consider a different context where we employ $\mathtt{D} = 100$ direct observations, $\mathcal{D} = \{25.4, 23.6, 24.1, \cdots, 24.3\}_{1 \times 100}$, and the prior is $p(x) = \mathcal{U}(x; 0, 50)$. The prior, likelihood, and posterior are presented in figure 6.22. Note that here, out of the 100 Monte Carlo samples displayed, none lie in the high probability region for the likelihood. As a result, it leads to a poor approximation of the evidence where the true value is $f(\mathcal{D}) = 1.19 \times 10^{-78}$ and where the approximation

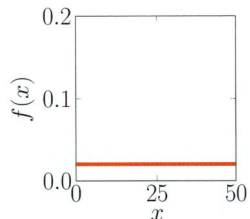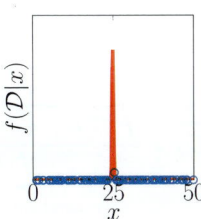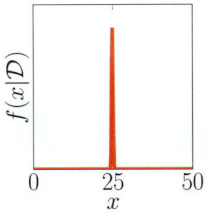

Figure 6.22: Example where drawing Monte Carlo samples from the prior leads to a poor estimate of the posterior because no sample falls in the high probability region.

is

$$\hat{f}(\mathcal{D}) = \frac{1}{S} \sum_{s=1}^{S} \left( \prod_{j=1}^{100} f(y_j|x_s) \right) = 1.95 \times 10^{-83} \; \triangle.$$

Although it is possible to solve this problem by increasing the number of samples, this approach is computationally inefficient and even becomes impossible for high-dimensional domains. Chapter 7 presents advanced Monte Carlo methods for solving this limitation.

### 6.5.3   Monte Carlo Sampling: Parameter Estimation

For D observations $\mathcal{D} = \{y_1, y_2, \cdots, y_D\}$, and samples $\boldsymbol{\theta}_s \sim h(\boldsymbol{\theta}) = f(\boldsymbol{\theta})$, $s \in \{1, 2, \cdots, S\}$ drawn from the prior PDF of parameters, we can estimate the evidence

$$\hat{f}(\mathcal{D}) = \frac{1}{S} \sum_{s=1}^{S} \left( \prod_{j=1}^{D} f(y_j|\boldsymbol{\theta}_s) \right) \tag{6.8}$$

as well as the posterior expected values and variance for parameters

$$\hat{\mathbb{E}}[\boldsymbol{\theta}|\mathcal{D}] = \frac{1}{S} \sum_{s=1}^{S} \left( \boldsymbol{\theta}_s \cdot \frac{\prod_{j=1}^{D} f(y_j|\boldsymbol{\theta}_s)}{f(\mathcal{D})} \right)$$

$$\hat{\mathrm{cov}}(\boldsymbol{\theta}|\mathcal{D}) = \frac{1}{S-1} \sum_{s=1}^{S} \left( (\boldsymbol{\theta}_s - \mathbb{E}[\boldsymbol{\theta}|\mathcal{D}])(\boldsymbol{\theta}_s - \mathbb{E}[\boldsymbol{\theta}|\mathcal{D}])^{\mathsf{T}} \cdot \frac{\prod_{j=1}^{D} f(y_j|\boldsymbol{\theta}_s)}{f(\mathcal{D})} \right).$$

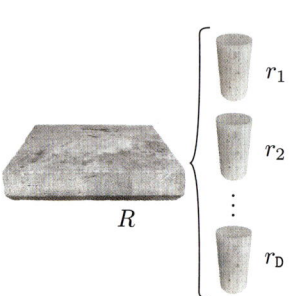

Figure 6.23: Example of samples taken from the same concrete batch.

*Example: Concrete strength*   We consider the case where we want to characterize the resistance $R$ of a given concrete batch, where the intrinsic variability is estimated by testing the resistance from several samples, as depicted in figure 6.23. We assume that the true yet unknown parameter values are $\check{\mu}_R = 42\,\mathrm{MPa}$ and $\check{\sigma}_R = 2\,\mathrm{MPa}$ and that the resistance $R$ is described by a log-normal random variable

$$R \sim \ln \mathcal{N}(r; \check{\mu}_{\ln R}, \check{\sigma}_{\ln R})\,\mathrm{MPa}.$$

The unknown parameters we want to learn about are $\boldsymbol{\theta} = [\mu_R \; \sigma_R]^{\mathsf{T}}$. Here, $R$ is a hidden random variable because it is not directly observed; only imprecise observations $y_i$ are available. The set of observations $\mathcal{D} = \{y_1, y_2, \cdots, y_D\}$ is employed to estimate the posterior PDF for the parameters $\boldsymbol{\theta}$ that are controlling the PDF of the hidden random variable $X$,

$$\overbrace{f(\mu_R, \sigma_R|\mathcal{D})}^{\text{Posterior PDF}} = \frac{\overbrace{f(\mathcal{D}|\mu_R, \sigma_R)}^{\text{Likelihood}} \cdot \overbrace{f(\mu_R, \sigma_R)}^{\text{Prior PDF}}}{\underbrace{f(\mathcal{D})}_{\text{Normalization constant}}}.$$

The joint prior knowledge for parameters $f(\mu_R, \sigma_R) = f(\mu_R) \cdot f(\sigma_R)$, $(\mu_R \perp\!\!\!\perp \sigma_R)$ is obtained from the product of the marginal PDFs,

$$\underbrace{f(\mu_R) = \ln\mathcal{N}(\mu_R; \mu_{\ln\mu_R}, \sigma_{\ln\mu_R})}_{\substack{\mu_{\mu_R} = 45\,\text{MPa} \\ \sigma_{\mu_R} = 7\,\text{MPa}}}, \quad \underbrace{f(\sigma_R) = \ln\mathcal{N}(\sigma_R; \mu_{\ln\sigma_R}, \sigma_{\ln\sigma_R})}_{\substack{\mu_{\sigma_R} = 5\,\text{MPa} \\ \sigma_{\sigma_R} = 2\,\text{MPa}}}.$$

The prior PDF for $\mu_R$ and $\sigma_R$ are themselves described by log-normal PDFs with respective mean $\{\mu_{\mu_R}, \mu_{\sigma_R}\}$ and variance $\{\sigma_{\mu_R}^2, \sigma_{\sigma_R}^2\}$.

The observation model is $\ln y_i = \ln r_i + v_i$, where the observation errors $v_i : V \sim \mathcal{N}(v; 0, 0.01^2)$ MPa. With these assumptions regarding the observation model, the marginal likelihood for each observation is

$$f(y|\boldsymbol{\theta}_{\ln}) = f(y|\mu_{\ln R}, \sigma_{\ln R}) = \ln\mathcal{N}(y; \mu_{\ln R}, (\sigma_{\ln R}^2 + \overbrace{\sigma_V^2}^{=0.01^2})^{1/2}),$$

and the joint likelihood for the entire set of observations is

$$f(\mathcal{D}|\overbrace{\mu_{\ln R}, \sigma_{\ln R}}^{\boldsymbol{\theta}_{\ln}=\text{fct}(\boldsymbol{\theta})}) = \underbrace{\prod_{j=1}^{D} \frac{1}{y_j\sqrt{2\pi}\sqrt{\sigma_{\ln R}^2 + 0.01^2}} \exp\left(-\frac{1}{2}\left(\frac{\ln y_j - \mu_{\ln R}}{\sqrt{\sigma_{\ln R}^2 + 0.01^2}}\right)^2\right)}_{\text{Log-Normal PDF}}.$$

Note that the marginal and joint likelihood are defined as a function of the log-space parameters. In order to evaluate the likelihood for the parameters $\{\mu_R, \sigma_R\}$, we employ the transformation functions given in equation 4.4. Note that the likelihood $f(\mathbf{y}|\boldsymbol{\theta})$ describes a probability density only for $\mathbf{y}$ and not for $\boldsymbol{\theta}$. Therefore this re-parametrization $\boldsymbol{\theta}_{\ln} = \text{fct}(\boldsymbol{\theta})$ of the likelihood does not require using the change of variable rule we saw in §3.4.

Figure 6.24 presents the prior PDF for parameters, the likelihood, the posterior, and the posterior predictive for (a) $D = 1$, (b) $D = 5$, and (c) $D = 100$ observations. We see that, as the number of observations $D$ increases, the likelihood and posterior for $\boldsymbol{\theta}$ become concentrated around the true values, and the posterior predictive for $R$ tends toward the true PDF with parameters $\mu_R \to \check{\mu}_R = 42\,\text{MPa}$ and $\sigma_R \to \check{\sigma}_R = 2\,\text{MPa}$.

## 6.6  Conjugate Priors

The previous section went through the process of using numerical approximation methods for estimating the posterior PDF for the parameters of a hidden random variable $X$. This complex process

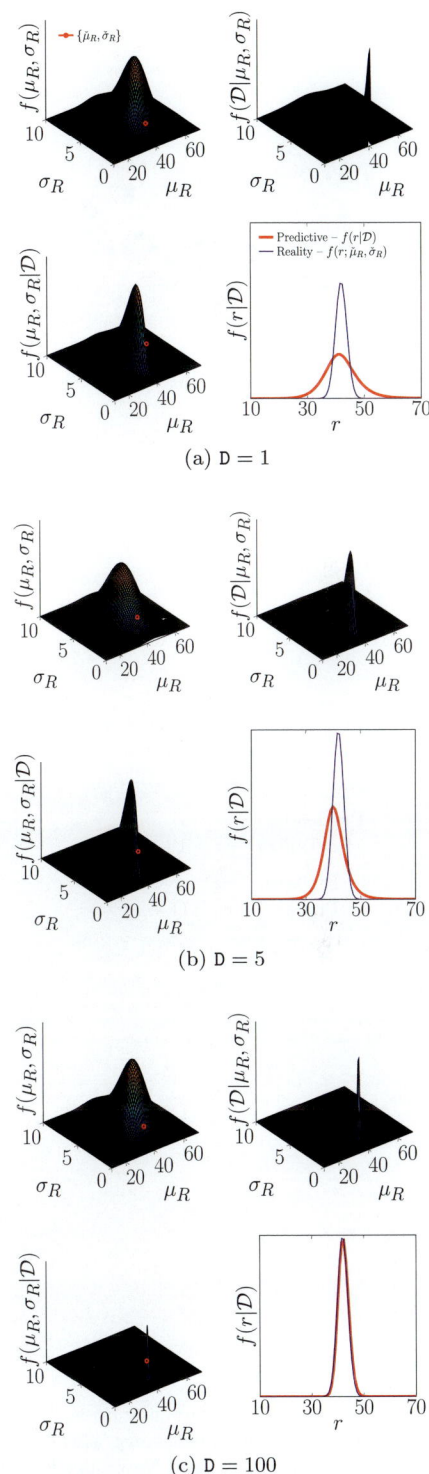

Figure 6.24: Examples of Bayesian estimation as a function of the number of observations employed, $D = \{1, 5, 100\}$.

Figure 6.25: Thumbtacks described by a random variable with sample space $\mathcal{S} = \{\text{Head}\,(\text{H}), \text{Tail}(\text{T})\}$. (Photo: Michel Goulet)

$$
\begin{aligned}
\Pr(\text{Head}|x) &= x \\
\Pr(\text{Tail}|x) &= 1 - x \\
\Pr(\mathcal{D}_\text{H}|x) &= x^{\text{D}_\text{H}} \\
\Pr(\mathcal{D}_\text{T}|x) &= (1 - x)^{\text{D}_\text{T}} \\
\Pr(\mathcal{D}|x) &= x^{\text{D}_\text{H}}(1 - x)^{\text{D}_\text{T}}
\end{aligned}
$$

[5] Wikipedia (2017). Conjugate prior. URL https://en.wikipedia.org/wiki/conjugate_prior. Accessed November 8, 2019.

[6] Murphy, K. P. (2007). Conjugate Bayesian analysis of the Gaussian distribution. Citeseer.

is necessary when no analytical formulation exists to describe the posterior.

For specific combinations of *prior* distribution and *likelihood* function, the *posterior* PDF follows the same type of distribution as the prior PDF. These specific combinations are referred to as *conjugate priors*. For conjugate priors, analytic formulations are available to describe the posterior PDF of parameters $f(\boldsymbol{\theta}|\mathcal{D})$ as well as the posterior predictive $f(x|\mathcal{D})$.

One classic example of conjugate prior is the *Beta-Binomial*. Take, for example, the probability that a thumbtack (see figure 6.25) lands on either *Head* or *Tail*, $\mathcal{S} = \{\text{Head}\,(\text{H}), \text{Tail}\,(\text{T})\}$. We can employ the Beta-Binomial conjugate prior to describe the posterior PDF for $x = \Pr(\text{Head}|\mathcal{D}) \in (0, 1)$, given a set of observations $\mathcal{D}$. It is suited in this case to assume that the prior PDF for $x = \Pr(\text{Head})$ follows a Beta PDF,

$$
f(x) = \mathcal{B}(x; \alpha, \beta) = \frac{x^{\alpha-1}(1 - x)^{\beta-1}}{\mathbf{B}(\alpha, \beta)},
$$

where $\mathbf{B}(\alpha, \beta)$ is the normalization constant. In this case, observations are binary, so for a set containing $\text{D} = \text{D}_\text{H} + \text{D}_\text{T}$ observations, heads are assigned the outcome H: $\mathcal{D}_\text{H} = \{\text{H}, \text{H}, \cdots, \text{H}\}_{1 \times \text{D}_\text{H}}$ and tails are assigned the outcome T: $\mathcal{D}_\text{T} = \{\text{T}, \text{T}, \cdots, \text{T}\}_{1 \times \text{D}_\text{T}}$. In that case, the likelihood describing the probability of observing $\text{D}_\text{H}$ heads and $\text{D}_\text{T}$ tails follows a Binomial distribution,

$$
p(\mathcal{D}|x) \propto x^{\text{D}_\text{H}}(1 - x)^{\text{D}_\text{T}}.
$$

The multiplication of the prior and the likelihood remains analytically tractable so that

$$
f(x|\mathcal{D}) \propto x^{\text{D}_\text{H}}(1 - x)^{\text{D}_\text{T}} \cdot \frac{x^{\alpha-1}(1 - x)^{\beta-1}}{\mathbf{B}(\alpha, \beta)},
$$

and by recalculating the normalization constant, we find that the posterior also follows a Beta PDF with parameters $\alpha + \text{D}_\text{H}$ and $\beta + \text{D}_\text{T}$,

$$
f(x|\mathcal{D}) = \frac{x^{\alpha+\text{D}_\text{H}-1}(1 - x)^{\beta+\text{D}_\text{T}-1}}{\mathbf{B}(\alpha + \text{D}_\text{H}, \beta + \text{D}_\text{T})} = \mathcal{B}(x; \alpha + \text{D}_\text{H}, \beta + \text{D}_\text{T}).
$$

The analytic formulation for conjugate priors with their parameters, likelihood, posterior, and posterior predictive can be found online[5] or in specialized publications such as the compendium by Murphy.[6]

*Example: Traffic safety* We take as another example an intersection where the number of users is $u = 3 \times 10^5$/yr and where 20 accidents occurred over the last 5 years. Note that for the purpose of simplicity, the number of passages without accidents is taken as the total number of users rather than this number minus the number of accidents. This current information can be summarized as

$$\mathcal{D}_{\text{old}} : \begin{cases} \mathcal{D}_{\mathsf{N}} = 5 \cdot (3 \times 10^5) & \text{(passages without accident, } \mathsf{N}) \\ \mathcal{D}_{\mathsf{A}} = 20 & \text{(accidents, } \mathsf{A}). \end{cases}$$

Our prior knowledge for $x$, the probability of accident per user, is described by the Beta PDF $f(x) = \mathcal{B}(x; 1, 1)$, where the parameters are $\alpha_0 = \beta_0 = 1$, which corresponds to a uniform distribution over the interval $(0, 1)$. The posterior PDF describing the daily probability of accident given the observations $\mathcal{D}_{\text{old}}$ is thus also described by the Beta PDF,

$$f(x|\mathcal{D}_{\text{old}}) = \mathcal{B}(x; \alpha_0 + \mathcal{D}_{\mathsf{A}}, \beta_0 + \mathcal{D}_{\mathsf{N}}).$$

Figure 6.27 presents the posterior PDF transformed in the $\log_{10}$-space using the change of variable rule (see §3.4),

$$\begin{aligned} f(\log_{10}(x)|\mathcal{D}_{\text{old}}) &= f(x|\mathcal{D}_{\text{old}}) \cdot \left| \frac{d\log_{10}(x)}{dx} \right|^{-1} \\ &= f(x|\mathcal{D}_{\text{old}}) \cdot x \ln 10. \end{aligned}$$

Now, we consider that following refection works made on the intersection in order to decrease the probability of accidents, we observe $a = 1$ accident over a period of one year, that is, $3 \times 10^5$ users without accident. In order to estimate the new probability of accident, we assume that our previous posterior knowledge $f(x|\mathcal{D}_{\text{old}})$ is an upper-bound for $X$, because the refection works made the intersection safer. The new posterior then becomes

$$f(x|\mathcal{D}) \propto \underbrace{x^1 \cdot (1-x)^{3 \times 10^5}}_{\text{likelihood}} \cdot \underbrace{(1 - F(x|\mathcal{D}_{\text{old}}))}_{\text{prior}}.$$

Note that because the prior is now described by the complement of the CDF of $X|\mathcal{D}_{\text{old}}$, which is not a conjugate prior, the posterior does not have an analytic solution anymore, and the evidence $f(\mathcal{D}_{\text{new}})$ must be approximated using numerical integration. Figure 6.28 presents the prior, $1 - F(x|\mathcal{D}_{\text{old}})$ and posterior PDF, $f(x|\mathcal{D})$, for the probability of an accident after the intervention. The posterior predictive probability of an accident before the refection works was $1.3 \times 10^{-5}$ compared to $5.8 \times 10^{-6}$ after.

Figure 6.26: Example of application of conjugate priors on traffic data. (Photo: John Matychuk)

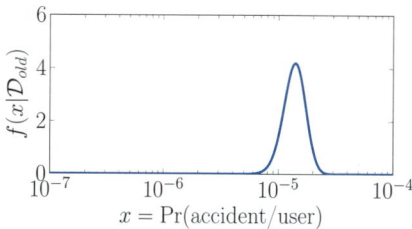

Figure 6.27: Posterior PDF for the probability of an accident *before* refection works.

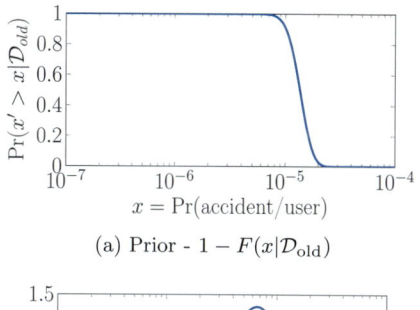

(a) Prior - $1 - F(x|\mathcal{D}_{\text{old}})$

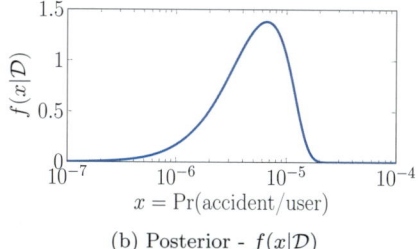

(b) Posterior - $f(x|\mathcal{D})$

Figure 6.28: Prior and posterior PDF for the probability of an accident *after* refection works.

## 6.7  Approximating the Posterior

For complex machine learning models such as those that will be covered in the next chapters, conjugate priors are often not applicable, and it can be computationally prohibitive to rely on sampling methods for performing Bayesian estimation. In those cases, we may have to employ *point estimates* rather than a probability density function for describing the posterior PDF of model parameters.

### 6.7.1  Maximum Likelihood and Posterior Estimates

Instead of estimating the posterior using a Bayesian approach, a common approximation is to find a single set of optimal parameter values $\boldsymbol{\theta}^*$ that maximizes the likelihood (MLE) or the product of the likelihood and the prior (MAP) where

$$
\boldsymbol{\theta}^* = \begin{cases}
\arg\max_{\boldsymbol{\theta}} f(\mathcal{D}|\boldsymbol{\theta}) & \text{Maximum likelihood estimate (MLE)} \\
\arg\max_{\boldsymbol{\theta}} f(\mathcal{D}|\boldsymbol{\theta}) \cdot f(\boldsymbol{\theta}) & \text{Maximum a-posteriori (MAP).}
\end{cases}
$$

Optimization methods covered in chapter 5 can be employed to identify $\boldsymbol{\theta}^*$. Note that it is common to perform the optimization using a log-transformed objective function, for example, $\ln(f(\mathcal{D}|\boldsymbol{\theta}))$, for numerical stability reasons (see equation 6.5). Note that it is equivalent to optimize either the original or the log-transformed objective function,

$$
\arg\max_{\boldsymbol{\theta}} f(\mathcal{D}|\boldsymbol{\theta}) \equiv \arg\max_{\boldsymbol{\theta}} \ln f(\mathcal{D}|\boldsymbol{\theta}).
$$

Estimating only the most likely value with respect to either the MAP or MLE criterion can lead to a negligible approximation error when applied to *identifiable* problems (see §6.3.4), where the amount of independent data is large; in that case, the posterior is concentrated around a single value $f(\boldsymbol{\theta}|\mathcal{D}) \to \delta(\boldsymbol{\theta}^*)$. When the set of observations is small, in the case of a multimodal posterior, or for *non-identifiable* problems, estimating only $\boldsymbol{\theta}^*$ is no longer a good approximation of the posterior. The definition of what is a *small* or a *large* data set is problem dependent. In simple cases, data sets containing fewer than 10 observations will typically lead to significant epistemic uncertainty associated with unknown parameter values. Cases involving complex models, such as those treated in chapter 12, may need tens of thousands ($D \propto 10^5$) of observations to reach the point where the epistemic uncertainty associated with unknown parameter values is negligible.

We saw in §5.5 that it is common to resort to parameter-space transformations in order to perform optimization in an unconstrained domain. For an MLE, the quantity maximized is the likelihood of observations $\mathbf{y}$, given the parameters $\boldsymbol{\theta}$. Because $f(\mathbf{y}|\boldsymbol{\theta})$ describes a probability density only for $\mathbf{y}$ and not for $\boldsymbol{\theta}$, a parameter-space transformation for an MLE does not require using the change of variable rule we saw in §3.4. On the other hand, the MAP seeks to maximize the product of the likelihood and the prior, $f(\mathbf{y}|\boldsymbol{\theta}) \cdot f(\boldsymbol{\theta})$. Here, the prior is a probability density, so the change of variable rule applies when we employ space transformations. Nevertheless, as we saw in figure 3.19, when a PDF is subject to a nonlinear transformation, the modes in the original and transformed spaces do not coincide. Therefore, the MLE is invariant to parameter-space transformations and the MAP is not.[7] Note that Bayesian estimation is invariant to parameter-space transformations, given that we apply the change of variable rule. We will apply parameter-space transformations in the context of Bayesian estimation in chapter 7.

[7] Jermyn, I. H. (2005). Invariant Bayesian estimation on manifolds. *The Annals of Statistics 33*(2), 583–605.

### 6.7.2   Laplace Approximation

Another method for approximating a posterior PDF for the unknown parameters of a hidden random variable is the *Laplace approximation*. This method approximates the posterior as a multivariate Normal with a posterior mean vector equal to either the MLE or MAP estimate $\boldsymbol{\theta}^*$, and a posterior covariance $\boldsymbol{\Sigma}_{\boldsymbol{\theta}^*}$,

$$\underbrace{f(\boldsymbol{\theta}|\mathcal{D})}_{\text{posterior}} \approx \underbrace{\mathcal{N}(\boldsymbol{\theta}; \overbrace{\boldsymbol{\theta}^*}^{\text{MLE/MAP estimate}}, \boldsymbol{\Sigma}_{\boldsymbol{\theta}^*})}_{\text{Laplace approximation}}.$$

When we choose the MLE approximation to describe the posterior mode, the posterior covariance is approximated by the inverse Hessian matrix of the negative log-likelihood evaluated at $\boldsymbol{\theta}^*$,

$$
\begin{aligned}
\boldsymbol{\Sigma}_{\boldsymbol{\theta}^*} &= \mathbf{H}[-\overbrace{\ln f(\mathcal{D}|\boldsymbol{\theta}^*)}^{\text{log-likelihood}}]^{-1} \\
&= \left[ -\frac{\partial^2 \ln f(\mathcal{D}|\boldsymbol{\theta})}{\partial \boldsymbol{\theta} \partial \boldsymbol{\theta}^{\mathsf{T}}} \right]^{-1}_{\boldsymbol{\theta}^*}.
\end{aligned}
$$

Details regarding the numerical estimation of the Hessian are presented in §5.4. In order to include prior knowledge, we have to estimate the Laplace approximation using $\ln(f(\mathcal{D}|\boldsymbol{\theta}^*) \cdot f(\boldsymbol{\theta}^*))$ instead of the log-likelihood $\ln f(\mathcal{D}|\boldsymbol{\theta}^*)$ alone.

The relation between the second derivative of the log-likelihood evaluated at $\boldsymbol{\theta}^*$ and the covariance $\boldsymbol{\Sigma}_{\boldsymbol{\theta}^*}$ can be explained by looking at the multivariate Normal with probability density function described by

$$f(\boldsymbol{\theta}) = \frac{1}{(2\pi)^{\mathsf{P}/2}(\det \boldsymbol{\Sigma_\theta})^{1/2}} \exp\left(-\frac{1}{2}(\boldsymbol{\theta}-\boldsymbol{\mu_\theta})^\mathsf{T}\boldsymbol{\Sigma_\theta}^{-1}(\boldsymbol{\theta}-\boldsymbol{\mu_\theta})\right),$$

where P is the number of parameters $\boldsymbol{\theta} = [\theta_1\ \theta_2\ \cdots\ \theta_\mathsf{P}]^\mathsf{T}$. By taking the natural logarithm of $f(\boldsymbol{\theta})$, we obtain

$$\ln f(\boldsymbol{\theta}) = -\frac{\mathsf{P}}{2}\ln 2\pi - \frac{1}{2}\ln\det\boldsymbol{\Sigma_\theta} - \frac{1}{2}(\boldsymbol{\theta}-\boldsymbol{\mu_\theta})^\mathsf{T}\boldsymbol{\Sigma_\theta}^{-1}(\boldsymbol{\theta}-\boldsymbol{\mu_\theta}),$$

where the derivative of $\ln f(\boldsymbol{\theta})$ with respect to $\boldsymbol{\theta}$ is

$$
\begin{aligned}
\frac{\partial \ln f(\boldsymbol{\theta})}{\partial \boldsymbol{\theta}} &= -\frac{1}{2}\frac{\partial\left((\boldsymbol{\theta}-\boldsymbol{\mu_\theta})^\mathsf{T}\boldsymbol{\Sigma_\theta}^{-1}(\boldsymbol{\theta}-\boldsymbol{\mu_\theta})\right)}{\partial\boldsymbol{\theta}} \\
&= -\frac{1}{2}\left(2\boldsymbol{\Sigma_\theta}^{-1}(\boldsymbol{\theta}-\boldsymbol{\mu_\theta})\right) \\
&= -\boldsymbol{\Sigma_\theta}^{-1}(\boldsymbol{\theta}-\boldsymbol{\mu_\theta}).
\end{aligned}
$$

If we differentiate this last expression a second time with respect to $\boldsymbol{\theta}$, we obtain

$$\frac{\partial^2 \ln f(\boldsymbol{\theta})}{\partial\boldsymbol{\theta}\partial\boldsymbol{\theta}^\mathsf{T}} = -\boldsymbol{\Sigma_\theta}^{-1}.$$

This development shows how the Laplace approximation allows estimating the covariance matrix using the posterior PDF's curvatures.

In the case of a Gaussian log-posterior, the second-order derivatives are constant, so the curvatures can be evaluated anywhere in the domain. In practice, we employ the Laplace approximation when the posterior is not Gaussian; in order to minimize the approximation error, we evaluate the second-order derivatives of the log-posterior at the MLE or MAP $\boldsymbol{\theta}^*$, where the probability density is the highest.

Figure 6.29 presents an example of an application of the Laplace approximation, where a generic posterior PDF for two parameters is approximated by a bivariate Normal PDF. The Hessian matrix evaluated at the mode $\boldsymbol{\theta}^* = [-1.47\ 0.72]^\mathsf{T}$ is

$$\mathbf{H}[-\ln f(\mathcal{D}|\boldsymbol{\theta}^*)] = \begin{bmatrix} 0.95 & 0.02 \\ 0.02 & 0.95 \end{bmatrix},$$

so the Laplace approximation of the posterior covariance is described by

$$\boldsymbol{\Sigma}_{\boldsymbol{\theta}^*} = \mathbf{H}[-\ln f(\mathcal{D}|\boldsymbol{\theta}^*)]^{-1} = \begin{bmatrix} 1.05 & -0.02 \\ -0.02 & 1.05 \end{bmatrix}.$$

The closer the posterior is to a multivariate Normal, the better the Laplace approximation is. Note that for problems involving several parameters, computing the Hessian matrix using numerical derivatives becomes computationally prohibitive. One possible

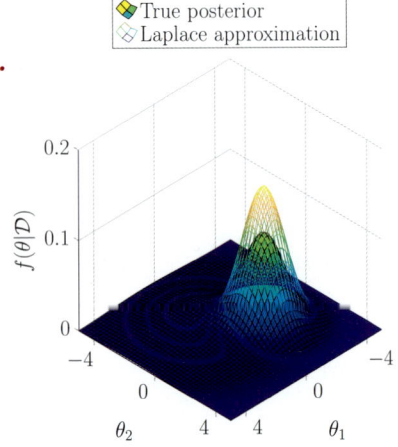

Figure 6.29: Example of an application of the Laplace approximation to describe a posterior PDF.

approximation is to estimate only the main diagonal of the Hessian. The main diagonal of the Hessian contains information about the variances of the posterior with respect to each variable without providing information about the covariances (i.e., the pairwise linear dependencies between variables).

In addition to being employed to approximate the posterior PDF, the Laplace approximation can be used to estimate the evidence $f(\mathcal{D})$. With the Laplace approximation, the posterior is

$$
\begin{aligned}
f(\boldsymbol{\theta}|\mathcal{D}) &\approx \mathcal{N}(\boldsymbol{\theta};\boldsymbol{\theta}^*,\boldsymbol{\Sigma}_{\boldsymbol{\theta}^*}) \\
&= \underbrace{\frac{1}{(2\pi)^{\mathrm{P}/2}(\det\boldsymbol{\Sigma}_{\boldsymbol{\theta}^*})^{1/2}}}_{(a)} \underbrace{\exp\left(-\frac{1}{2}(\boldsymbol{\theta}-\boldsymbol{\theta}^*)^{\mathsf{T}}\boldsymbol{\Sigma}_{\boldsymbol{\theta}^*}^{-1}(\boldsymbol{\theta}-\boldsymbol{\theta}^*)\right)}_{(b)}.
\end{aligned}
$$

$$(6.9)$$

In equation 6.9, (a) corresponds to the normalization constant for the exponential term indicated in (b), for which the value is $\exp(0) = 1$ when evaluated at the mode $\boldsymbol{\theta}^*$ (see §4.1). When estimating $\tilde{f}(\boldsymbol{\theta}^*)$ using either an MAP $\tilde{f}(\boldsymbol{\theta}^*) = f(\mathcal{D}|\boldsymbol{\theta}^*) \cdot f(\boldsymbol{\theta}^*)$ or an MLE $\tilde{f}(\boldsymbol{\theta}^*) = f(\mathcal{D}|\boldsymbol{\theta}^*)$, its value is not going to equal 1. Therefore, the normalization constant in (a) must be multiplied by the value of the unnormalized posterior evaluated at the mode $\boldsymbol{\theta}^*$ so that

$$
\hat{f}(\mathcal{D}) = \tilde{f}(\boldsymbol{\theta}^*) \cdot (2\pi)^{\mathrm{P}/2}(\det\boldsymbol{\Sigma}_{\boldsymbol{\theta}^*})^{1/2}.
\tag{6.10}
$$

The reader interested in the formal derivation of this result from a Taylor-series expansion should consult MacKay.[8]

[8] MacKay, D. J. C. (2003). *Information theory, inference, and learning algorithms.* Cambridge University Press.

## 6.8   Model Selection

In the concrete strength estimation example presented in §6.5.3, it is known that the data was generated from a log-normal distribution. In real-life cases, the PDF form for the hidden random variable $X$ may either remain unknown or not correspond to an analytically tractable probability density function. In these cases, it is required to compare the performance of several *model classes* at explaining the observations $\mathcal{D}$.

Given a set of M *model classes* $m \in \mathcal{M} = \{1, 2, \cdots, \mathtt{M}\}$, the posterior probability of each class given $\mathcal{D}$ is described using Bayes rule,

$$
\underbrace{p(m|\mathcal{D})}_{\text{model class posterior}} = \frac{\overbrace{f(\mathcal{D}|m)}^{\text{model likelihood}} \cdot \overbrace{p(m)}^{\text{model class prior}}}{\underbrace{f(\mathcal{D}^{\{1:\mathtt{M}\}})}_{\text{evidence}}},
\tag{6.11}
$$

where $p(m|\mathcal{D})$ and $p(m)$ are respectively the posterior and prior

**Note:** With Bayesian model selection, the objective is not to identify the single best model; it allows quantifying the posterior probability of each model, which can be employed for performing subsequent predictions.

probability mass functions for the model classes. The likelihood of a model class $m$ is obtained by marginalizing the model parameters from the joint distribution of observations and parameters,

$$f(\mathcal{D}|m) = \int f(\mathcal{D}|\boldsymbol{\theta}^{\{m\}}) \cdot f(\boldsymbol{\theta}^{\{m\}}) d\boldsymbol{\theta}, \qquad (6.12)$$

where $\boldsymbol{\theta}^{\{m\}}$ is the set of parameters associated with the model class $m$. The likelihood of a model class in equation 6.12 corresponds to its evidence, which also corresponds to the normalization constant from the posterior PDF for its parameters,

$$f(\mathcal{D}|m) \equiv \overbrace{f(\mathcal{D}^{\{m\}})}^{\text{normalization constant for model class } m}. \qquad (6.13)$$

The normalization constant for the model comparison problem presented in equation 6.11 is obtained by summing the likelihood of each model class times its prior probability,

$$f(\mathcal{D}^{\{1:M\}}) = \sum_{m \in \mathcal{M}} f(\mathcal{D}|m) \cdot p(m).$$

Note that using the importance-sampling procedure presented in equation 6.8 for estimating the normalization constants in equation 6.13 is only a viable option for simple problems involving a small number of parameters and a small number of observations. Section 7.5 further discusses the applicability of more advanced sampling methods for estimating $f(\mathcal{D})$.

Here, we revisit the concrete strength example presented in §6.5.3. We compare the posterior probability for two model classes $m \in \mathcal{M} = \{L, N\}$ (i.e., a log-normal and a Normal PDF); note that we know the correct hypothesis is $m = L$ because simulated data were generated using a log-normal PDF.

For $D = 5$ observations, the evidence of each model is

$$\begin{aligned} f(\mathcal{D}|L) &= 8.79 \times 10^{-7} \\ f(\mathcal{D}|N) &= 6.44 \times 10^{-7}. \end{aligned}$$

With the hypothesis that the prior probability of each model is equal, $p(L) = p(N) = 0.5$, the posterior probability for model classes $m$ are

$$\begin{aligned} p(L|\mathcal{D}) &= \frac{8.79 \times 10^{-7} \times 0.5}{8.79 \times 10^{-7} \times 0.5 + 6.44 \times 10^{-7} \times 0.5} \\ &= 0.58 \end{aligned}$$

$$\begin{aligned} p(N|\mathcal{D}) &= 1 - p(L|\mathcal{D}) \\ &= \frac{6.44 \times 10^{-7} \times 0.5}{8.79 \times 10^{-7} \times 0.5 + 6.44 \times 10^{-7} \times 0.5} \\ &= 0.42. \end{aligned}$$

This result indicates that both models predict with almost equal performance the observations $\mathcal{D}$. Therefore, if we want to make predictions for unobserved quantities, it is possible to employ both models where the probability of each of them is quantified by its posterior probability $p(m|\mathcal{D})$.

In practice, because estimating the evidence $f(\mathcal{D}|m)$ is a computationally demanding task, it is common to resort to approximations such as the *Akaike information criterion* (AIC) and *Bayesian information criterion* (BIC) for comparing model classes. Note that the BIC is a special case derived from the Laplace approximation in equation 6.10 for estimating $f(\mathcal{D})$. Other advanced approximate methods for comparing model classes are the *Deviance information criterion*[9] (DIC) and the *Watanabe-Akaike information criterion*[10] (WAIC). The reader interested in these methods should refer to specialized textbooks.[11] Another common model selection technique widely employed in machine learning is *cross-validation*; the main idea consists in employing a first subset of data to estimate the model parameter and then test the predictive performance of the model using a second independent data set. The details regarding cross-validation are presented in §8.1.2, where the method is employed to discriminate among model structures in the context of regression problems.

[9] Spiegelhalter, D. J., N. G. Best, B. P. Carlin, and A. van der Linde (2002). Bayesian measures of model complexity and fit. *Journal of the Royal Statistical Society: Series B (Statistical Methodology) 64* (4), 583–639.

[10] Watanabe, S. (2010). Asymptotic equivalence of Bayes cross validation and widely applicable information criterion in singular learning theory. *Journal of Machine Learning Research 11*, 3571–3594.

[11] Burnham, K. P. and D. R. Anderson (2002). *Model selection and multimodel inference: A practical information-theoretic approach* (2nd ed.). Springer; and Gelman, A., J. B. Carlin, H. S. Stern, and D. B. Rubin (2014). *Bayesian data analysis* (3rd ed.). CRC Press.

*Exercises*

**P6.1** Explain the difference between a posterior, a prior, and a likelihood function.

**P6.2** Explain what a non-informative prior is.

**P6.3** Explain why it is hard to compute the normalization constant $f(\mathcal{D})$.

**P6.4** Explain in what circumstances the prior plays an important role and when it has a negligible role.

**P6.5** Explain what a censored observation is.

**P6.6** Explain why, in the context of importance sampling, when the number of observations D is large, it is typically not efficient to draw samples from the prior in order to estimate $f(\mathcal{D})$.

**P6.7** Explain what a posterior predictive PDF is.

**P6.8** Explain what a non-identifiable problem is and what we can do about it.

**P6.9** Explain what the main difficulty is when using Bayesian model selection.

**P6.10** Explain what a conjugate prior is and why it is computationally efficient.

**P6.11** Given the prior failure probability of a pipe $\Pr(\mathsf{F}) = 0.001$ and an observation $y \in \{+, -\}$ from a non-destructive test with a likelihood

$$\begin{array}{llll} \Pr(+|\mathsf{F}) &=& 0.9 & \Pr(-|\mathsf{F}) &=& 0.05 \\ \Pr(+|\overline{\mathsf{F}}) &=& 0.05 & \Pr(-|\overline{\mathsf{F}}) &=& 0.9 \end{array},$$

compute the posterior probabilities of a pipe failure for both measurement outcomes $\Pr(\mathsf{F}|+)$ and $\Pr(\mathsf{F}|-)$, and explain why $\Pr(\mathsf{F}|+) \neq \Pr(\overline{\mathsf{F}}|-)$.

**P6.12** For the observation models below, formulate the likelihood for a set of observations $\mathcal{D} = \{y_1, y_2, \cdots, y_{\mathsf{D}}\}$.

a) $y = x + v$, $\begin{cases} x = \text{constant} \\ v : V \sim \mathcal{N}(v; 0, \sigma_V^2), \ V_i \perp\!\!\!\perp V_j, \forall i \neq j \end{cases}$

b) $y > x + v$, $\begin{cases} x = \text{constant} \\ v : V \sim \mathcal{N}(v; 0, \sigma_V^2), \ V_i \perp\!\!\!\perp V_j, \forall i \neq j \end{cases}$

c) $y = x + v$, $\begin{cases} x : X \sim \mathcal{N}(x; \mu_X, \sigma_X^2), \ X_i \perp\!\!\!\perp X_j, \forall i \neq j \\ v : V \sim \mathcal{N}(v; 0, \sigma_V^2), \ V_i \perp\!\!\!\perp V_j, \forall i \neq j \end{cases}$

# 7
# *Markov Chain Monte Carlo*

The *importance sampling* method was introduced in chapter 6 for estimating the normalization constant of a posterior probability density function (PDF) as well as its expected values and covariance. A limitation of this approach is related to the difficulties associated with the choice of an efficient sampling distribution $h(\cdot)$. The sampling methods covered in this chapter address this limitation by providing ways of sampling directly from the posterior using Markov chain Monte Carlo (MCMC) methods.

*Markov chain Monte Carlo* takes its name from the *Markov property*, which states:

> *Given the present, the future is independent from the past.*

It means that if we know the current state of the system, we can predict future states without having to consider past states. Starting from $X_t \sim p(x_t)$, we can transition to $x_{t+1}$ using a transition probability $p(x_{t+1}|x_t)$ that only depends on $x_t$. It implicitly conveys that the conditional probability depending on $x_t$ is equivalent to the one depending on $x_{1:t}$,

$$p(x_{t+1}|x_t) = p(x_{t+1}|x_1, x_2, \cdots, x_t).$$

A *Markov chain* defines the joint distribution for random variables by combining the chain rule (see §3.3.4) and the Markov property so that

$$p(x_{1:\mathrm{T}}) = p(x_1)p(x_2|x_1)p(x_3|x_2)\cdots p(x_\mathrm{T}|x_{\mathrm{T}-1}) = p(x_1)\prod_{t=2}^{\mathrm{T}} p(x_t|x_{t-1}).$$

The idea behind MCMC is to construct a Markov chain for which the stationary distribution is the posterior. Conceptually, it corresponds to *randomly walking* through the parameter space so that the *fraction of steps spent* at exploring each part of the domain is proportional to the posterior density. Figure 7.1 presents an ex-

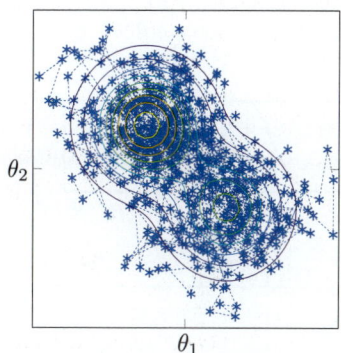

$\theta_2$

$\theta_1$

Figure 7.1: Example of random walk generated with MCMC.

ample of a random walk generated with MCMC that follows the density described by the underlying contour plot.

Several MCMC methods exist: Metropolis-Hastings, Gibbs sampling, slice sampling, Hamiltonian Monte Carlo, and more. This chapter covers only the Metropolis and Metropolis-Hastings methods because they are the most accessible. The reader interested in advanced methods should refer to dedicated textbooks such as the one by Brooks et al.[1]

[1] Brooks, S., A. Gelman, G. Jones, and X.-L. Meng (2011). *Handbook of Markov Chain Monte Carlo*. CRC Press.

## 7.1   Metropolis

The *Metropolis* algorithm[2] was developed during the Second World War while working on the Manhattan project (i.e., the atomic bomb) at Los Alamos, New Mexico. Metropolis is not the most efficient sampling algorithm, yet it is a simple one allowing for an easy introduction to MCMC methods.

[2] Metropolis, N., A. W. Rosenbluth, M. N. Rosenbluth, A. H. Teller, and E. Teller (1953). Equation of state calculations by fast computing machines. *The Journal of Chemical Physics 21*(6), 1087–1092.

The Metropolis algorithm requires defining an *initial state* for a set of P parameters $\boldsymbol{\theta}_0 = [\theta_1 \ \theta_2 \ \cdots \ \theta_P]_0^\mathsf{T}$, a target distribution $\tilde{f}(\boldsymbol{\theta}) = f(\mathcal{D}|\boldsymbol{\theta}) \cdot f(\boldsymbol{\theta})$ corresponding to the unnormalized posterior we want to sample from, and a *proposal distribution*, $q(\boldsymbol{\theta}'|\boldsymbol{\theta})$, which describes where to move next given the current parameter values. The proposal must have a *nonzero probability to transit from the current state to any state* supported by the *target distribution* and must be *symmetric*, that is,

**Notation**
Initial state: $\boldsymbol{\theta}_0$
Target distribution: $\tilde{f}(\boldsymbol{\theta})$
Proposal distribution: $q(\boldsymbol{\theta}'|\boldsymbol{\theta})$

$$q(\boldsymbol{\theta}'|\boldsymbol{\theta}) = q(\boldsymbol{\theta}|\boldsymbol{\theta}').$$

The *Normal distribution* (see §4.1) is a common general-purpose proposal distribution,

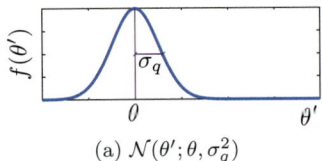

(a) $\mathcal{N}(\theta'; \theta, \sigma_q^2)$

$$q(\boldsymbol{\theta}'|\boldsymbol{\theta}) = \mathcal{N}(\boldsymbol{\theta}'; \boldsymbol{\theta}, \boldsymbol{\Sigma}_q),$$

where the mean is defined by the current position and the probability to move in a region around the current position is controlled by the covariance matrix $\boldsymbol{\Sigma}_q$. Figure 7.2 presents an example of 1-D and 2-D Normal proposal distributions.

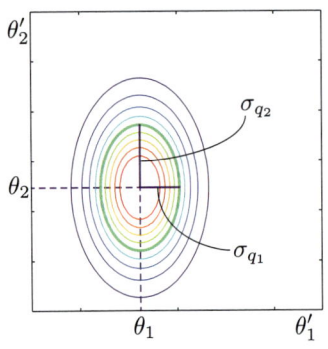

(b) $\mathcal{N}\left(\boldsymbol{\theta}'; [\theta_1 \ \theta_2]^\mathsf{T}, \mathrm{diag}([\sigma_{q_1}^2 \ \sigma_{q_2}^2])\right)$

Figure 7.2: Examples of 1-D and 2-D proposal distributions, $q(\boldsymbol{\theta}'|\boldsymbol{\theta})$.

The Metropolis algorithm is recursive so at the $s^\text{th}$ step, given a current position in the parameter space $\boldsymbol{\theta} = \boldsymbol{\theta}_s$, we employ $q(\boldsymbol{\theta}'|\boldsymbol{\theta})$ to propose moving to a new position $\boldsymbol{\theta}'$. If the target distribution evaluated at the proposed location is greater than or equal to the current one, that is, $\tilde{f}(\boldsymbol{\theta}') \geq \tilde{f}(\boldsymbol{\theta})$—we accept the proposed location. If the proposed location has a target value that is lower than the current one, we accept moving to the proposed location with a probability equal to the *acceptance ratio* $\frac{\tilde{f}(\boldsymbol{\theta}')}{\tilde{f}(\boldsymbol{\theta})}$. In the case where

**Note:** In order to accept a proposed location with a probability equal to $r = \tilde{f}(\boldsymbol{\theta}')/\tilde{f}(\boldsymbol{\theta})$, we compare it with a sample $u$ taken from $\mathcal{U}(0, 1)$. If $u \leq r$, we accept the move; otherwise, we reject it.

the proposed location is rejected, we stay at the current location and $\boldsymbol{\theta}_{s+1} = \boldsymbol{\theta}_s$. Each step from the Metropolis sampling method is formalized in algorithm 4.

---

**Algorithm 4:** Metropolis sampling

| | | |
|---|---|---|
| 1 | define $\tilde{f}(\boldsymbol{\theta})$ | (target distribution) |
| 2 | define $q(\boldsymbol{\theta}'|\boldsymbol{\theta})$ | (proposal distribution) |
| 3 | define S | (number of samples) |
| 4 | initialize $\mathcal{S} = \emptyset$ | (set of samples) |
| 5 | initialize $\boldsymbol{\theta}_0$ | (initial starting location) |
| 6 | **for** $s \in \{0, 1, 2, \cdots, S-1\}$ **do** | |
| 7 | $\quad$ define $\boldsymbol{\theta} = \boldsymbol{\theta}_s$ | |
| 8 | $\quad$ sample $\boldsymbol{\theta}' \sim q(\boldsymbol{\theta}'|\boldsymbol{\theta})$ | |
| 9 | $\quad$ compute $\alpha = \frac{\tilde{f}(\boldsymbol{\theta}')}{\tilde{f}(\boldsymbol{\theta})}$ | (acceptance ratio) |
| 10 | $\quad$ compute $r = \min(1, \alpha)$ | |
| 11 | $\quad$ sample $u \sim \mathcal{U}(0,1)$ | |
| 12 | $\quad$ **if** $u \leq r$ **then** | |
| 13 | $\quad\quad$ $\boldsymbol{\theta}_{s+1} = \boldsymbol{\theta}'$ | |
| 14 | $\quad$ **else** | |
| 15 | $\quad\quad$ $\boldsymbol{\theta}_{s+1} = \boldsymbol{\theta}_s$ | |
| 16 | $\quad$ $\mathcal{S} \leftarrow \{\mathcal{S} \cup \{\boldsymbol{\theta}_{s+1}\}\}$ | (add to the set of samples) |

---

If we apply this recursive algorithm over S iterations, the result is that the fraction of the steps spent exploring each part of the domain is proportional to the density of the target distribution $\tilde{f}(\boldsymbol{\theta})$. Note that this last statement is valid under some conditions regarding the chain starting location and number of samples S that we will further discuss in §7.3. If we select a target distribution that is an unnormalized posterior, it implies that each sample $\boldsymbol{\theta}_s$ is a realization of that posterior, $\boldsymbol{\theta}_s : \boldsymbol{\theta} \sim f(\mathcal{D}|\boldsymbol{\theta}) \cdot f(\boldsymbol{\theta})$.

Figure 7.3 presents a step-by-step application of algorithm 4 for sampling a given target density $\tilde{f}(\theta)$. The proposal distribution employed in this example has a standard deviation $\sigma_q = 1$. At step $s = 1$, the target value at the proposed location $\theta'$ is greater than at the current location $\theta$, so the move is accepted. For the second step, $s = 2$, the target value at the proposed location is smaller than the value at the current location. The move is nevertheless accepted because the random number $u$ drawn from $\mathcal{U}(0,1)$ turned out to be smaller than the ratio $\frac{\tilde{f}(\theta')}{\tilde{f}(\theta)}$. At step $s = 3$, the target value at the proposed location $\theta'$ is greater than the value at the current location $\theta$, so the move is accepted. Figure 7.3d presents the chain containing a total of S = 2400 samples. The superposition of the sample's histogram and the target density confirms that the Metropolis algorithm is sampling from $\tilde{f}(\boldsymbol{\theta})$.

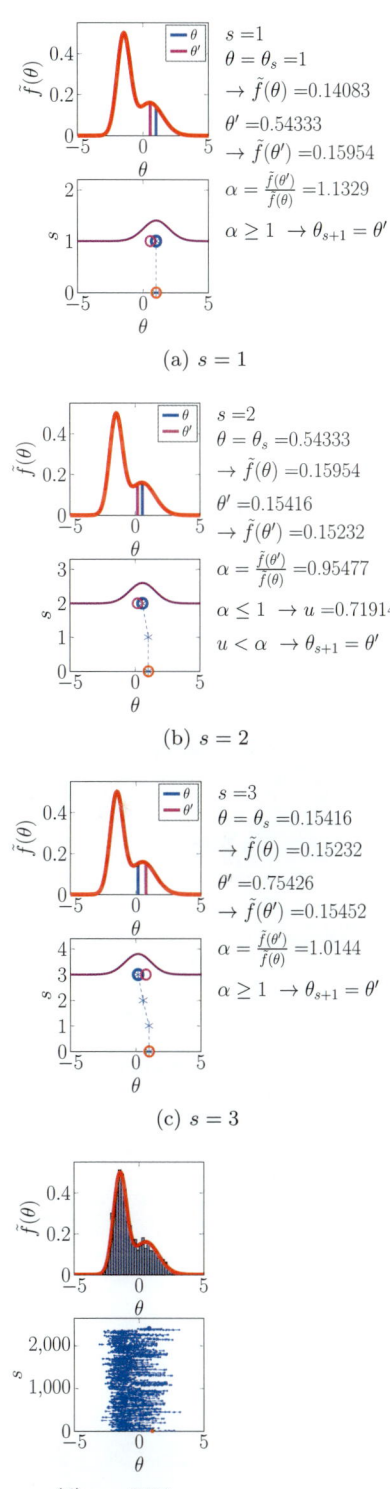

(a) $s = 1$

(b) $s = 2$

(c) $s = 3$

(d) $s = 2400$

Figure 7.3: Step-by-step example of 1-D Metropolis sampling.

[3] Hastings, W. K. (1970). Monte Carlo sampling methods using Markov chains and their applications. *Biometrika* 57(1), 97–109.

## 7.2   Metropolis-Hastings

The *Metropolis-Hastings* algorithm[3] is identical to the Metropolis algorithm except that it allows for *nonsymmetric transition probabilities* where

$$q(\boldsymbol{\theta}'|\boldsymbol{\theta}) \neq q(\boldsymbol{\theta}|\boldsymbol{\theta}').$$

The main change with Metropolis-Hastings is that when the proposed location $\boldsymbol{\theta}'$ has a target value that is lower than the current location $\boldsymbol{\theta}$, we accept the proposed location with a probability equal to the ratio $\frac{\tilde{f}(\boldsymbol{\theta}')}{\tilde{f}(\boldsymbol{\theta})}$ times the ratio $\frac{q(\boldsymbol{\theta}|\boldsymbol{\theta}')}{q(\boldsymbol{\theta}'|\boldsymbol{\theta})}$, that is, the ratio of the probability density of going from the proposed location to the current one, divided by the probability density of going from the current location to the proposed one.

Applying Metropolis-Hastings only requires replacing line 9 in algorithm 4 with the new acceptance ratio,

$$\alpha = \frac{\tilde{f}(\boldsymbol{\theta}')}{\tilde{f}(\boldsymbol{\theta})} \cdot \frac{q(\boldsymbol{\theta}|\boldsymbol{\theta}')}{q(\boldsymbol{\theta}'|\boldsymbol{\theta})}.$$

In the particular case where transition probabilities are symmetric, that is, $q(\boldsymbol{\theta}'|\boldsymbol{\theta}) = q(\boldsymbol{\theta}|\boldsymbol{\theta}')$, then Metropolis-Hastings is equivalent to Metropolis. Note that when we employ a Normal proposal density for parameters defined in the unbounded real domain, $\boldsymbol{\theta} \in \mathbb{R}^{\mathsf{P}}$, then there is no need for Metropolis-Hastings. On the other hand, using Metropolis-Hastings for sampling bounded domains requires modifying the proposal density at each step. Figure 7.4 illustrates why if we employ a truncated Normal PDF as a proposal, $q(\boldsymbol{\theta}'|\boldsymbol{\theta}) \neq q(\boldsymbol{\theta}|\boldsymbol{\theta}')$ because the normalization constant needs to be recalculated at each iteration. Section 7.4 shows how to leverage transformation functions $\theta^{\mathrm{tr}} = g(\theta)$ in order to transform a bounded domain into an unbounded one $\theta^{\mathrm{tr}} \in \mathbb{R}$ so that the Metropolis method can be employed.

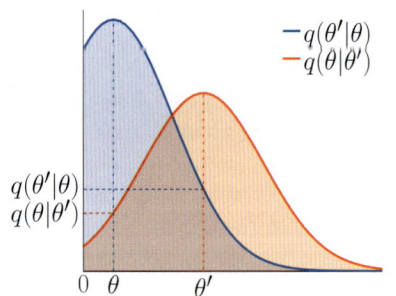

Figure 7.4: Example illustrating how, when using a truncated Normal PDF as a proposal, $q(\theta'|\theta) \neq q(\theta|\theta')$.

## 7.3   Convergence Checks

So far, the presentation of the Metropolis and Metropolis-Hastings algorithms overlooked the notions of convergence. For that, we need to further address two aspects: the *burn-in phase* and *convergence metrics*.

### 7.3.1   Burn-In Phase

For samples taken with an MCMC method, in order to be considered as realizations from the stationary distribution describing the target PDF, a chain must have *forgotten where it started from*. Depending on the choice of initial state $\theta_0$ and transition PDF, the sampling procedure may initially stay trapped in a part of the domain so that the sample's density is not representative of the target density. As depicted in figure 7.5, this issue requires discarding samples taken before reaching the stationary distribution. These discarded samples are called the *burn-in phase*. In practice, it is common to discard the first half of each chain as the burn-in phase and then perform the convergence check that will be further detailed in §7.3.2.

Figure 7.6 presents an example adapted from Murphy,[4] where given a current location $x \in \{0, 1, \cdots, 20\}$, the transition model is defined so there is an equal probability of moving to the nearest neighbor either on the left or on the right. If we apply this random transition an infinite number of times, we will reach the stationary distribution $p^{(\infty)}(x) = 1/21, \forall x$. Each graph in figure 7.6 presents the probability $p^{(n)}(x)$ of being in any state $x$ after $n$ transitions from the initial state $x_0 = 17$. Here, we see that even after 100 transitions, the chain has not yet forgotten where it started from because it is still skewed toward $x = 17$. After 400 transitions, the initial state has been forgotten, because in this graph we can no longer infer the chain's initial value.

### 7.3.2 Monitoring Convergence

Monitoring convergence means assessing whether or not the MCMC samples belong to a stationary distribution. For one-dimensional problems, it is possible to track the convergence by plotting the sample numbers versus their values and identifying visually whether or not they belong to a stationary distribution (e.g., see figure 7.3). When sampling in several dimensions, this visual check is limited. Instead, the solution is to generate samples from multiple chains (e.g., 3–5), each having a different starting location $\theta_0$. The stationarity of these chains can be quantified by comparing the variance within and between chains using the *estimated potential scale reduction* (EPSR). Figure 7.7 illustrates the notation employed to describe samples from multiple chains, where $\theta_{s,c}$ identifies the $s^{\text{th}}$ sample out of S, from the $c^{\text{th}}$ chains out of C. Note that because the final number of samples desired is S, a quantity equal to 2S samples must be generated in order to account for those discarded during the burn-in period.

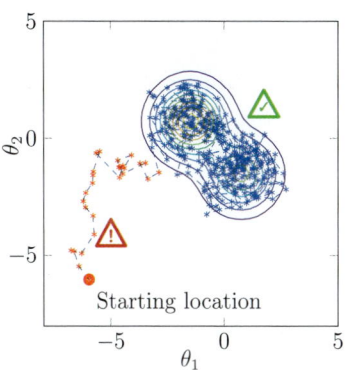

Figure 7.5: The impact of the initial starting location on MCMC sampling.

[4] Murphy, K. P. (2012). *Machine learning: A probabilistic perspective*. MIT Press.

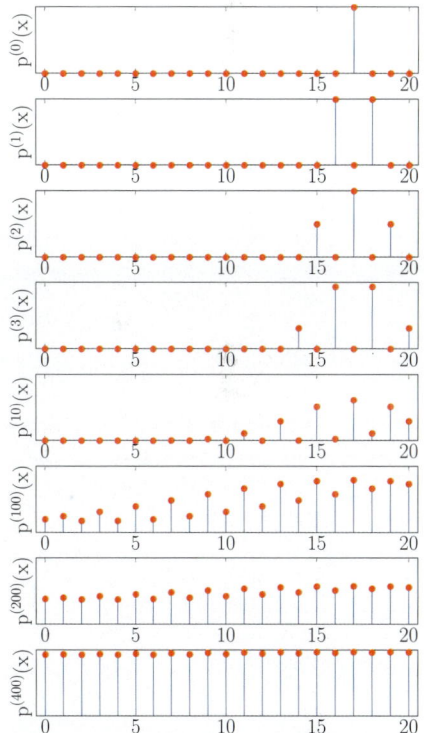

Figure 7.6: The impact of the initial starting location using a stochastic process.

Figure 7.7: The notation for samples taken from multiple chains.

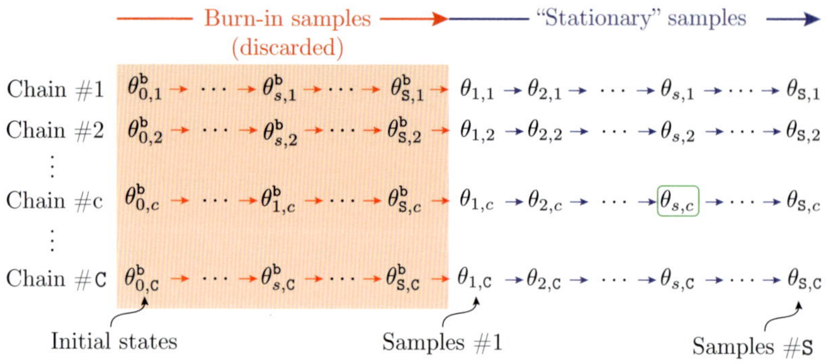

$$
\begin{array}{ll}
\text{Chain \#1} & \theta^b_{0,1} \to \cdots \to \theta^b_{s,1} \to \cdots \to \theta^b_{S,1} \;\; \theta_{1,1} \to \theta_{2,1} \to \cdots \to \theta_{s,1} \to \cdots \to \theta_{S,1} \\
\text{Chain \#2} & \theta^b_{0,2} \to \cdots \to \theta^b_{s,2} \to \cdots \to \theta^b_{S,2} \;\; \theta_{1,2} \to \theta_{2,2} \to \cdots \to \theta_{s,2} \to \cdots \to \theta_{S,2} \\
& \vdots \\
\text{Chain \#c} & \theta^b_{0,c} \to \cdots \to \theta^b_{1,c} \to \cdots \to \theta^b_{S,c} \;\; \theta_{1,c} \to \theta_{2,c} \to \cdots \to \boxed{\theta_{s,c}} \to \cdots \to \theta_{S,c} \\
& \vdots \\
\text{Chain \#C} & \theta^b_{0,C} \to \cdots \to \theta^b_{s,C} \to \cdots \to \theta^b_{S,C} \;\; \theta_{1,C} \to \theta_{2,C} \to \cdots \to \theta_{s,C} \to \cdots \to \theta_{S,C}
\end{array}
$$

Burn-in samples (discarded) — "Stationary" samples

Initial states — Samples #1 — Samples #S

### 7.3.3  Estimated Potential Scale Reduction

The *estimated potential scale reduction*[5] metric denoted by $\hat{R}$ is computed from two quantities: the within-chains and between-chains variances. The *within-chains* mean $\overline{\theta}_{\cdot c}$ and variance $W$ are estimated using

$$
\overline{\theta}_{\cdot c} = \frac{1}{S} \sum_{s=1}^{S} \theta_{s,c}
$$

and

$$
W = \frac{1}{C} \sum_{c=1}^{C} \left( \frac{1}{S-1} \sum_{s=1}^{S} (\theta_{s,c} - \overline{\theta}_{\cdot c})^2 \right).
$$

The property of the within-chains variance is that it underestimates the true variance of samples. The *between-chains* mean $\overline{\theta}_{\cdot\cdot}$ is estimated using

$$
\overline{\theta}_{\cdot\cdot} = \frac{1}{C} \sum_{c=1}^{C} \overline{\theta}_{\cdot c},
$$

and the variance between the means of chains is given by

$$
B = \frac{S}{C-1} \sum_{c=1}^{C} (\overline{\theta}_{\cdot c} - \overline{\theta}_{\cdot\cdot})^2.
$$

Contrarily to the within-chain variance $W$, the between-chain estimate

$$
\hat{V} = \frac{S-1}{S} W + \frac{1}{S} B
$$

overestimates the variance of samples. The metric $\hat{R}$ is defined as the square root of the ratio between $\hat{V}$ and $W$,

$$
\hat{R} = \sqrt{\frac{\hat{V}}{W}}.
$$

Because $\hat{V}$ overestimates and $W$ underestimates the variance, $\hat{R}$ should be greater than one. As illustrated in figure 7.8, a value

[5] Gelman, A. and D. B. Rubin (1992). Inference from iterative simulation using multiple sequences. *Statistical Science* 7(4), 457–472.

$\hat{R} \approx 1$ indicates that the upper and lower bounds have converged to the same value. Otherwise, if $\hat{R} > 1$, it is an indication that convergence is not reached and the number of samples needs to be increased. In practice, convergence can be deemed to be met when $\hat{R} < 1 + \epsilon \approx 1.1$. When generating MCMC samples for $\boldsymbol{\theta} = [\theta_1 \ \theta_2 \ \cdots \ \theta_P]^\mathsf{T}$, we have to compute the EPSR $\hat{R}_i$ for each dimension $i \in \{1, 2, \cdots, P\}$.

Figures 7.9a–c compare the EPSR convergence metric $\hat{R}$ for the Metropolis method applied to a 2-D target distribution using a different number of samples. Figures 7.9a–c are using an isotropic bivariate Normal proposal PDF with $\sigma_q = 1$. From the three tests, only the one employing $\mathsf{S} = 10^4$ samples meets the criterion that we have set here to $\hat{R} < 1.01$.

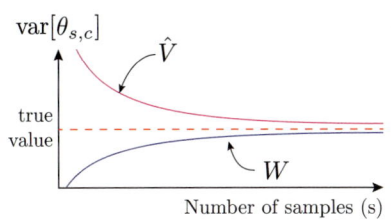

Figure 7.8: The EPSR metric to check for convergence.

Figure 7.9: Comparison of the ESPR convergence metric $\hat{R}$ for 100, 1,000, and 10,000 MCMC samples.

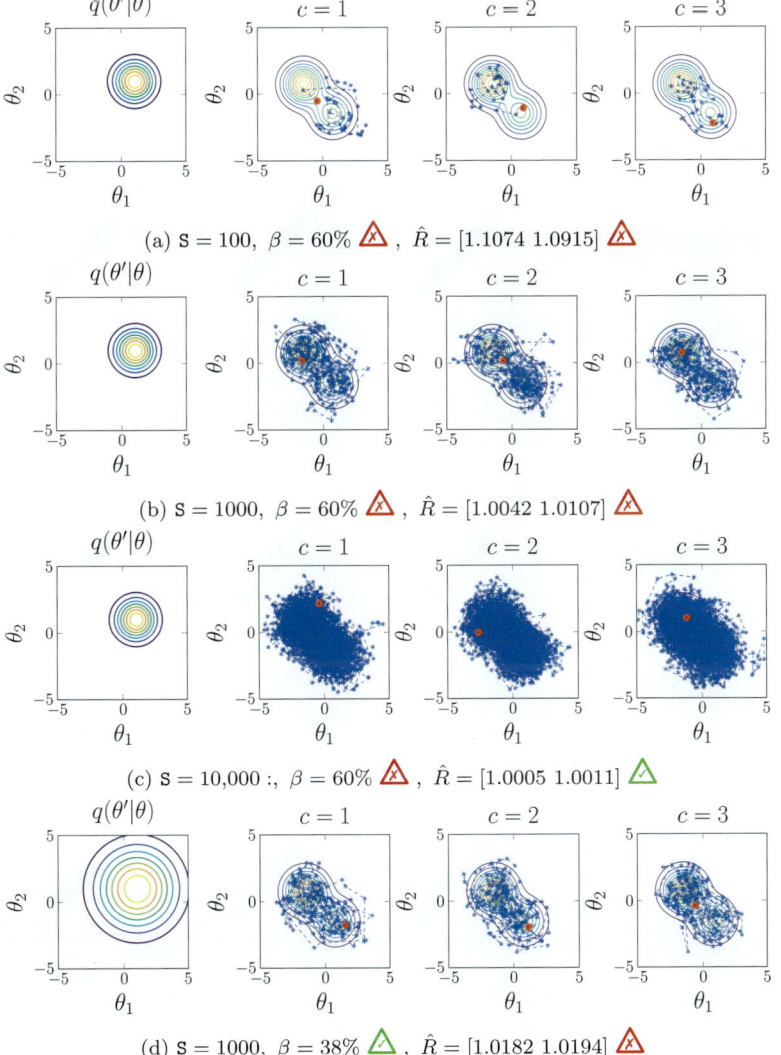

(a) $\mathsf{S} = 100$, $\beta = 60\%$ , $\hat{R} = [1.1074 \ 1.0915]$

(b) $\mathsf{S} = 1000$, $\beta = 60\%$ , $\hat{R} = [1.0042 \ 1.0107]$

(c) $\mathsf{S} = 10,000$ :, $\beta = 60\%$ , $\hat{R} = [1.0005 \ 1.0011]$

(d) $\mathsf{S} = 1000$, $\beta = 38\%$ , $\hat{R} = [1.0182 \ 1.0194]$

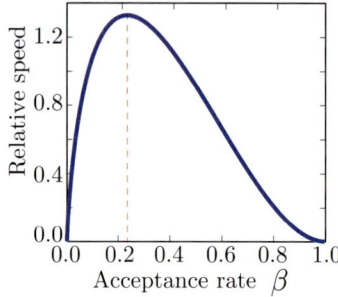

Figure 7.10: Relative convergence speed of MCMC for $P \geq 5$ as a function of the acceptance rate $\beta$.
(Adapted from Rosenthal (2011).)

[6] Rosenthal, J. S. (2011). Optimal proposal distributions and adaptive MCMC. In *Handbook of Markov Chain Monte Carlo*, 93–112. CRC Press.

Figure 7.11: Comparison of the EPSR convergence metric $\hat{R}$ for 1,000 MCMC samples using different proposal distributions.

### 7.3.4   Acceptance Rate

The *acceptance rate* $\beta$ is the ratio between the number of steps where the proposed move is accepted and the total number of steps. Figures 7.9a–c have an acceptance rate of 60 percent.

In the case presented in figure 7.9d, the standard deviation of the proposal was doubled to $\sigma_q = 2$. It has the effect of reducing the acceptance rate to 38 percent. The convergence speed of MCMC methods is related to the acceptance rate. For ideal cases involving Normal random variables, the optimal acceptance rate is approximately 23 percent for parameter spaces having five dimensions or more (i.e., $P \geq 5$), and approximately 44 percent for cases in one dimension ($P = 1$).[6] Figure 7.10 presents the relative convergence speed as a function of the acceptance rate for a case involving five dimensions or more ($P \geq 5$). Note that there is a wide range of values for $\beta$ with similar efficiency, so the optimal values should not be sought strictly.

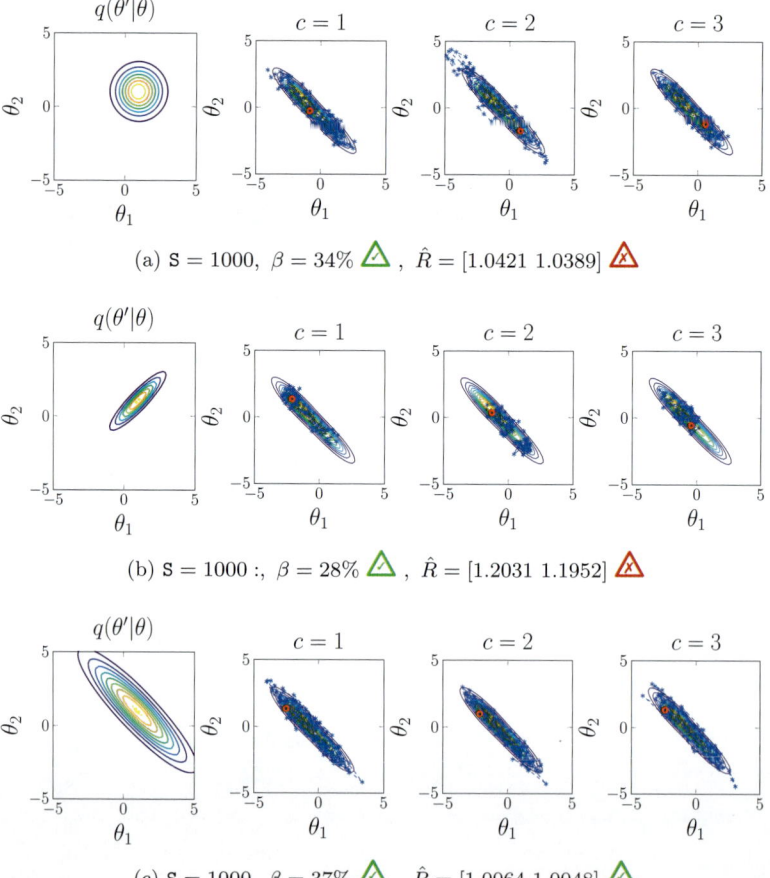

(a) $S = 1000$, $\beta = 34\%$ , $\hat{R} = [1.0421\ 1.0389]$

(b) $S = 1000 :$, $\beta = 28\%$ , $\hat{R} = [1.2031\ 1.1952]$

(c) $S = 1000$, $\beta = 37\%$ , $\hat{R} = [1.0064\ 1.0048]$

Figure 7.11 presents another set of examples applied to a target distribution having a high (negative) correlation between the dimensions $\theta_1$ and $\theta_2$, and using a fixed number of samples equal to $S = 1000$. In comparison with the case presented in (a), the case in (b) displays worse $\hat{R}$ values. This difference is due to the poor choice of correlation coefficient for the proposal PDF in (b); the correlation coefficient of the proposal has a sign that is the opposite of the one for the target PDF. If, like in figure 7.11c, the proposal is well selected for the target distribution, the convergence speed will be higher.

For trivial cases, it may be possible to infer efficient parameters for proposals from trials and errors or from heuristics. However, for more complex cases involving a large number of dimensions, manual tuning falls short. The next section presents a method for automatically tuning the proposal covariance matrix.

### 7.3.5 Proposal Tuning

One generic way to define the proposal PDF is using a multivariate Normal centered on the current state $\boldsymbol{\theta}$ and with covariance matrix $\gamma^2 \boldsymbol{\Sigma}_{\boldsymbol{\theta}^*}$,

$$q(\boldsymbol{\theta}'|\boldsymbol{\theta}) = \mathcal{N}(\boldsymbol{\theta}'; \boldsymbol{\theta}, \gamma^2 \boldsymbol{\Sigma}_{\boldsymbol{\theta}^*}), \quad \gamma^2 = \frac{2.4^2}{P},$$

where the scaling factor $\gamma$ depends on the number of parameters P. The covariance matrix $\boldsymbol{\Sigma}_{\boldsymbol{\theta}^*}$ is an approximation of the posterior covariance obtained using the *Laplace approximation* (see §6.7.2) calculated for the *maximum a-posteriori* value (MAP), $\boldsymbol{\theta}^*$. The MAP value can be estimated using gradient-based optimization techniques such as those presented in §5.2.

Algorithm 5 presents a simple implementation of the Metropolis algorithm with convergence checks and with a tuning procedure for the scaling factor $\gamma$. Note that this algorithm is a minimal example intended to help the reader understand the algorithm flow. Implementations with state-of-the-art efficiency include several more steps.

### 7.4 Space Transformation

When dealing with parameters that are not defined in the unbounded real space, one solution is to employ the Metropolis-Hastings algorithm to account for the non-reversibility of the transitions caused by the domain constraints, as illustrated in figure 7.4.

---

**Algorithm 5:** Convergence check and scaling parameter tuning

---

1  define $\tilde{f}(\boldsymbol{\theta})$, S, C, $\epsilon$

2  initialize $\boldsymbol{\theta}_{0,c}$, $i = 0$

3  compute MAP: $\boldsymbol{\theta}^*$          (e.g., gradient-based optimization)

4  compute $\boldsymbol{\Sigma}_{\boldsymbol{\theta}^*} = \mathbf{H}[-\ln \tilde{f}(\boldsymbol{\theta}^*)]^{-1}$      (Laplace approximation)

5  define $q(\boldsymbol{\theta}'|\boldsymbol{\theta}) = \mathcal{N}(\boldsymbol{\theta}'; \boldsymbol{\theta}, \gamma^2 \boldsymbol{\Sigma}_{\boldsymbol{\theta}^*})$,  $\gamma^2 = \frac{2.4^2}{\mathtt{P}}$

6  **for** *chains* $c \in \{1, 2, \cdots, \mathtt{C}\}$ **do**

7  $\quad$ $\mathcal{S}_c = \emptyset$

8  $\quad$ **for** *samples* $s \in \{0, 1, 2, \cdots, \mathtt{S} - 1\}$ **do**

9  $\quad\quad$ Metropolis algorithm: $\mathcal{S}_c \leftarrow \{\mathcal{S}_c \cup \{\boldsymbol{\theta}_{s+1,c}\}\}$

10 **if** $i = 0$ *(Pilot run)* **then**

11 $\quad$ **for** $c \in \{1, 2, \cdots, \mathtt{C}\}$ **do**

12 $\quad\quad$ $\mathcal{S}_c = \{\boldsymbol{\theta}_{\mathtt{S}/2,c}, \cdots, \boldsymbol{\theta}_{\mathtt{S},c}\}$          (discard burn-in samples)

13 $\quad$ compute $\beta$          (acceptance rate)

14 $\quad$ **if** $\beta < 0.15$ **then**

15 $\quad\quad$ $\gamma^2 = \gamma^2/2$, Goto 5          (decrease scaling factor)

16 $\quad$ **else if** $\beta > 0.50$ **then**

17 $\quad\quad$ $\gamma^2 = 2\gamma^2$, Goto 5          (increase scaling factor)

18 compute $\hat{R}_p, \forall p \in \{1, 2, \cdots, \mathtt{P}\}$          (EPSR)

19 **if** $\hat{R}_p > 1 + \epsilon$ *for any p* **then**

20 $\quad$ $\boldsymbol{\theta}_{0,c} - \boldsymbol{\theta}_{\mathtt{S},c}$          (restart using the last sample)

21 $\quad$ $\mathtt{S} = 2^i \times \mathtt{S}$          (increase the number of samples)

22 $\quad$ $i = i + 1$, Goto 6

23 converged

---

Another solution is to transform each constrained parameter in the real space in order to employ the Metropolis algorithm. Transformation functions $\theta^{\mathrm{tr}} = g(\theta_i)$ suited for this purpose are described in §3.4. When working with transformed parameters, the proposal distribution has to be defined in the transformed space,

$$q(\boldsymbol{\theta}^{\mathrm{tr}'}|\boldsymbol{\theta}^{\mathrm{tr}}) = \mathcal{N}(\boldsymbol{\theta}^{\mathrm{tr}'}; \boldsymbol{\theta}^{\mathrm{tr}}, \gamma^2 \boldsymbol{\Sigma}_{\boldsymbol{\theta}^*}^{\mathrm{tr}}),$$

where $\boldsymbol{\Sigma}_{\boldsymbol{\theta}^*}^{\mathrm{tr}}$ is estimated using the Laplace approximation, which is itself defined in the transformed space. The target probability can be evaluated in the transformed space using the inverse transformation function $g^{-1}(\boldsymbol{\theta}^{\mathrm{tr}})$, its gradient $|\nabla g(\boldsymbol{\theta})|_{\boldsymbol{\theta}^{\mathrm{tr}}}^{-1}$ evaluated at $\boldsymbol{\theta}^{\mathrm{tr}}$, and the change of variable rule,

**Change of variable rule**
$f_{\mathbf{Y}}(\mathbf{y}) = f_{\mathbf{X}}(\mathbf{x}) |\det \mathbf{J}_{\mathbf{y},\mathbf{x}}|^{-1}$
(see §3.4)

$$\tilde{f}(\boldsymbol{\theta}^{\mathrm{tr}}) = \tilde{f}(\overbrace{g^{-1}(\boldsymbol{\theta}^{\mathrm{tr}})}^{=\boldsymbol{\theta}}) \cdot \prod_{i=1}^{\mathtt{P}} \frac{1}{|\nabla_i g(\boldsymbol{\theta})|_{\boldsymbol{\theta}^{\mathrm{tr}}}}.$$

Because each transformation function $g(\theta_i)$ depends on a single parameter, the determinant of the diagonal Jacobian matrix required for the transformation $g(\boldsymbol{\theta})$ simplifies to the product of the inverse absolute value of the gradient of each transformation function evaluated at $\theta_i = g^{-1}(\theta_i^{\text{tr}})$. Algorithm 6 presents a simple implementation of the Metropolis algorithm applied to a transformed space.

**Determinant of a diagonal matrix**
$\det(\text{diag}(\mathbf{x})) = \prod_{i=1}^{\text{X}} x_i$

---

**Algorithm 6:** Metropolis with transformed space

1  define $\tilde{f}(\boldsymbol{\theta})$                      (target distribution)

2  define $g(\boldsymbol{\theta}) \to \boldsymbol{\theta}^{\text{tr}}$, $g^{-1}(\boldsymbol{\theta}^{\text{tr}}) \to \boldsymbol{\theta}$    (transformation functions)

3  define $\nabla g(\boldsymbol{\theta})$               (transformation gradient)

4  define $\mathsf{S}$                    (number of samples)

5  initialize $\mathcal{S} = \emptyset$               (set of samples)

6  initialize $\boldsymbol{\theta}_0$             (initial starting location)

7  compute MAP: $\boldsymbol{\theta}^{*\text{tr}}$     (e.g., gradient-based optimization)

8  compute $\boldsymbol{\Sigma}_{\boldsymbol{\theta}^*}^{\text{tr}} = \mathbf{H}[-\ln \tilde{f}(\boldsymbol{\theta}^{*\text{tr}})]^{-1}$    (Laplace approximation)

9  **for** $s \in \{0, 1, 2, \cdots, \mathsf{S} - 1\}$ **do**

10     define $\boldsymbol{\theta}^{\text{tr}} = g(\boldsymbol{\theta}_s)$

11     sample $\boldsymbol{\theta}'^{\text{tr}} \sim \mathcal{N}(\boldsymbol{\theta}'^{\text{tr}}; \boldsymbol{\theta}^{\text{tr}}, \gamma^2 \boldsymbol{\Sigma}_{\boldsymbol{\theta}^*}^{\text{tr}})$

12     compute $\alpha = \dfrac{\tilde{f}(g^{-1}(\boldsymbol{\theta}'^{\text{tr}}))}{\tilde{f}(g^{-1}(\boldsymbol{\theta}^{\text{tr}}))} \cdot \displaystyle\prod_{i=1}^{\mathsf{P}} \dfrac{|\nabla_i g(\boldsymbol{\theta})|_{\boldsymbol{\theta}^{\text{tr}}}}{|\nabla_i g(\boldsymbol{\theta}')|_{\boldsymbol{\theta}'^{\text{tr}}}}$

13     compute $r = \min(1, \alpha)$

14     sample $u \sim \mathcal{U}(0, 1)$

15     **if** $u \leq r$ **then**

16         $\boldsymbol{\theta}_{s+1} = g^{-1}(\boldsymbol{\theta}'^{\text{tr}})$

17     **else**

18         $\boldsymbol{\theta}_{s+1} = \boldsymbol{\theta}_s$

19     $\mathcal{S} \leftarrow \{\mathcal{S} \cup \{\boldsymbol{\theta}_{s+1}\}\}$

---

## 7.5 Computing with MCMC Samples

When employing an MCMC method, each sample $\boldsymbol{\theta}_s$ is a realization of the target distribution $\tilde{f}(\boldsymbol{\theta})$. When we are interested in performing a Bayesian estimation for parameters using a set of observations $\mathcal{D} = \{y_1, y_2, \cdots, y_{\mathsf{D}}\}$, the unnormalized target distribution is defined as the product of the likelihood and the prior,

$$\tilde{f}(\boldsymbol{\theta}) = f(\mathcal{D}|\boldsymbol{\theta}) \cdot f(\boldsymbol{\theta}).$$

The posterior expected values and covariance for $f(\boldsymbol{\theta}|\mathcal{D})$ are then obtained by computing the empirical average and covariance of the samples $\boldsymbol{\theta}_s$,

$$
\begin{aligned}
\mathbb{E}[\boldsymbol{\theta}|\mathcal{D}] &= \int \boldsymbol{\theta} \cdot f(\boldsymbol{\theta}|\mathcal{D})d\boldsymbol{\theta} \\
&\approx \frac{1}{\mathsf{S}} \sum_{s=1}^{\mathsf{S}} \boldsymbol{\theta}_s \\
\mathrm{cov}(\boldsymbol{\theta}|\mathcal{D}) &= \mathbb{E}\big[(\boldsymbol{\theta} - \mathbb{E}[\boldsymbol{\theta}|\mathcal{D}])(\boldsymbol{\theta} - \mathbb{E}[\boldsymbol{\theta}|\mathcal{D}])^{\mathsf{T}}\big] \\
&\approx \frac{1}{\mathsf{S}-1} \sum_{s=1}^{\mathsf{S}} (\boldsymbol{\theta}_s - \mathbb{E}[\boldsymbol{\theta}|\mathcal{D}])(\boldsymbol{\theta}_s - \mathbb{E}[\boldsymbol{\theta}|\mathcal{D}])^{\mathsf{T}}.
\end{aligned}
$$

Given $f_X(x; \boldsymbol{\theta})$, a probability density function defined by the parameters $\boldsymbol{\theta}$. The posterior predictive density of $X$ given observations $\mathcal{D}$ is obtained by marginalizing the uncertainty of $\boldsymbol{\theta}$,

$$
X|\mathcal{D} \sim f(x|\mathcal{D}) = \int f_X(x; \boldsymbol{\theta}) \cdot \overbrace{f(\boldsymbol{\theta}|\mathcal{D})}^{\boldsymbol{\theta}_s} d\boldsymbol{\theta}.
$$

Predictive samples $x_s$ can be generated by using samples $\boldsymbol{\theta}_s$ and evaluating them in $f_X(x; \boldsymbol{\theta}_s)$, so

$$
\underbrace{x_s : \ X|\boldsymbol{\theta}_s \sim f_X(x; \boldsymbol{\theta}_s)}_{\text{sample from } X \text{ using MCMC samples } \boldsymbol{\theta}_s}.
$$

The posterior predictive expected value and variance for $X|\mathcal{D} \sim f(x|\mathcal{D})$ can then be estimated as

$$
\begin{aligned}
\mathbb{E}[X|\mathcal{D}] &= \int x \cdot f(x|\mathcal{D})dx \\
&\approx \frac{1}{\mathsf{S}} \sum_{s=1}^{\mathsf{S}} x_s \\
\mathrm{var}[X|\mathcal{D}] &= \mathbb{E}[(X - \mathbb{E}[X|\mathcal{D}])^2] \\
&\approx \frac{1}{\mathsf{S}-1} \sum_{s=1}^{\mathsf{S}} (x_s - \mathbb{E}[X|\mathcal{D}])^2.
\end{aligned}
$$

Given a function that depends on parameters $\boldsymbol{\theta}$ such that $z = g(\boldsymbol{\theta})$, posterior predictive samples from that function can be obtained by evaluating it for randomly selected samples $\boldsymbol{\theta}_s$,

$$
\underbrace{z_s : \ Z|\boldsymbol{\theta}_s = g(\boldsymbol{\theta}_s)}_{\text{sample from } Z \text{ using MCMC samples } \boldsymbol{\theta}_s}.
$$

The posterior predictive expected value and variance for the function output are thus once again

$$
\begin{aligned}
\mathbb{E}[Z|\mathcal{D}] &= \int z \cdot f(z|\mathcal{D})dz \\
&\approx \frac{1}{\mathsf{S}} \sum_{s=1}^{\mathsf{S}} z_s \\
\mathrm{var}[Z|\mathcal{D}] &= \mathbb{E}[(Z - \mathbb{E}[Z|\mathcal{D}])^2] \\
&\approx \frac{1}{\mathsf{S}-1} \sum_{s=1}^{\mathsf{S}} (z_s - \mathbb{E}[Z|\mathcal{D}])^2.
\end{aligned}
$$

As we saw in §6.8, performing Bayesian model selection requires computing the evidence

$$
f(\mathcal{D}) = \int f(\mathcal{D}|\boldsymbol{\theta}) \cdot f(\boldsymbol{\theta})d\boldsymbol{\theta}.
$$

It is theoretically possible to estimate this normalization constant from Metropolis-Hastings samples using the *harmonic mean of the likelihood*.[7] This method is not presented here because it is known for its poor performance in practice.[8] The *annealed importance sampling*[9] (AIS) is a method combining annealing optimization methods, importance sampling, and MCMC sampling. Despite being one of the efficient methods for estimating $f(\mathcal{D})$, we have to keep in mind that estimating the evidence is intrinsically difficult when the number of parameters is large and when the posterior is multimodal or when it displays nonlinear dependencies. Therefore, we should always be careful when estimating $f(\mathcal{D})$ because no perfect black-box solution is currently available for estimating it.

[7] Newton, M. A. and A. E. Raftery (1994). Approximate Bayesian inference with the weighted likelihood bootstrap. *Journal of the Royal Statistical Society. Series B (Methodological) 56*(1), 3–48.

[8] Neal, R. (2008). The harmonic mean of the likelihood: Worst Monte Carlo method ever. URL https://radfordneal.wordpress.com/2008/08/17/the-harmonic-mean-of-the-likelihood-worst-monte-carlo-method-ever/. Accessed November 8, 2019.

[9] Neal, R. M. (2001). Annealed importance sampling. *Statistics and Computing 11*(2), 125–139.

*Example: Concrete tests* We are revisiting the example presented in §6.5.3, where Bayesian estimation is employed to characterize the resistance $R$ of a concrete mix. The resistance is now modeled as a Normal random variable with unknown mean and variance, $R \sim \mathcal{N}(r; \mu_R, \sigma_R^2)$. The observation model is $y = r + v$, $v : V \sim \mathcal{N}(y; 0, 0.01^2)$ MPa, where $V$ describes the observation errors that are independent of each other, that is, $V_i \perp\!\!\!\perp V_j, \forall i \neq j$. Our goal is to estimate the posterior PDF for the resistance's mean $\mu_R$ and standard deviation $\sigma_R$,

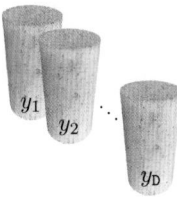

$$
\overbrace{f(\mu_R, \sigma_R|\mathcal{D})}^{\text{Posterior PDF}} = \frac{\overbrace{f(\mathcal{D}|\mu_R, \sigma_R)}^{\text{Likelihood}} \cdot \overbrace{f(\mu_R, \sigma_R)}^{\text{Prior PDF}}}{\underbrace{f(\mathcal{D})}_{\text{Evidence}}}.
$$

In order to reflect an absence of prior knowledge, we employ non-informative priors (see §6.4.1) for both parameters, so that

$f(\mu_R) \propto 1$, $f(\sigma_R) \propto 1/\sigma_R$, and thus $f(\mu_R, \sigma_R) \propto 1/\sigma_R$. The likelihood of an observation $y$, given a set of parameters $\boldsymbol{\theta}$, is $f(y|\boldsymbol{\theta}) = \mathcal{N}(y; \mu_R, \sigma_R^2 + \sigma_V^2 = 0.01^2)$. Because of the conditional independence assumption for the observations, the joint likelihood for D observations is obtained from the product of the marginals,

$$f(\mathcal{D}|\mu_R, \sigma_R) = \prod_{i=1}^{D} \underbrace{\frac{1}{\sqrt{2\pi}\sqrt{\sigma_R^2 + \sigma_V^2}} \exp\left[-\frac{1}{2}\left(\frac{y_i - \mu_R}{\sqrt{\sigma_R^2 + \sigma_V^2}}\right)^2\right]}_{\text{Normal PDF}}.$$

In this example, we assume we only have three observations $\mathcal{D} = \{43.3, 40.4, 44.8\}$ MPa.

The parameter $\sigma_R \in (0, \infty) = \mathbb{R}^+$ is constrained to positive real numbers. Therefore, we perform the Bayesian estimation in the transformed space $\boldsymbol{\theta}^{\text{tr}}$,

$$\begin{aligned} \boldsymbol{\theta}^{\text{tr}} &= [\theta_1^{\text{tr}} \; \theta_2^{\text{tr}}]^\mathsf{T} = [\mu_R \; \ln(\sigma_R)]^\mathsf{T} \\ \boldsymbol{\theta} &= [\mu_R \; \sigma_R]^\mathsf{T} = [\theta_1^{\text{tr}} \; \exp(\theta_2^{\text{tr}})]^\mathsf{T}. \end{aligned}$$

Note that no transformation is applied for $\mu_R$ because it is already defined over $\mathbb{R}$.

The Newton-Raphson algorithm is employed in order to identify the set of parameters maximizing the target PDF defined in the transformed space,

$$\tilde{f}(\boldsymbol{\theta}^{\text{tr}}) = f(\mathcal{D}|\boldsymbol{\theta}^{\text{tr}}) \cdot f(\boldsymbol{\theta}^{\text{tr}}).$$

The optimal values found are

$$\begin{aligned} \boldsymbol{\theta}^{*\text{tr}} &= [43.0 \; 0.6]^\mathsf{T} \\ \boldsymbol{\theta}^* &= [43.0 \; 1.8]^\mathsf{T}. \end{aligned}$$

The Laplace approximation evaluated at $\boldsymbol{\theta}^{*\text{tr}}$ is employed to estimate the posterior covariance

$$\boldsymbol{\Sigma}_{\boldsymbol{\theta}^*}^{\text{tr}} = \mathbf{H}[-\ln \tilde{f}(\boldsymbol{\theta}^{*\text{tr}})]^{-1} = \begin{bmatrix} 1.11 & 0 \\ 0 & 0.17 \end{bmatrix}.$$

This estimation is employed with a scaling factor $\gamma^2 = 2.88$ in order to initialize the proposal PDF covariance. A total of C $= 4$ chains are generated, where each contains S $= 10^4$ samples. The scaling factor obtained after tuning is $\gamma^2 = 1.44$, which leads to an acceptance rate of $\beta = 0.29$ and EPSR convergence metrics $\hat{R} = [1.0009 \; 1.0017]$.

The posterior mean and posterior covariance are estimated from the MCMC samples $\boldsymbol{\theta}_s$,

$$\hat{\mathbb{E}}[\boldsymbol{\theta}|\mathcal{D}] = \frac{1}{\mathsf{S}}\sum_{s=1}^{\mathsf{S}}\boldsymbol{\theta}_s = [42.9\ 3.8]^{\mathsf{T}} \quad \text{(Posterior mean)}$$

$$\hat{\text{cov}}(\boldsymbol{\theta}|\mathcal{D}) = \frac{1}{\mathsf{S}-1}\sum_{s=1}^{\mathsf{S}}(\boldsymbol{\theta}_s - \mathbb{E}[\boldsymbol{\theta}|\mathcal{D}])(\boldsymbol{\theta}_s - \mathbb{E}[\boldsymbol{\theta}|\mathcal{D}])^{\mathsf{T}}$$

$$= \begin{bmatrix} 8.0 & 2.5 \\ 2.5 & 20.6 \end{bmatrix} \quad \text{(Posterior covariance)}.$$

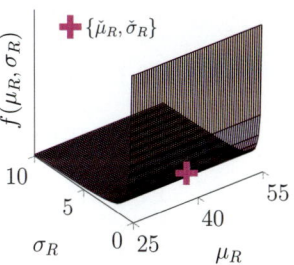

The posterior predictive mean and variance for the concrete resistance, which include the epistemic uncertainty about the parameters $\mu_R$ and $\sigma_R$, are

$$\hat{\mathbb{E}}[R|\mathcal{D}] = \frac{1}{\mathsf{S}}\sum_{s=1}^{\mathsf{S}}r_s = 42.9$$

$$\hat{\text{var}}[R|\mathcal{D}] = \frac{1}{\mathsf{S}-1}\sum_{s=1}^{\mathsf{S}}(r_s - \mathbb{E}[R|\mathcal{D}])^2 = 40.5,$$

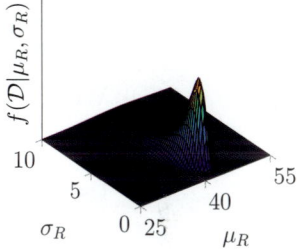

where samples $r_s : R \sim \mathcal{N}(r; \mu_{R,s}, \sigma_{R,s}^2)$ are generated using the MCMC samples $\boldsymbol{\theta}_s = [\mu_{R,s}\ \sigma_{R,s}]^{\mathsf{T}}$.

Figure 7.12 presents the prior, likelihood, posterior, and posterior predictive PDFs. Samples $\boldsymbol{\theta}_s$ are represented by dots on the posterior. Note how the posterior predictive does not exactly correspond with the true PDF for $R$, which has for true values $\check{\mu}_R = 42$ MPa and $\check{\sigma}_R = 2$ MPa. This discrepancy is attributed to the fact that only 3 observations are available, so epistemic uncertainty remains in the estimation of parameters $\{\mu_R, \sigma_R\}$. In figure 7.12, this uncertainty translates in heavier tails for the posterior predictive than for the true PDF.

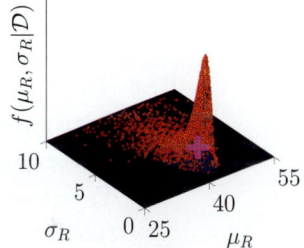

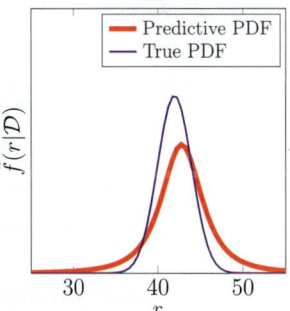

Figure 7.12: Example of application of MCMC sampling for the estimation of the resistance of a concrete mix. The cross indicates the true (unknown) parameter values.

*Exercises*

**P7.1** What is the Markov hypothesis, and what is its role in MCMC methods?

**P7.2** In the MCMC sampling context, what is the role of the burn-in phase?

**P7.3** With MCMC sampling, how can we quantify the convergence?

**P7.4** Explain what factors affect the sampling efficiency of the Metropolis algorithm.

**P7.5** What case is best suited to employ MCMC sampling rather than an MAP or MLE point approximation for a posterior?

**P7.6** With MCMC, what is the purpose of performing the sampling in a transformed space?

**P7.7** Use the Metropolis algorithm to sample the target function

$$\tilde{f}(\theta_1, \theta_2) = 0.6 \cdot \mathcal{N}\left(\theta_1; [-1.5 \ 0.75]^\mathsf{T}, \mathrm{diag}([1 \ 1]^2)\right) + 0.4 \cdot \mathcal{N}\left(\theta_2; [0.75 \ -1.5]^\mathsf{T}, \mathrm{diag}([1 \ 1]^2)\right).$$

Check your results with figure 7.9.

**P7.8** Three soil specimens are tested to estimate the parameters $\{\mu_x, \sigma_x\}$ of the probability distribution function describing the friction angle $X \sim \mathcal{N}(x; \mu_x, \sigma_x^2) = f(x; \mu_x, \sigma_x)$. Observations $y_i$ are realizations of the observation model $y : Y = X + V$, where $V_i \sim \mathcal{N}(v; 0, 1^2)$ is a random variable describing the measurement device's uncertainty (note: $V_i \perp\!\!\!\perp V_j, \forall i \neq j$). The set of observations available is $\mathcal{D} = \{y_1 = 30°, y_2 = 35°, y_3 = 27°\}$. The prior knowledge $f(\mu_x) = \mathcal{U}(\mu_x; 20, 35)$ follows a uniform PDF, $f(\sigma_x) \propto 1/\sigma_x$ follows a non-informative prior, and $\mu_x \perp\!\!\!\perp \sigma_x$.

 a. In the context of MCMC sampling, write down the formulation for the target distribution $\tilde{f}(\mu_x, \sigma_x)$.

 b. Implement your own MCMC algorithm to draw samples from the unnormalized posterior $f(\mu_x, \sigma_x | \mathcal{D})$. Use the proposal model $f(\mathbf{s}_t | \mathbf{s}_{t-1}) = \mathcal{N}\left(\mathbf{s}_t; \mathbf{s}_{t-1}, \mathrm{diag}([4 \ 4])\right)$. Report the acceptance rate and present a 2-D contour plot of your posterior overlaid with your MCMC samples.

 c. Compute the marginal posterior means and standard deviations $\mathbb{E}[\mu_x | \mathcal{D}]$, $\mathbb{E}[\sigma_x | \mathcal{D}]$, $\sigma[\mu_x | \mathcal{D}]$, $\sigma[\sigma_x | \mathcal{D}]$.

 d. Using results from (b), generate samples from the posterior predictive distribution

$$f(x | \mathcal{D}) = \int \int f(x; \mu_x, \sigma_x) \cdot f(\mu_x, \sigma_x | \mathcal{D}) d\mu_x d\sigma_x$$

and plot a histogram for these samples.

# Part III

# Supervised Learning

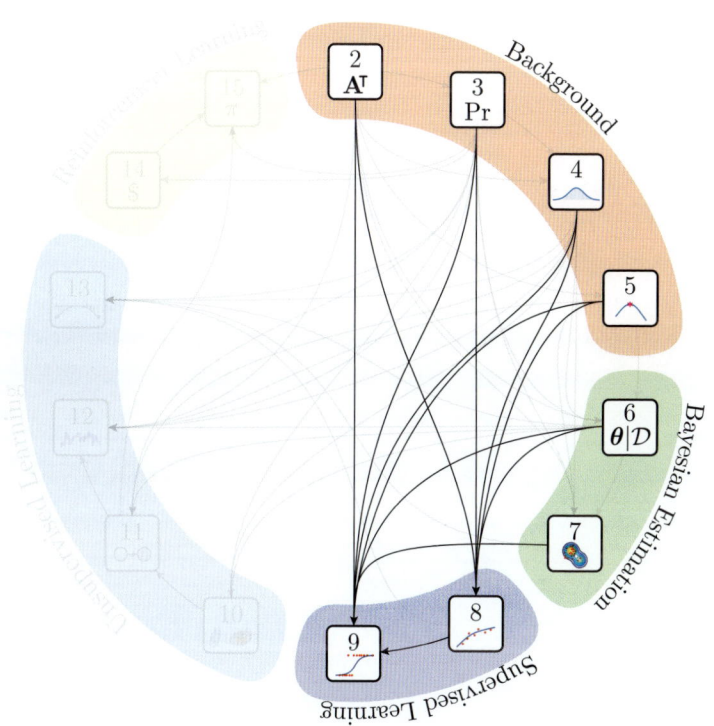

# 8
# Regression

Regression consists in the task of modeling the relationships between a *system's* responses and the covariates describing the properties of this system. Here, the generic term *system* is employed because regression can be applied to many types of problems. Figure 8.1 presents examples of applications in the field of civil engineering; for instance, one could model the relationship between (a) the number of pathogens and chlorine concentration, (b) the number of accidents at a road intersection as a function of the hour of the day, and (c) concrete strength as a function of the water-cement ratio. These examples are all functions of a single covariate; later in this chapter we will see how regression is generalized to model the responses of a system as a function of any number of covariates.

In regression problems, the data sets are described by $D$ pairs of observations $\mathcal{D} = \{(\mathbf{x}_i, y_i), \forall i \in \{1 : D\}\}$ where $\mathbf{x}_i = [x_1 \ x_2 \ \cdots \ x_X]_i^{\mathsf{T}} \in \mathbb{R}^X$ are *covariates*, also called *attributes* or *regressors*, and $y_i \in \mathbb{R}^1$ is an observed system response. A regression model $g(\mathbf{x}) \equiv \text{fct}(\mathbf{x}) \in \mathbb{R}^1$ is thus a $X \rightarrow 1$ function, that is, a function of $X$ covariates leading to one output. This chapter covers three regression methods for building $g(\mathbf{x})$: *linear regression, Gaussian process regression,* and *neural networks.*

## 8.1   Linear Regression

*Linear regression* is one of the easiest methods for approaching regression problems. Contrarily to what its name may lead us to think, linear regression is not limited to cases where the relationships between the covariates and system responses are linear. The term *linear* refers to the linear dependence of the model with respect to its parameters. In the case involving a single covariate $x$,

**Data**

$\mathcal{D} = \{(\mathbf{x}_i, y_i), \forall i \in \{1 : D\}\}$

$\mathbf{x}_i \in \mathbb{R}^X : \begin{cases} \text{Covariate} \\ \text{Attribute} \\ \text{Regressor} \end{cases}$

$y_i \in \mathbb{R} : \text{Observation}$

**Model**

$g(x) \equiv \text{fct}(x)$

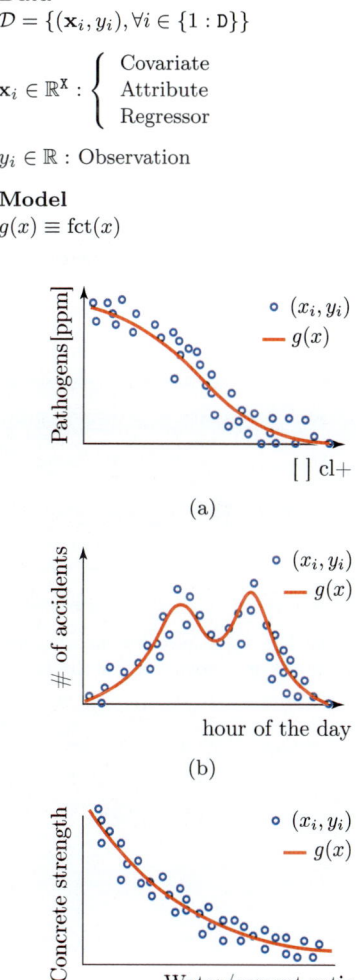

Figure 8.1: Examples of applications of regression analysis in the context of civil engineering.

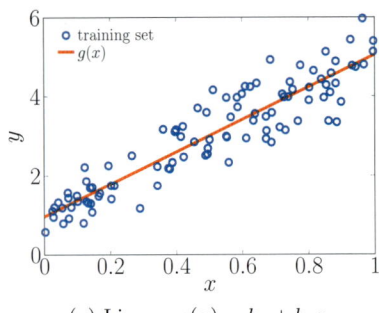

(a) Linear: $g(x) = b_0 + b_1 x$

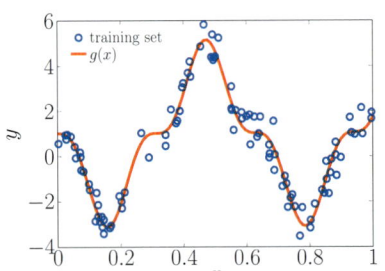

(b) Nonlinear: $g(x) = b_0 + b_1 \sin(10x)^3$

Figure 8.2: Examples of linear regression applied to linear and nonlinear relationships.

**Note:** The 1 in $\begin{bmatrix} 1 \\ x \end{bmatrix}$ stands for the absence of dependence on a covariate for the bias parameter $b_0$.

the model employed in linear regression takes the form

$$g(x) = b_0 + b_1\phi_1(x) + b_2\phi_2(x) + \ldots + b_{\mathsf{B}}\phi_{\mathsf{B}}(x),$$

where $\boldsymbol{\theta} = \mathbf{b} = [b_0 \ b_1 \ \cdots \ b_{\mathsf{B}}]^{\mathsf{T}}$ is a vector of model parameters and $\phi_i(x)$ is a *basis function* applied on the covariate $x$. Basis functions are nonlinear functions employed when we want to apply linear regression to describe nonlinear relationships. Figure 8.2 presents two examples of linear regression applied to (a) linear and (b) nonlinear relationships.

### 8.1.1   Mathematical Formulation

For the simplest case where there is only one covariate, the data set consists in D pairs of covariates and system responses $\mathcal{D} = \{(x_i, y_i), \forall i \in \{1 : \mathsf{D}\}\}$, where a covariate $x_i \in \mathbb{R}$, and its associated observation $y_i \in \mathbb{R}$. The observation model is defined as

$$\underbrace{y}_{\text{Observation}} = \overbrace{g(x)}^{\text{Model}} + \underbrace{v, \ v : V \sim \mathcal{N}(v; 0, \sigma_V^2)}_{\text{Observation error}},$$

where $v$ describes the observation errors that are assumed to be independent of each other, $V_i \perp\!\!\!\perp V_j, \forall i \neq j$. Note that observed covariates $x$ are assumed to be exact and free from any observation errors. Let us consider the special case of a linear basis function $\phi(x) = x$ where the model takes the form

$$g(x) = b_0 + b_1 x = [b_0 \ b_1]\begin{bmatrix} 1 \\ x \end{bmatrix}. \tag{8.1}$$

We can express the model for the entire data set using the following matrix notation,

$$\mathbf{b} = \begin{bmatrix} b_0 \\ b_1 \end{bmatrix}, \ \mathbf{y} = \begin{bmatrix} y_1 \\ y_2 \\ \vdots \\ y_{\mathsf{D}} \end{bmatrix}, \ \mathbf{X} = \begin{bmatrix} 1 & x_1 \\ 1 & x_2 \\ \vdots & \vdots \\ 1 & x_{\mathsf{D}} \end{bmatrix}, \ \mathbf{v} = \begin{bmatrix} v_1 \\ v_2 \\ \vdots \\ v_{\mathsf{D}} \end{bmatrix}.$$

With this matrix notation, we reformulate the observation model as

$$\mathbf{y} = \mathbf{X}\mathbf{b} + \mathbf{v}, \ \mathbf{v} : \mathbf{V} \sim \mathcal{N}(\mathbf{v}; \mathbf{0}, \sigma_V^2 \cdot \mathbf{I}),$$

where $\mathbf{I}$ is the identity matrix indicating that all measurement errors are independent from each other and have the same variance. If we separate the set of observations $\mathcal{D} = \{\mathcal{D}_x, \mathcal{D}_y\}$ by isolating covariates $\mathcal{D}_x$ and system responses $\mathcal{D}_y$, the joint likelihood of observed system responses given the set of observed covariates and model parameters $\boldsymbol{\theta} = \mathbf{b}$ is expressed as

$$
\begin{aligned}
f(\mathcal{D}_y | \mathcal{D}_x, \boldsymbol{\theta}) &\equiv f(\mathcal{D}_y = \mathbf{y} | \mathcal{D}_x = \mathbf{x}, \boldsymbol{\theta} = \mathbf{b}) \\
&= \prod_{i=1}^{D} \mathcal{N}(y_i; g(x_i), \sigma_V^2) \\
&\propto \prod_{i=1}^{D} \exp\left(-\frac{1}{2} \frac{\left(y_i - g(x_i)\right)^2}{\sigma_V^2}\right) \\
&= \exp\left(-\frac{1}{2} \sum_{i=1}^{D} \frac{\left(y_i - g(x_i)\right)^2}{\sigma_V^2}\right) \\
&= \exp\left(-\frac{1}{2\sigma_V^2} (\mathbf{y} - \mathbf{Xb})^\mathsf{T} (\mathbf{y} - \mathbf{Xb})\right).
\end{aligned}
$$

The goal is to estimate optimal parameters $\boldsymbol{\theta}^* = \mathbf{b}^*$, which maximize the likelihood $f(\mathcal{D}_y | \mathcal{D}_x, \boldsymbol{\theta})$, or equivalently, which maximize the log-likelihood,

$$
\begin{aligned}
\ln f(\mathcal{D}_y | \mathcal{D}_x, \boldsymbol{\theta}) &= -\frac{1}{2\sigma_V^2} (\mathbf{y} - \mathbf{Xb})^\mathsf{T} (\mathbf{y} - \mathbf{Xb}) \\
&\propto -\frac{1}{2} (\mathbf{y} - \mathbf{Xb})^\mathsf{T} (\mathbf{y} - \mathbf{Xb}).
\end{aligned}
$$

**Note:** Because we have assumed that $\sigma_V$ is the same for all observations, optimal parameters values $\boldsymbol{\theta}^*$ are independent of the observation-error standard deviation.

The set of data $\mathcal{D}$ employed to obtain the maximum likelihood estimate (MLE) for parameters $\boldsymbol{\theta} = \mathbf{b}$ is called the *training set*. In the context of linear regression, the objective of maximizing $\ln f(\mathcal{D}_y | \mathcal{D}_x, \boldsymbol{\theta})$ is often replaced by the equivalent objective of *minimizing* a *loss function* $J(\mathbf{b})$ corresponding to the sum of the squares of the prediction errors,

**Note:**
$$
\arg\max_{\mathbf{b}} \ln f(\mathcal{D}_y | \mathcal{D}_x, \boldsymbol{\theta}) \equiv \arg\min_{\mathbf{b}} J(\mathbf{b})
$$

$$
\begin{aligned}
J(\mathbf{b}) &= \frac{1}{2} \sum_{i=1}^{D} \left(y_i - g(x_i)\right)^2 \\
&= \frac{1}{2} (\mathbf{y} - \mathbf{Xb})^\mathsf{T} (\mathbf{y} - \mathbf{Xb}) \\
&= -\ln f(\mathcal{D}_y | \mathcal{D}_x, \boldsymbol{\theta}).
\end{aligned}
$$

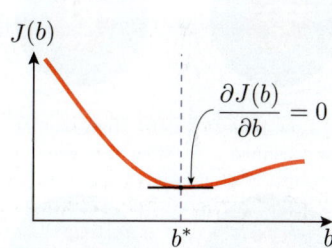

Figure 8.3: The minimum of a continuous function corresponds to the location where its derivative equals zero.

Figure 8.3 presents a loss function $J(b)$ as a function of the model parameters $b$. Optimal parameter values $\boldsymbol{\theta}^* = \mathbf{b}^*$ are defined as

$$
\mathbf{b}^* = \arg\min_{\mathbf{b}} \frac{1}{2} (\mathbf{y} - \mathbf{Xb})^\mathsf{T} (\mathbf{y} - \mathbf{Xb}).
$$

As for other continuous functions, the minimum of $J(\mathbf{b}^*)$ coincides with the point where its derivative equals zero,

$$
\frac{\partial J(\mathbf{b}^*)}{\partial \mathbf{b}} = 0 = \mathbf{X}^\mathsf{T} \mathbf{Xb}^* - \mathbf{X}^\mathsf{T} \mathbf{y}.
$$

By reorganizing the terms from the last equation, optimal parameters are defined as

$$
\mathbf{b}^* = (\mathbf{X}^\mathsf{T} \mathbf{X})^{-1} \mathbf{X}^\mathsf{T} \mathbf{y}. \tag{8.2}
$$

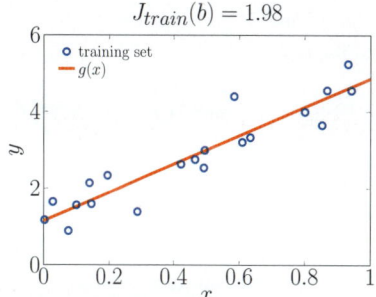

Figure 8.4: Example of linear regression minimizing the performance metric $J$ computed on the training set.

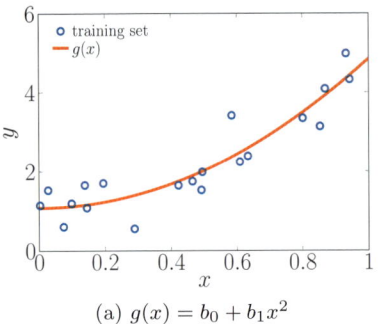

(a) $g(x) = b_0 + b_1 x^2$

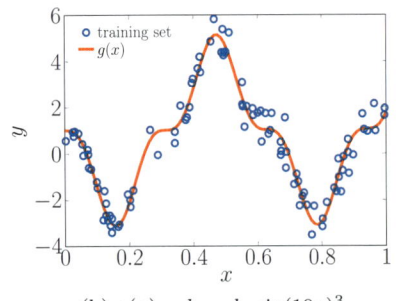

(b) $g(x) = b_0 - b_2 \sin(10x)^3$

Figure 8.5: Examples of linear regression applied to nonlinear relationships.

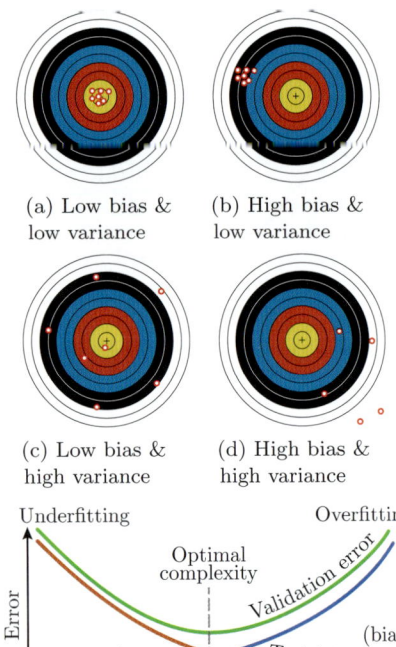

(a) Low bias & low variance

(b) High bias & low variance

(c) Low bias & high variance

(d) High bias & high variance

(e) Bias-variance tradeoff

Figure 8.6: Illustration of the *bias* and *variance* when talking about predictive performance and model complexity.

Figure 8.4 presents an example of linear regression employing a linear basis function

$$g(x) = b_0 + b_1 x.$$

The optimal parameters found using the training set are $\mathbf{b}^* = [1.14\ 3.71]^{\mathsf{T}}$, for which $J_{train}(\mathbf{b}^*) = 1.98$.

We can extend the formulation presented above for the general case where the relationship between $x$ and $y$ is modeled by the summation of multiple basis functions,

$$g(x) = b_0 + b_1 \phi_1(x) + b_2 \phi_2(x) + \cdots + b_{\mathsf{B}} \phi_{\mathsf{B}}(x) = \mathbf{X} \mathbf{b},$$

where $\phi_i(x)$ can be any linear or nonlinear function, for example: $\phi_i(x) = x^2$, $\phi_i(x) = \sin(x), \cdots$. In a matrix form, this general model is expressed as

$$\mathbf{b} = \begin{bmatrix} b_0 \\ b_1 \\ \vdots \\ b_{\mathsf{B}} \end{bmatrix},\ \mathbf{y} = \begin{bmatrix} y_1 \\ y_2 \\ \vdots \\ y_{\mathsf{D}} \end{bmatrix},\ \mathbf{X} = \begin{bmatrix} 1 & \phi_1(x_1) & \phi_2(x_1) & \cdots & \phi_{\mathsf{B}}(x_1) \\ 1 & \phi_1(x_2) & \phi_2(x_2) & \cdots & \phi_{\mathsf{B}}(x_2) \\ \vdots & \vdots & \vdots & \ddots & \vdots \\ 1 & \phi_1(x_{\mathsf{D}}) & \phi_2(x_{\mathsf{D}}) & \cdots & \phi_{\mathsf{B}}(x_{\mathsf{D}}) \end{bmatrix}.$$

Remember that with linear regression, the model is necessarily linear with respect to $\phi_i(x)$ and $\mathbf{b}$, yet not $x$. Figure 8.5 presents two examples of linear regression applied to nonlinear relationships between a system response and a covariate.

### 8.1.2   Overfitting and Cross-Validation

One of the main challenges with linear regression, as with many other machine learning methods, is *overfitting*. Overfitting happens when you select a model class or model structure that represents the data in your training set well, yet it fails to provide good predictions for other covariates that were not seen during training. A highly complex model capable of adapting to the training data is said to have a large *capacity*.

Overfitting with a model that has too much capacity corresponds to the case (c) in figure 8.6, where prediction errors are displaying a low bias when evaluated with the data employed to train the model; however, when validated with the true objective of predicting unobserved data, there is a high variance. Case (b) represents the situation where a model underfits the data because its low capacity causes a large bias and a small variance. As depicted in (e), we want the optimal *model complexity* offering a tradeoff between the prediction bias and variance by minimizing the validation error. This is known as the *bias-variance tradeoff*. The concept illustrated in figure 8.6 will be revisited at the end of this section.

In the context of the linear regression, overfitting arises when one chooses basis functions $\phi(x)$ that have too much capacity and that will be able to fit the data in a training set but not the data of an independent validation set. Figure 8.7a presents an example where data is generated from a linear basis function and where the regression model also employs a linear basis function to fit the data. Figure 8.7b presents the case where the same data is fitted with a model using polynomial basis functions of order 5. Notice how when more complex basis functions are introduced in the model, it becomes more flexible, and it can better fit the data in the training set. Here the fitting ability is quantified by the loss function $J$ that is lower for the polynomial model than for the linear one.

One solution to this challenge is to replace the MLE procedure for optimizing the parameters $\mathbf{b}$ (see equation 8.2) with the Bayesian estimation procedure as described in §6.3. The appropriate model complexity should then be selected by employing Bayesian model selection as presented in §6.8. In practice, this is seldom performed because calculating the evidence $f(\mathcal{D}|m_i)$ for a model is computationally demanding. An alternative is to perform *ridge regression*,[1] which is a special case of Bayesian linear regression that employs *regularization* to constrain the parameters (see §8.3.3).

Another general procedure to identify the optimal model complexity is to employ *cross-validation*. The idea is to avoid employing the same data to (1) estimate optimal parameters, (2) discriminate between model classes, and (3) test the final model performance using the loss function. To start with, the data is separated into a training set and a test set. In order to maximize the utilization of the data available, we can rely on the *N-fold cross-validation* procedure, where the training set is again divided into N subsets called *folds*; one at a time, each subset is employed as an independent validation set while the training is performed on the remaining folds. The total loss is obtained by summing the loss obtained for each validation subset. The model class selected is the one with the lowest aggregated loss. Once a model class is selected as the best one, its parameters can be retrained using the entire training set, and its predictive performance is evaluated on the independent test set that was not employed to estimate parameters or to discriminate between model classes.

Figure 8.8 presents an example of a 3-fold cross-validation where the green-shaded boxes describe the part of the data set employed to train the model, the blue boxes describe the part of the data set employed to compare the predictive performance while choosing

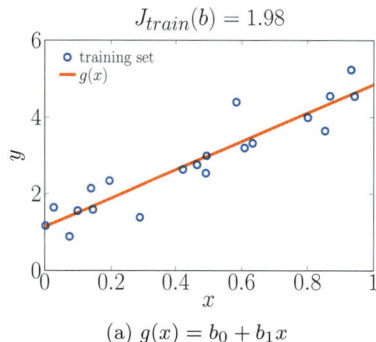

(a) $g(x) = b_0 + b_1 x$

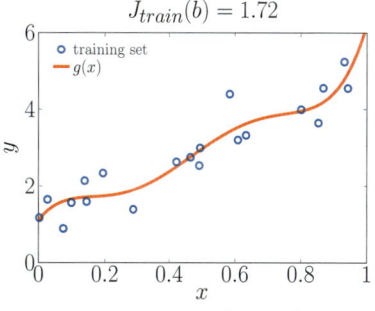

(b) $g(x) = b_0 + b_1 x + b_2 x^2 + b_3 x^3 + b_4 x^4 + b_5 x^5$

Figure 8.7: Example of (a) a model with correct complexity, and (b) a model with too much capacity.

[1] Hoerl, A. E. and R. W. Kennard (1970). Ridge regression: Biased estimation for nonorthogonal problems. *Technometrics 12*(1), 55–67.

Figure 8.8: The *N*-fold cross-validation procedure.

$J_{train}(b) = 0.43 \mid J_{validation}(b) = 1.56$

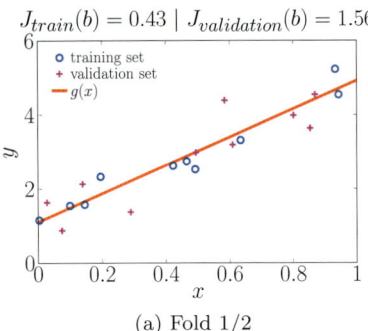

(a) Fold 1/2

$J_{train}(b) = 1.53 \mid J_{validation}(b) = 0.46$

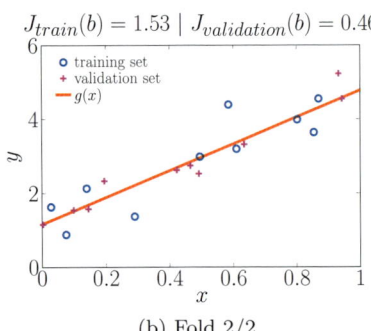

(b) Fold 2/2

Figure 8.9: Example of 2-fold cross-validation for the model $g_1(x) = b_0 + b_1x$.

$J_{train}(b) = 0.7 \mid J_{validation}(b) = 5.42$

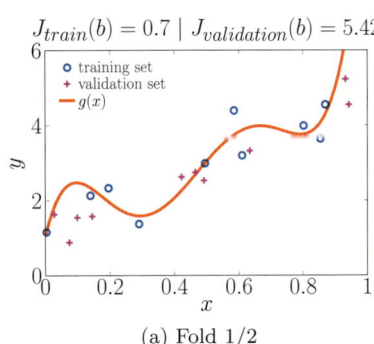

(a) Fold 1/2

$J_{train}(b) = 0.11 \mid J_{validation}(b) = 15.47$

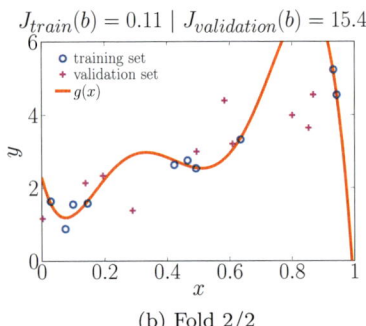

(b) Fold 2/2

Figure 8.10: Example of 2-fold cross-validation for the model $g_2(x) = b_0 + b_1x + b_2x^2 + b_3x^3 + b_4x^4 + b_5x^5$.

between model classes, and the red box represents the independent test set that is only employed to quantify the model performance on unobserved data. It is essential that for each fold, the training and validation sets, and the test set are kept separated. Typically, we would employ 80 percent of the data for the training and validation sets and 20 percent for the test set.

It is common to employ 10-fold cross-validation because it offers a balance between the computational effort required and the amount of data available to estimate the model parameters. For N folds, the model needs to be trained N times, which may involve a substantial computational effort. On the other hand, using few folds may lead to poor parameter estimates. For example, if we only employ 2 folds, for each of them the model is only trained on half the training set. When $N = D_{train}$, the procedure is called *leave-one-out cross-validation* (LOOCV). This approach leads to the best estimation of validation errors because more data is employed during training, yet it is the most computationally demanding. Algorithm 7 summarizes the $N$-fold cross-validation procedure for choosing between M model classes.

---

**Algorithm 7:** *N*-fold cross-validation

1  define $\{\mathbf{x}^{\text{train}}, \mathbf{y}^{\text{train}}\}$        (training set with $D_{train}$ observations)
2  define $\{\mathbf{x}^{\text{test}}, \mathbf{y}^{\text{test}}\}$                                    (test set)
3  define $g_m(\mathbf{x}), \forall m \in \{1 : \mathtt{M}\}$                                 (models)
4  define N                                                          (N-folds)
5  $\mathbf{d} = \text{randperm}(\mathtt{D}_{train})$        (random permutation of $\lfloor 1 : \mathtt{D}_{train}\rfloor^\mathsf{T}$)
6  $n = \text{round}(\mathtt{D}_{train}/\mathtt{N})$                        (# observations per fold)
7  **for** $m \in \{1, 2, \cdots, \mathtt{M}\}$ **do**
8      **for** $i \in \{1, 2, \cdots, \mathtt{N}\}$ **do**
9          $\mathbf{x}_t = \mathbf{x}^{\text{train}}; \mathbf{y}_t = \mathbf{y}^{\text{train}}$                     (initialize training set)
10          $\mathbf{k} = \mathbf{d}\big((n(i-1)+1) : \min(i \cdot n, \mathtt{D}_{train})\big)$   (val. set indices)
11          $\mathbf{x}_v = \mathbf{x}^{\text{train}}(\mathbf{k}, :); \mathbf{y}_v = \mathbf{y}^{\text{train}}(\mathbf{k}, :)$          (validation set)
12          $\mathbf{x}_t(\mathbf{k}, :) = [\ ]; \mathbf{y}_t(\mathbf{k}, :) = [\ ]$            (remove validation set)
13          $\mathbf{X} = \text{fct}(\mathbf{x}_t)$                   (training model-matrix)
14          $\mathbf{b}_i^* = (\mathbf{X}^\mathsf{T}\mathbf{X})^{-1}\mathbf{X}^\mathsf{T}\mathbf{y}_t$             (estimate parameters)
15          $\mathbf{X} = \text{fct}(\mathbf{x}_v)$               (validation model-matrix)
16          $J_i = J(\mathbf{b}_i^*, \mathbf{y}_v)$                   ($i^{\text{th}}$ validation-set loss)
17      $J_{\text{validation}}^m = \sum_{i=1}^{\mathtt{N}} J_i$       (aggregated loss for model $m$)
18  $m^* = \arg\min_m J_{\text{validation}}^m$                     (select best model)
19  $\mathbf{X} = \text{fct}(\mathbf{x}^{\text{train}})$               (training model-matrix for $m^*$)
20  $\mathbf{b}^* = (\mathbf{X}^\mathsf{T}\mathbf{X})^{-1}\mathbf{X}^\mathsf{T}\mathbf{y}^{\text{train}}$             (estimate parameters)
21  $\mathbf{X} = \text{fct}(\mathbf{x}^{\text{test}})$                     (test-set model matrix)
22  $J_{\text{test}}^{m^*} = J(\mathbf{b}^*, \mathbf{y}^{\text{test}})$                     (test-set loss for $m^*$)

Figures 8.9 and 8.10 present a 2-fold cross-validation procedure employed to choose between a linear model $g_1(x)$ and a fifth-order polynomial one $g_2(x)$,

$$
\begin{aligned}
g_1(x) &= b_0 + b_1 x && \text{(linear)} \\
g_2(x) &= b_0 + b_1 x + b_2 x^2 + b_3 x^3 + b_4 x^4 + b_5 x^5 && \text{($5^{\text{th}}$ order)}.
\end{aligned}
$$

For the linear model, the sum of the losses for the two training and validation sets is

$$
\left.
\begin{aligned}
\sum J_{train} &= 1.97 \\
\sum J_{validation} &= 2.02
\end{aligned}
\right\} g_1(x) \; \triangle,
$$

and for the fifth-order polynomial model, the sum of the losses for the two training and validation sets is

$$
\left.
\begin{aligned}
\sum J_{train} &= 0.81 \\
\sum J_{validation} &= 20.89
\end{aligned}
\right\} g_2(x) \; \triangle.
$$

These results indicate that the fifth-order polynomial function is better at fitting the training set; however, it fails at predicting data that was not seen in the training set. This is reflected in the loss function for the validation set, which is lower for the linear model having the correct complexity, compared to the fifth-order polynomial function that is overcomplex.

We now come back to the targets presented in figure 8.6; the model $g_1(x)$ presented in figure 8.9 corresponds to the target in figure 8.6a, where the complexity is optimal so there is both a low bias and a low variance. The model $g_2(x)$ presented in figure 8.10 corresponds to figure 8.6c, where because the capacity is too high, the model closely fits the training data in each fold (low bias), yet the model is drastically different (i.e., high variance) for each fold. Here, the high model capacity allowed to overfit the noise contained in the data. Finally, the target in figure 8.6b corresponds to the new case illustrated in figure 8.11, where the model does not have enough capacity to fit the data. The result is that the model is highly biased, yet there is almost no variability between the models obtained for each fold.

### 8.1.3 Mathematical Formulation $>$1-D

Linear regression can be employed for cases involving more than one covariate. The data set is then $\mathcal{D} = \{(\mathbf{x}_i, y_i), \forall i \in \{1 : \mathtt{D}\}\}$, where $\mathbf{x}_i = [x_1 \; x_2 \; \cdots \; x_{\mathtt{X}}]_i^{\mathsf{T}} \in \mathbb{R}^{\mathtt{X}}$ is an $\mathtt{X}$ by one vector associated with an observation $y_i \in \mathbb{R}^1$. The model is an $\mathtt{X} \to 1$ function, $g(\mathbf{x}) \equiv \mathrm{fct}(\mathbf{x}) \in \mathbb{R}^1$, which is defined using basis functions so that

$$
g(\mathbf{x}) = b_0 + b_1 \phi_1(x_1) + b_2 \phi_2(x_2) + \cdots + b_{\mathtt{B}} \phi_{\mathtt{B}}(x_{\mathtt{X}}) = \mathbf{X}\mathbf{b}.
$$

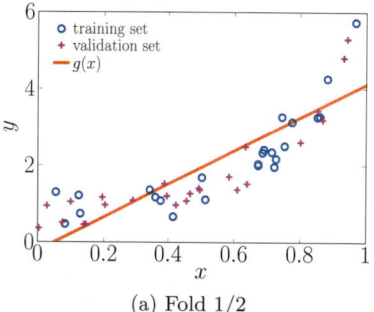

(a) Fold 1/2

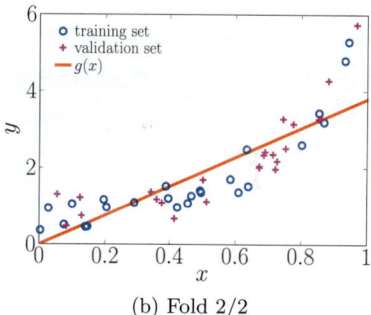

(b) Fold 2/2

Figure 8.11: Example of 2-fold cross-validation for the model $g_3(x) = b_0 + b_1 x$, where the data follows the real model $\check{g}(x) = 1 + 5x^4$.

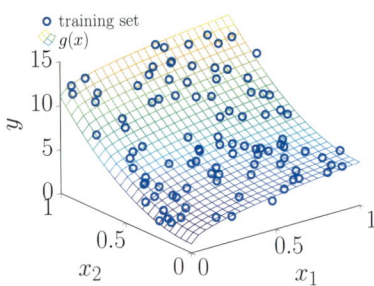

Figure 8.12: Example of application of linear regression to a 2-D data set.

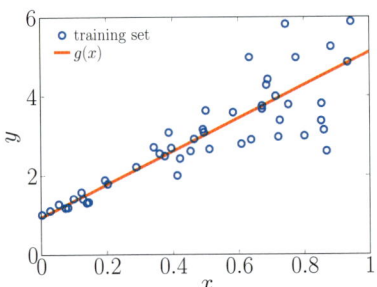

(a) Example of heteroscedasticity where the variance of errors is a function of the covariate $x$

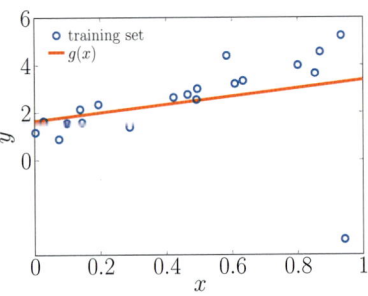

(b) Example of a bias introduced by an outlier on the bottom right corner

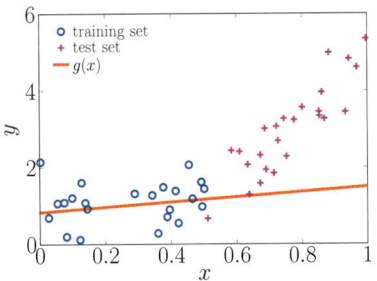

(c) Example of biased extrapolation for unobserved covariates

Figure 8.13: Illustrations of some of the limitations of the linear regression method presented in §8.1.1.

The matrix notation for the model defined over the entire data set is

$$\mathbf{b} = \begin{bmatrix} b_0 \\ b_1 \\ \vdots \\ b_B \end{bmatrix}, \ \mathbf{y} = \begin{bmatrix} y_1 \\ y_2 \\ \vdots \\ y_D \end{bmatrix}, \ \mathbf{X} = \begin{bmatrix} 1 & \phi_1(x_{1,1}) & \phi_2(x_{1,2}) & \cdots & \phi_B(x_{1,X}) \\ 1 & \phi_1(x_{2,1}) & \phi_2(x_{2,2}) & \cdots & \phi_B(x_{2,X}) \\ \vdots & \vdots & \vdots & \ddots & \vdots \\ 1 & \phi_1(x_{D,1}) & \phi_2(x_{D,2}) & \cdots & \phi_B(x_{D,X}) \end{bmatrix},$$

where $[\mathbf{x}_i]_j \equiv x_{i,j}$ refers to the $j^{\text{th}}$ component of the $\mathbf{x}$ vector for the $i^{\text{th}}$ observation. Figure 8.12 presents an example of multivariate linear regression for $\mathbf{x} = [x_1 \ x_2]^{\mathsf{T}}$ and for which the model is

$$g(\mathbf{x}) = b_0 + b_1 x_1^{1/2} + b_2 x_2^2.$$

Note that even if the model now involves multiple covariates, the optimal model parameters $\mathbf{b}^*$ are still estimated using equation 8.2.

### 8.1.4   Limitations

Some of the limitations of the linear regression method presented in §8.1.1 are illustrated in figure 8.13: (a) it assumes *homoscedastic* errors and is thus not suited for the cases where the variance of errors depends on the covariate $x$, that is, *heteroscedastic* cases, (b) it is sensitive to *outliers*, so a single biased observation may severely affect the model, and (c) it does not differentiate between *interpolation* and *extrapolation*. Interpolation refers to predictions made between covariates for which observations are available to train the model, and extrapolation refers to predictions made beyond the domain the model has been trained for. In (c), the predictions performed with $g(x)$ are good at interpolating between data points in the training set, yet they offer a poor performance for extrapolating beyond them on the test data. A main limitation with the linear regression formulation presented here is that it does not provide uncertainty estimates indicating that its predictive performance degrades for covariate values that are far from those available in the training set.

Another key limitation not illustrated in figure 8.13 is that the performance of linear regression depends on our ability to hand-pick the correct basis functions. For cases involving one or two covariates, this task is feasible from a simple visual representation of the data set. This task becomes difficult when the number of covariates $X > 2$. In short, linear regression is simple and parameter estimation is quick, yet it is outperformed by modern techniques such as *Gaussian process regression* and *neural networks*, which will be covered in the next sections. Note that the cross-validation technique covered here is not limited to linear regression and can be employed with a wide range of machine learning methods.

## 8.2   Gaussian Process Regression

Gaussian process regression (GPR) works by describing the prior knowledge of system responses over its covariate domain using a joint multivariate Normal probability density function (PDF). Then, it employs the properties of the multivariate Normal (see §4.1.3) in order to update the joint prior PDF using empirical observations.

*Pipeline example*   Take for example the 100 km long pipeline illustrated in figure 8.14, for which we are interested in quantifying the temperature at any coordinate $x \in (0, 100)$ km. Given that we know the pipeline temperature to be $y = 8°C$ at $x = 30$ km, *what is the temperature at $x = 30.1$ km?* Because of the proximity between the two locations, it is intuitive that the temperature has to be close to $8°C$. Then, *what is the temperature at $x = 70$ km?* We cannot say for sure that it is going to be around $8°C$ because over a 40 km distance, the change in temperature can reach a few degrees. This example illustrates how we can take advantage of the underlying dependencies between system responses for different observed covariates $x$, for predicting the responses for unobserved covariates.

Let us consider system responses defined by a single covariate $x_i$ so that

$$\underbrace{g_i}_{\text{observation}} = \overbrace{g(x_i)}^{\text{model}}.$$

The data set is defined as $\mathcal{D} = \{(x_i, g_i), \forall i \in \{1 : \text{D}\}\} = \{\mathcal{D}_x, \mathcal{D}_g\}$, where for D observations,

$$\mathcal{D}_g = \{\mathbf{g}\} = \{[g_1 \ g_2 \ \cdots \ g_\text{D}]^\mathsf{T}\}, \text{ and } \mathcal{D}_x = \{\mathbf{x}\} = \{[x_1 \ x_2 \ \cdots \ x_\text{D}]^\mathsf{T}\},$$

so for each observed $g_i$, there is an associated covariate $x_i$. A *Gaussian process* is a multivariate Normal random variable defined over a domain described by covariates. In the case of Gaussian process regression, a set of system responses is assumed to be a realization from a Gaussian process $g(\mathbf{x}) : \mathbf{G}(\mathbf{x}) \sim \mathcal{N}(g(\mathbf{x}); \boldsymbol{\mu_\mathbf{G}}, \boldsymbol{\Sigma_\mathbf{G}})$, where $\boldsymbol{\mu_\mathbf{G}}$ is the mean vector $[\boldsymbol{\mu_\mathbf{G}}]_i = \mu_\mathbf{G}(x_i)$ and the covariance matrix $\boldsymbol{\Sigma_\mathbf{G}}$ is

$$[\boldsymbol{\Sigma_\mathbf{G}}]_{ij} = \rho(x_i, x_j) \cdot \sigma_G(x_i) \cdot \sigma_G(x_j), \quad \rho(x_i, x_j) = \text{fct}(x_i - x_j).$$

$\boldsymbol{\Sigma_\mathbf{G}}$ defines the covariance between a discrete set of Gaussian random variables for which the pairwise correlation $\rho(x_i, x_j)$ between

$x = 0 \qquad\qquad\qquad x = 100$
$[km]$

Figure 8.14: Example of a 100 km long pipeline.

**Note:** As for linear regression (see §8.1), covariates $x$ are assumed to be deterministic constants that are free from observation errors.

$G(x_i)$ and $G(x_j)$ is a function of the distance between the covariates $x_i$ and $x_j$. Note that $\boldsymbol{\mu_G}$ and $\boldsymbol{\Sigma_G}$ depend implicitly on the covariates $\mathbf{x}$.

Figure 8.15: Representation of the prior knowledge for the temperature along a pipeline. The dashed horizontal red line represents the prior mean, and the pink-shaded area is the $\pm 2\sigma_G$ confidence region. Overlaid colored curves represent five realizations of the Gaussian process, $\mathbf{g}_i : \mathbf{G} \sim \mathcal{N}(g(\mathbf{x}); \boldsymbol{\mu_G}, \boldsymbol{\Sigma_G})$.

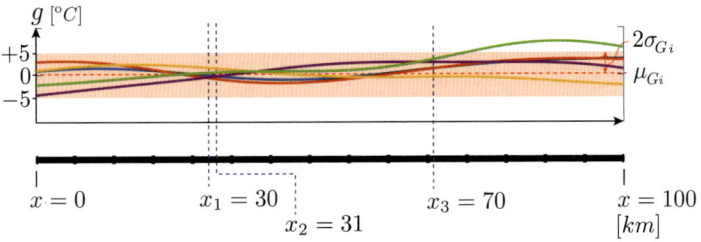

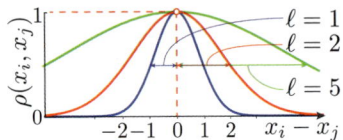

Figure 8.16: Square-exponential correlation function.

*Pipeline example (continued)*   Figure 8.15 illustrates our prior knowledge of the temperature for coordinates $\mathbf{x} = [0\ 1\ \cdots\ 100]^\mathsf{T}$ km,

$$\underbrace{\mathbf{G} \sim \mathcal{N}(g(\mathbf{x}); \boldsymbol{\mu_G}, \boldsymbol{\Sigma_G})}_{\text{Prior knowledge}}, \quad [\boldsymbol{\Sigma_G}]_{ij} = \rho(x_i, x_j)\sigma_G^2,$$

where each colored curve represents a different realization of the Gaussian process $\mathbf{G}$. The prior means and standard deviations are assumed to be equal for all locations $x$, so $\mu_G(x) = \mu_G = 0°C$ and $\sigma_G(x_i) = \sigma_G = 2.5°C$. The correlation between the temperature at two locations $G(x_i)$ and $G(x_j)$ is defined here by a *square exponential* correlation function

$$\rho(x_i, x_j) = \exp\left(-\frac{1}{2}\frac{(x_i - x_j)^2}{\ell^2}\right),$$

where $\ell$ is the *length-scale* parameter. Figure 8.16 presents the square-exponential correlation function plotted for parameter values $\ell = \{1, 2, 5\}$.

In figure 8.15, for a length-scale of $\ell = 25$ km, the correlation for the three pairs of coordinates $x = \{30, 31, 70\}$ is

$$\left.\begin{array}{l}\rho(x_1 = 30\,\text{km}, x_2 = 31\,\text{km}) = 0.999 \\ \rho(x_1 = 30\,\text{km}, x_3 = 70\,\text{km}) = 0.278 \\ \rho(x_2 = 31\,\text{km}, x_3 = 70\,\text{km}) = 0.296\end{array}\right\} \text{for } \ell = 25\,\text{km}.$$

Figure 8.17 presents examples of realizations for the same Gaussian process using different length-scale parameters.

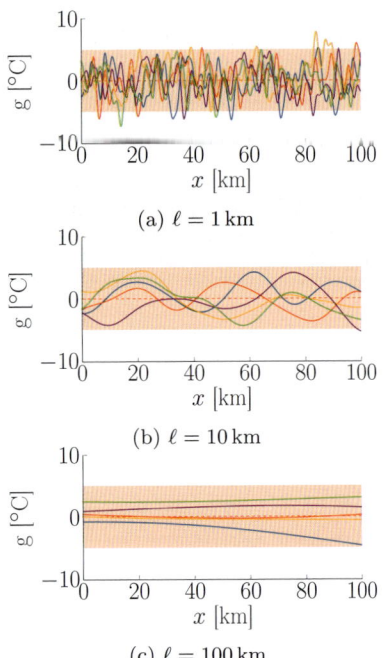

(a) $\ell = 1$ km

(b) $\ell = 10$ km

(c) $\ell = 100$ km

Figure 8.17: Examples of Gaussian process realizations for different length-scale parameter values. The dashed horizontal red line represents the prior mean, and the pink-shaded area is the $\pm 2\sigma_G$ confidence region.

## 8.2.1   Updating a GP Using Exact Observations

Let us consider $\mathcal{D} = \{(x_i, g_i),\ i \in \{1 : \mathtt{D}\}\}$, a set of D exact observations (i.e., without observation errors), and $\mathbf{x}_* = [x_{*1}\ x_{*2}\ \cdots\ x_{*\mathtt{P}}]^\mathsf{T}$, a vector of P covariates corresponding to either observed or unobserved locations where we want to obtain predictions. The goal is to obtain the posterior PDF $f(\mathbf{g}_*|\mathbf{x}_*, \mathcal{D})$. Before obtaining this posterior PDF, we have to define the joint prior knowledge for the system responses at both the observed $\mathbf{x}$ and prediction $\mathbf{x}_*$ locations,

$$\underbrace{\left\{ \begin{array}{c} \mathbf{G} \\ \mathbf{G}_* \end{array} \right\}, \ \boldsymbol{\mu} = \left\{ \frac{\boldsymbol{\mu}_{\mathbf{G}}}{\boldsymbol{\mu}_*} \right\}, \ \boldsymbol{\Sigma} = \left[ \begin{array}{c|c} \boldsymbol{\Sigma}_{\mathbf{G}} & \boldsymbol{\Sigma}_{\mathbf{G}*} \\ \hline \boldsymbol{\Sigma}_{\mathbf{G}*}^{\mathsf{T}} & \boldsymbol{\Sigma}_* \end{array} \right],}_{\text{Prior knowledge}}$$

where $\boldsymbol{\mu}_{\mathbf{G}}$ and $\boldsymbol{\mu}_*$ are, respectively, the prior means at observed and at prediction locations; $[\boldsymbol{\Sigma}_{\mathbf{G}}]_{ij} = \rho(x_i, x_j)\sigma_G^2$ and $[\boldsymbol{\Sigma}_*]_{ij} = \rho(x_{*i}, x_{*j})\sigma_G^2$ are their respective covariances. The covariance between observed and prediction locations is described by $[\boldsymbol{\Sigma}_{\mathbf{G}*}]_{ij} = \rho(x_i, x_{*j})\sigma_G^2$. The posterior probability of $\mathbf{G}_*$ at prediction locations $\mathbf{x}_*$, conditioned on the observations $\mathcal{D}$, is given by $f_{\mathbf{G}_*|\mathbf{x}_*, \mathcal{D}}(\mathbf{g}_*|\mathbf{x}_*, \mathcal{D}) = \mathcal{N}(\mathbf{g}_*; \boldsymbol{\mu}_{*|\mathcal{D}}, \boldsymbol{\Sigma}_{*|\mathcal{D}})$, where the posterior mean vector and covariance matrix are

$$\boldsymbol{\mu}_{*|\mathcal{D}} = \boldsymbol{\mu}_{\mathbf{G}*} + \boldsymbol{\Sigma}_{\mathbf{G}*}^{\mathsf{T}} \boldsymbol{\Sigma}_{\mathbf{G}}^{-1} (\mathbf{g} - \boldsymbol{\mu}_{\mathbf{G}})$$
$$\underbrace{\boldsymbol{\Sigma}_{*|\mathcal{D}} = \boldsymbol{\Sigma}_* - \boldsymbol{\Sigma}_{\mathbf{G}*}^{\mathsf{T}} \boldsymbol{\Sigma}_{\mathbf{G}}^{-1} \boldsymbol{\Sigma}_{\mathbf{G}*}.}_{\text{Posterior knowledge}}$$

These two equations are analogous to those presented §4.1.3 for the conditional distribution of a multivariate Normal ($\equiv$ Gaussian).

*Strength and limitations*   One of the main differences between the Gaussian process regression and other regression methods that encode the relationships contained in the data set through a parametric model is that GPR is a *nonparametric approach*, for which predictions depend explicitly on the observations available in the data set. The upside, in comparison with other methods such as linear regression, is that there is no requirement to define basis functions that are compatible with the system responses. GPR thus allows modeling highly nonlinear responses using few parameters. Another key feature of GPR is that it is an intrinsically probabilistic method that allows differentiating between the prediction quality in interpolation and extrapolation situations. For example, figures 8.19c and 8.19d show how interpolating between observation leads to a narrower confidence region than when extrapolating.

One of the main downsides of Gaussian process regression is that when the data set is large ($D > 1000$), building the matrices describing the joint prior knowledge becomes computationally demanding and it is then necessary to employ approximate methods.[2]

**Nomenclature**

$\mathbf{x} = [x_1 \ x_2 \ \cdots \ x_D]^{\mathsf{T}}$: Covariates for D observed locations

$\mathbf{x}_* = [x_{*1} \ x_{*2} \ \cdots \ x_{*P}]^{\mathsf{T}}$: Covariates for P prediction locations

$\mathbf{g} = [g_1 \ g_2 \ \cdots \ g_D]^{\mathsf{T}}$: System responses for D observed locations

$\mathbf{g}_* = [g_{*1} \ g_{*2} \ \cdots \ g_{*P}]^{\mathsf{T}}$: Model predictions for P locations

$\boldsymbol{\Sigma}_{\mathbf{G}}$: Prior covariance of $\mathbf{G}$ for observed locations $\mathbf{x}$

$\boldsymbol{\Sigma}_*$: Prior covariance of $\mathbf{G}_*$ for prediction locations $\mathbf{x}_*$

$\boldsymbol{\Sigma}_{\mathbf{G}*}$: Prior covariance of $\mathbf{G}$ between observed and prediction locations, $\mathbf{x}$ and $\mathbf{x}_*$, respectively

[2] Quiñonero-Candela, J. and C. E. Rasmussen (2005). A unifying view of sparse approximate Gaussian process regression. *Journal of Machine Learning Research 6*, 1939–1959.

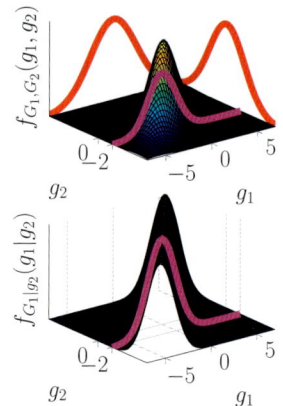

Figure 8.18: Joint Gaussian prior knowledge of the temperature at two locations and conditional PDF for the temperature $g_1$, given the observation $g_2 = -2$.

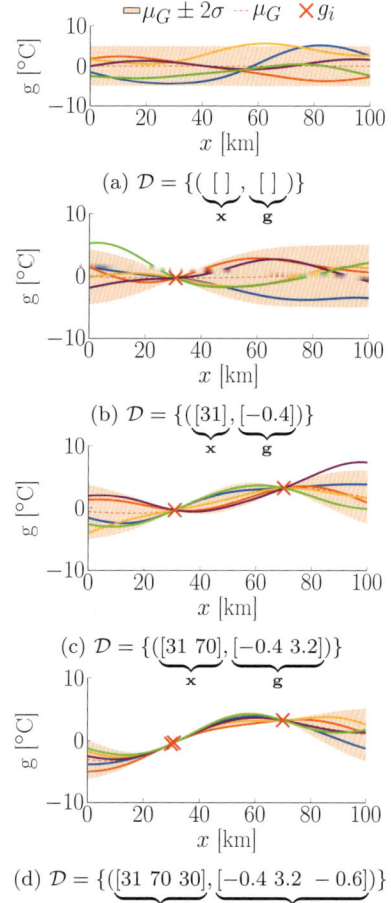

(a) $\mathcal{D} = \{(\underbrace{[\;]}_{\mathbf{x}}, \underbrace{[\;]}_{\mathbf{g}})\}$

(b) $\mathcal{D} = \{(\underbrace{[31]}_{\mathbf{x}}, \underbrace{[-0.4]}_{\mathbf{g}})\}$

(c) $\mathcal{D} = \{(\underbrace{[31\ 70]}_{\mathbf{x}}, \underbrace{[-0.4\ 3.2]}_{\mathbf{g}})\}$

(d) $\mathcal{D} = \{(\underbrace{[31\ 70\ 30]}_{\mathbf{x}}, \underbrace{[-0.4\ 3.2\ -0.6]}_{\mathbf{g}})\}$

Figure 8.19: Examples of how the Gaussian process describing the pipeline temperature evolves as it is updated with different data sets $\mathcal{D}$.

*Pipeline example (continued)*   We consider the case where the pipeline temperature $g_2 = -2°C$ is observed for the covariate $x_2$ and we want to use this information to obtain the posterior PDF describing the temperature $G_1$ at a location $x_1$. Given $\mu_G = 0°C$, $\sigma_G = 2.5°C$, and $\rho(x_1, x_2) = 0.8$, the joint PDF describing the prior probability of the temperature at locations $x_1$ and $x_2$ is defined by

$$f_{G_1 G_2}(g_1, g_2) = \mathcal{N}(\mathbf{g}; \boldsymbol{\mu_G}, \boldsymbol{\Sigma_G}) \begin{cases} \boldsymbol{\mu_G} = [0\ 0]^\mathsf{T} \\ \boldsymbol{\Sigma_G} = \begin{bmatrix} 2.5^2 & 0.8 \cdot 2.5^2 \\ 0.8 \cdot 2.5^2 & 2.5^2 \end{bmatrix} . \end{cases}$$

The conditional probability of $G_1$ given the observation $g_2$ is

$$\begin{aligned} f_{G_1|g_2}(g_1|g_2) &= \frac{f_{G_1 G_2}(g_1, g_2)}{f_{G_2}(g_2)} \\ &= \mathcal{N}(g_1; \mu_{1|2}, \sigma_{1|2}^2), \end{aligned}$$

where the posterior mean and standard deviation are

$$\begin{aligned} \mu_{1|2} &= \mu_1 + \rho \sigma_1 \frac{g_2 - \mu_2}{\sigma_2} &= 0 + 0.8 \times 2.5 \frac{-2-0}{2.5} &= -1.6°C \\ \sigma_{1|2} &= \sigma_1 \sqrt{1 - \rho^2} &= 2.5\sqrt{1 - 0.8^2} &= 1.5°C. \end{aligned}$$

Again, these last two equations are analogous to those presented in §4.1.3. The joint prior and posterior PDFs are presented in figure 8.18. Figure 8.19 presents the generalization of this update process for $\mathbb{P} = 101$ predicted locations $\mathbf{x}_\star = [0\ 1\ \cdots\ 100]^\mathsf{T}$ km and using, respectively, $\mathsf{D} = \{0, 1, 2, 3\}$ observations. Note how the uncertainty reduces to zero at locations where observations are available. Realizations of the updated Gaussian process are then required to pass by all observed values. The correlation embedded in the joint prior knowledge allows sharing the information carried by observations for predicting the system responses at adjacent unobserved covariate values.

### 8.2.2   Updating a GP Using Imperfect Observations

We now explore how to extend the formulation presented in the previous section to the case where observations are contaminated by noise. The observation model is now defined as

$$\underbrace{y}_{\text{observation}} = \overbrace{g(x)}^{\text{model}} + \underbrace{v}_{\text{measurement error}},$$

where $v : V \sim \mathcal{N}(v; 0, \sigma_V^2)$ describes zero-mean independent observation errors so that $V_i \perp\!\!\!\perp V_j \ \forall i \neq j$. Because $g(x) : \mathbf{G}(x)$ is now indirectly observed through $y$, we have to redefine our prior knowledge between the observations $\mathbf{Y}$ and the model responses $\mathbf{G}_*$ at prediction locations. The joint prior knowledge is described by

$$\underbrace{\left\{\begin{array}{c} \mathbf{Y} \\ \mathbf{G}_* \end{array}\right\}, \ \mathbf{m} = \left\{\begin{array}{c} \boldsymbol{\mu}_{\mathbf{Y}} \\ \boldsymbol{\mu}_{\mathbf{G}_*} \end{array}\right\}, \ \boldsymbol{\Sigma} = \left[\begin{array}{c|c} \boldsymbol{\Sigma}_{\mathbf{Y}} & \boldsymbol{\Sigma}_{\mathbf{Y}*} \\ \hline \boldsymbol{\Sigma}_{\mathbf{Y}*}^{\mathsf{T}} & \boldsymbol{\Sigma}_* \end{array}\right]}_{\text{prior knowledge}}.$$

In the case where the observation noise has a mean equal to zero, the mean vector for observations equals the prior mean for the system responses, $\boldsymbol{\mu}_{\mathbf{Y}} = \boldsymbol{\mu}_{\mathbf{G}}$. The global covariance matrix is then defined using

$$\begin{aligned} [\boldsymbol{\Sigma}_{\mathbf{Y}}]_{ij} &= \rho(x_i, x_j)\sigma_G^2 + \sigma_V^2 \delta_{ij}, \quad \text{where} \quad \delta_{ij} = 1 \text{ if } i = j \\ [\boldsymbol{\Sigma}_*]_{ij} &= \rho(x_{*i}, x_{*j})\sigma_G^2 \qquad\qquad\qquad \delta_{ij} = 0 \text{ if } i \neq j \\ [\boldsymbol{\Sigma}_{\mathbf{Y}*}]_{ij} &= \rho(x_i, x_{*j})\sigma_G^2. \end{aligned}$$

Given the prediction locations $\mathbf{x}_*$ and the observations $\mathcal{D}$, the posterior knowledge for the system responses $\mathbf{G}_*$ is

$$f_{\mathbf{G}_*|\mathbf{x}_*, \mathcal{D}}(\mathbf{g}_*|\mathbf{x}_*, \mathcal{D}) = \mathcal{N}(\mathbf{g}_*; \boldsymbol{\mu}_{*|\mathcal{D}}, \boldsymbol{\Sigma}_{*|\mathcal{D}}),$$

where the posterior mean vector $\boldsymbol{\mu}_{*|\mathcal{D}}$ and covariance matrix $\boldsymbol{\Sigma}_{*|\mathcal{D}}$ are

$$\underbrace{\begin{aligned} \boldsymbol{\mu}_{*|\mathcal{D}} &= \boldsymbol{\mu}_{\mathbf{G}_*} + \boldsymbol{\Sigma}_{\mathbf{Y}*}^{\mathsf{T}} \boldsymbol{\Sigma}_{\mathbf{Y}}^{-1}(\mathbf{y} - \boldsymbol{\mu}_{\mathbf{Y}}) \\ \boldsymbol{\Sigma}_{*|\mathcal{D}} &= \boldsymbol{\Sigma}_* - \boldsymbol{\Sigma}_{\mathbf{Y}*}^{\mathsf{T}} \boldsymbol{\Sigma}_{\mathbf{Y}}^{-1} \boldsymbol{\Sigma}_{\mathbf{Y}*}. \end{aligned}}_{\text{posterior knowledge}}$$

*Pipeline example (continued)* We continue our study of the pipeline example by considering noisy observations characterized by a measurement error standard deviation $\sigma_V = 1\,^\circ\text{C}$. Figure 8.20 presents the updated Gaussian process for $\mathtt{P} = 101$ prediction locations $\mathbf{x}_* = [0\ 1\ \cdots\ 100]^{\mathsf{T}}$ km, and using, respectively, $\mathtt{D} = \{0, 1, 2, 3\}$ noisy observations. Notice how, contrarily to results presented in figure 8.19, uncertainty now remains in the posterior at observed locations. The larger $\sigma_V$ is, the less information observations carry and the less they modify the prior knowledge.

### 8.2.3   Multiple Covariates

In the case where the regression is performed for multiple covariates $\mathbf{x} = [x_1\ x_2\ \cdots\ x_{\mathtt{X}}]^{\mathsf{T}}$, it is necessary to define a joint correlation function defining the correlation coefficient associated with a pair of covariate vectors. Typically, the joint correlation function is obtained from the product of a marginal correlation function $\rho([\mathbf{x}_i]_k, [\mathbf{x}_j]_k)$

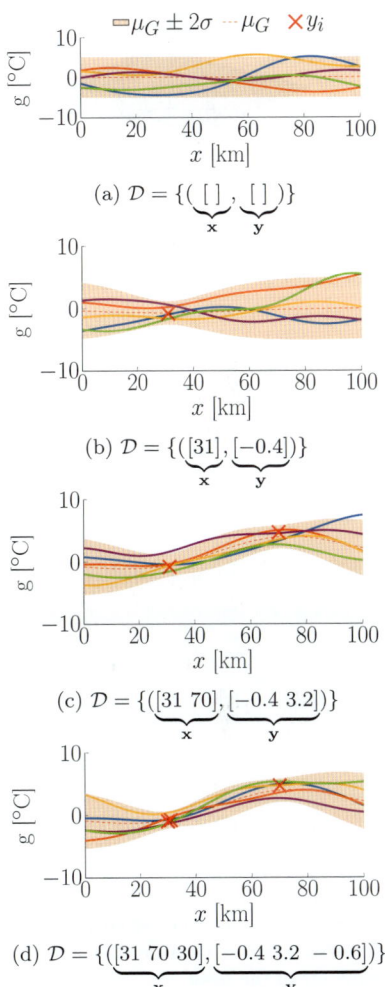

(a) $\mathcal{D} = \{(\underbrace{[\ ]}_{\mathbf{x}}, \underbrace{[\ ]}_{\mathbf{y}})\}$

(b) $\mathcal{D} = \{(\underbrace{[31]}_{\mathbf{x}}, \underbrace{[-0.4]}_{\mathbf{y}})\}$

(c) $\mathcal{D} = \{(\underbrace{[31\ 70]}_{\mathbf{x}}, \underbrace{[-0.4\ 3.2]}_{\mathbf{y}})\}$

(d) $\mathcal{D} = \{(\underbrace{[31\ 70\ 30]}_{\mathbf{x}}, \underbrace{[-0.4\ 3.2\ -0.6]}_{\mathbf{y}})\}$

Figure 8.20: Examples of how the Gaussian process describing the pipeline temperature evolves as it is updated with different data sets $\mathcal{D}$, with an observation standard deviation $\sigma_V = 1\,^\circ\text{C}$.

$$\underbrace{y_i}_{\text{observation}} = \underbrace{g(\mathbf{x}_i)}_{\text{model}} + \underbrace{v}_{\text{measurement error}}$$

$$\underbrace{\mathbf{x}_i = [x_1\ x_2\ \cdots\ x_{\mathtt{X}}]_i^{\mathsf{T}}}_{\text{covariates}}$$

$$\mathcal{D} = \{(\mathbf{x}_i, y_i), \forall i \in \{1 : \mathtt{D}\}\} = \{\mathcal{D}_x, \mathcal{D}_y\}$$

for each covariate,

$$\rho(\mathbf{x}_i, \mathbf{x}_j) = \prod_{k=1}^{\mathtt{X}} \rho\big([\mathbf{x}_i]_k, [\mathbf{x}_j]_k\big).$$

The multivariate formulation for the *square-exponential* correlation function is

$$\begin{aligned}
\rho(\mathbf{x}_i, \mathbf{x}_j) &= \prod_{k=1}^{\mathtt{X}} \exp\left(-\frac{1}{2}\frac{([\mathbf{x}_i]_k - [\mathbf{x}_j]_k)^2}{\ell_k^2}\right) \\
&\equiv \exp\left(-\frac{1}{2}(\mathbf{x}_i - \mathbf{x}_j)^\mathsf{T}\mathrm{diag}(\boldsymbol{\ell})^{-2}(\mathbf{x}_i - \mathbf{x}_j)\right),
\end{aligned}$$

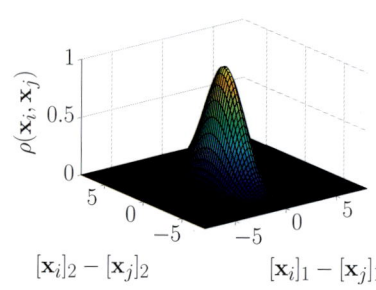

$\rho(\mathbf{x}_i, \mathbf{x}_j)$

$[\mathbf{x}_i]_2 - [\mathbf{x}_j]_2$     $[\mathbf{x}_i]_1 - [\mathbf{x}_j]_1$

Figure 8.21: Bivariate square-exponential correlation function with parameters $\ell_1 = 1$ and $\ell_2 = 1.5$.

where there is one parameter $\ell_k$ for each covariate. Figure 8.21 presents an example of a bivariate square-exponential correlation function. One relevant feature with this correlation function is that the length-scales $\ell_k$ are directly measuring the importance of each covariate for predicting the system responses. As a simple heuristic, we can say that if $\ell_k$ is greater than two times the empirical standard deviations of the observed covariates $\sigma_{X_k}$, then the $k^{\text{th}}$ covariate has a limited influence on the system response predictions. The reason is because the correlation coefficient remains close to one for any pairs of this $k^{\text{th}}$ covariate. In the opposite case, where $\ell_k$ is smaller than the empirical standard deviation $\sigma_{X_k}$, we can conclude that the $k^{\text{th}}$ covariate has a strong influence on the system response predictions.

### 8.2.4 Parameter Estimation

Up to now, we have assumed that we knew the parameters such as the observation standard deviation $\sigma_V$, the Gaussian process prior standard deviation $\sigma_G$, and the length-scales $\boldsymbol{\ell} = [\ell_1\ \ell_2\ \cdots\ \ell_{\mathtt{X}}]^\mathsf{T}$. In practice, we need to learn these *parameters* $\boldsymbol{\theta} = [\sigma_V\ \sigma_G\ \boldsymbol{\ell}^\mathsf{T}]^\mathsf{T}$ from the training data $\mathcal{D} = \{\mathcal{D}_x, \mathcal{D}_y\} = \{(\mathbf{x}_i, y_i),\ i \in \{1 : \mathtt{D}\}\}$. Following Bayes rule, the posterior PDF for parameters given data is

$\boldsymbol{\theta} = [\sigma_V\ \sigma_G\ \boldsymbol{\ell}^\mathsf{T}]^\mathsf{T}$ can be referred to as *hyperparameters* because they are parameters of the prior.

$$\underbrace{f(\boldsymbol{\theta}|\mathcal{D})}_{\text{posterior}} = \frac{\overbrace{f(\mathcal{D}_y|\mathcal{D}_x, \boldsymbol{\theta})}^{\text{likelihood}} \cdot \overbrace{f(\boldsymbol{\theta})}^{\text{prior}}}{\underbrace{f(\mathcal{D}_y)}_{\text{constant}}}.$$

Often, evaluating $f(\boldsymbol{\theta}|\mathcal{D}_x, \mathcal{D}_y)$ is computationally expensive, so a maximum likelihood estimation approach (see §6.7) is typically employed. The optimal parameter values $\boldsymbol{\theta}^*$ are then

$$\boldsymbol{\theta}^* = \arg\max_{\boldsymbol{\theta}} \overbrace{f(\mathcal{D}_y|\mathcal{D}_x, \boldsymbol{\theta})}^{\text{likelihood}} \equiv \arg\max_{\boldsymbol{\theta}} \overbrace{\ln f(\mathcal{D}_y|\mathcal{D}_x, \boldsymbol{\theta})}^{\text{log-likelihood}}.$$

The *likelihood* function $f(\mathcal{D}_y|\mathcal{D}_x, \boldsymbol{\theta})$ describes the joint prior probability density of observations $\mathcal{D}_y$, given a set of parameters $\boldsymbol{\theta}$,

$$f(\mathcal{D}_y|\mathcal{D}_x, \boldsymbol{\theta}) = \int f(\mathcal{D}_y|\mathbf{g}) \cdot f(\mathbf{g}|\mathcal{D}_x, \boldsymbol{\theta})d\mathbf{g} = \mathcal{N}(\mathcal{D}_y; \boldsymbol{\mu}_{\mathbf{Y}}, \boldsymbol{\Sigma}_{\mathbf{Y}}).$$

The log-likelihood for a vector of observations $\mathbf{y} = [y_1 \ y_2 \ \cdots \ y_{\mathsf{D}}]^{\mathsf{T}}$ is given by

$$\ln f(\mathcal{D}_y|\mathcal{D}_x, \boldsymbol{\theta}) = \ln \mathcal{N}(\mathbf{y}; \boldsymbol{\mu}_{\mathbf{Y}}, \boldsymbol{\Sigma}_{\mathbf{Y}})$$
$$= -\tfrac{1}{2}(\mathbf{y} - \boldsymbol{\mu}_{\mathbf{Y}})^{\mathsf{T}} \boldsymbol{\Sigma}_{\mathbf{Y}}^{-1}(\mathbf{y} - \boldsymbol{\mu}_{\mathbf{Y}}) - \tfrac{1}{2}\ln|\boldsymbol{\Sigma}_{\mathbf{Y}}| - \tfrac{\mathsf{D}}{2}\ln 2\pi.$$

The optimal parameters $\boldsymbol{\theta}^*$ correspond to the parameter values for which the derivative of the likelihood equals zero, as illustrated in figure 8.22. For $\boldsymbol{\mu}_{\mathbf{Y}} = \mathbf{0}$, the derivative of the log-likelihood is

$$\frac{\partial}{\partial \theta_j} \ln f(\mathcal{D}_y|\mathcal{D}_x, \boldsymbol{\theta}) = \frac{1}{2}\mathbf{y}^{\mathsf{T}} \boldsymbol{\Sigma}_{\mathbf{Y}}^{-1} \frac{\partial \boldsymbol{\Sigma}_{\mathbf{Y}}}{\partial \theta_j} \boldsymbol{\Sigma}_{\mathbf{Y}}^{-1}\mathbf{y} - \frac{1}{2}\text{tr}\left( \boldsymbol{\Sigma}_{\mathbf{Y}}^{-1} \frac{\partial \boldsymbol{\Sigma}_{\mathbf{Y}}}{\partial \theta_j} \right).$$

Efficient gradient-based optimization algorithms for learning parameters by maximizing the log-likelihood are already implemented open-source packages such as the GPML[3], GPstuff[4], and pyGPs[5] toolboxes.

### 8.2.5 Example: Soil Contamination Characterization

This section presents an applied example of GPR for the characterization of contaminant concentration in soils. Take, for example, a set of observations $\mathbf{y}_i$ of soil contaminant concentrations,

$$\mathcal{D} = \{\mathcal{D}_x, \mathcal{D}_y\} = \{(\mathbf{l}_i, y_i), \ i \in \{1 : 116\}\},$$

where $\mathbf{l}_i$ is a covariate vector associated with each concentration observation. These covariates denote the geographic coordinates (longitude, latitude, and depth) of each observation,

$$\underbrace{\mathbf{l}_i = [x \ y \ z]_i^{\mathsf{T}}}_{\text{geodesic coordinates}} \quad .$$

Each observation is represented in figure 8.23 by a circle where its size is proportional to the contaminant concentration. The goal is to employ Gaussian process regression in order to model the contaminant concentration across the whole soil volume, given the available observations for 116 discrete locations.

Because a concentration must be a positive number, and because several of the observations available are close to zero, the observation model is defined in the log-space,

$$\ln \underbrace{y}_{\text{observation}} = \overbrace{g(\mathbf{l})}^{\text{true contaminant [] in log space}} + \underbrace{v}_{\text{meas. error in log space}},$$

Theoretically, here, we should refer to $f(\mathcal{D}_y|\mathcal{D}_x, \boldsymbol{\theta})$ as the *marginal likelihood* because the hidden variables $\mathbf{g}$ are marginalized.

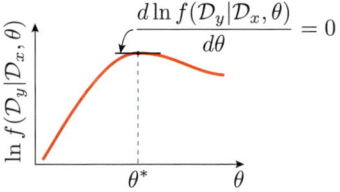

Figure 8.22: Log-likelihood maximization.

[3] Rasmussen, C. E. and H. Nickisch (2010). Gaussian processes for machine learning (GPML) toolbox. *The Journal of Machine Learning Research 11*, 3011–3015.

[4] Vanhatalo, J., J. Riihimäki, J. Hartikainen, P. Jylänki, V. Tolvanen, and A. Vehtari (2013). GPstuff: Bayesian modeling with Gaussian processes. *Journal of Machine Learning Research 14*, 1175–1179.

[5] Neumann, M., S. Huang, D. E. Marthaler, and K. Kersting (2015). pyGPs: A python library for Gaussian process regression and classification. *The Journal of Machine Learning Research 16*(1), 2611–2616.

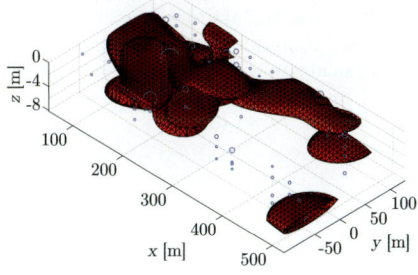

Figure 8.23: Example of application of GPR for soil contamination characterization. The iso-contours correspond to $\text{Pr}([] > 55\,\text{mg/kg}) = 0.5$, and the size of each circle is proportional to the observed contaminant concentration. (Adapted from Quach et al. (2017).)

where $g(\mathbf{l})$ models the natural logarithm of the true contaminant concentration. When transformed back to the original space using the exponential function, the model predictions are restricted to positive values. The observation errors are assumed to be described by the zero-mean independent Normal random variable $v : V \sim \mathcal{N}(v; 0, \sigma_V^2)$, $V_i \perp\!\!\!\perp V_j, \forall i \neq j$. The correlation between covariates is modeled using a square-exponential correlation function, and the prior mean at any location is assumed to be equal to one in the original space, which corresponds to zero in the log-transformed space.

The parameters to be estimated from the data are the length-scale along each direction, the Gaussian process standard deviation, and the observation noise standard deviation,

$$\boldsymbol{\theta} = \{\ell_x, \ell_y, \ell_z, \sigma_G, \sigma_V\}.$$

The maximum likelihood estimate (MLE) parameter values are

$$\boldsymbol{\theta}^* = \arg\max_{\boldsymbol{\theta}} \overbrace{\ln f(\mathcal{D}_y | \mathcal{D}_x, \boldsymbol{\theta})}^{\text{log-likelihood}} = \begin{cases} \ell_x = 56.4\,\text{m} \\ \ell_y = 53.7\,\text{m} \\ \ell_z = 0.64\,\text{m} \\ \sigma_G = 2.86 \\ \sigma_V = 0.18 \end{cases}.$$

**Covariate importance metric**

If $\ell_k \gg \sigma_{X_k}$ :   $\text{Imp}(X_k \to y) \approx 0$

If $\ell_k < \sigma_{X_k}$ :   $\text{Imp}(X_k \to y) \gg 0$

The optimal values for the length-scales $\{\ell_x, \ell_y, \ell_z\}$ are all smaller than the range of each covariate, as depicted in figure 8.23. This confirms that all three coordinates have an influence on the contaminant concentration. Figure 8.23 presents the iso-contours describing $\Pr([\,] > 55\,\text{mg/kg}) = 0.5$, where $\mathbf{l}^*$ are the covariates describing the $x, y, z$ geographic coordinate of a 3-D grid covering the entire soil volume for the site studied. Further details about this example are presented in the paper from Quach et al.[6]

[6] Quach, A. N. O., D. Tabor, L. Dumont, B. Courcelles, and J.-A. Goulet (2017). A machine learning approach for characterizing soil contamination in the presence of physical site discontinuities and aggregated samples. *Advanced Engineering Informatics 33*, 60–67.

### 8.2.6   Example: Metamodel

In civil engineering, it is common to employ computationally demanding models such as the finite element models presented in figure 8.24. When performing probabilistic studies related to structural or system reliability, it is often required to have millions of model realizations for different parameter values. In such a case, for the purpose of computational efficiency, it may be interesting to replace the finite element model with a regression model built from a set of simulated responses. Such a model of a model is called a *metamodel* or a *surrogate model*. Because observations are the output of a simulation, the observation model does not include any observation error and follows the formulation presented in §8.2.1.

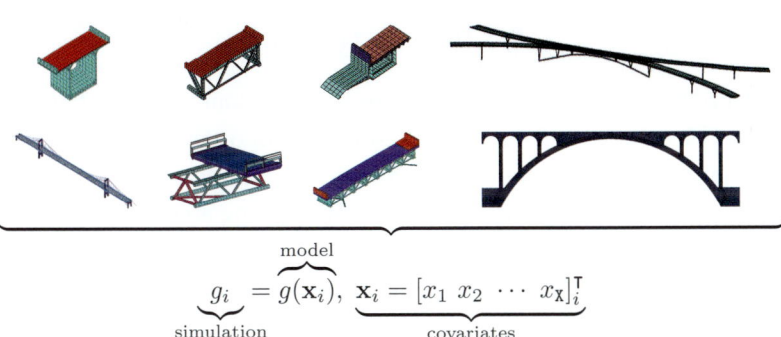

Figure 8.24: Examples of finite element structural models.

$$\underbrace{g_i}_{\text{simulation}} = \overbrace{g(\mathbf{x}_i)}^{\text{model}}, \ \underbrace{\mathbf{x}_i = [x_1 \ x_2 \ \cdots \ x_{\mathtt{X}}]_i^{\mathsf{T}}}_{\text{covariates}}$$

In this example, we built a metamodel for the Tamar Bridge finite element model.[7] Figure 8.25a presents an overview of the bridge finite element model, and figure 8.25b presents the modal displacement associated with its first natural frequency, which is the response modeled in this example. A set of D = 1000 model

[7] Westgate, R. J. and J. M. W. Brownjohn (2011). Development of a Tamar Bridge finite element model. In *Conference Proceedings of the Society for Experimental Mechanics*, Volume 5, pp. 13–20. Springer.

Figure 8.25: Tamar Bridge finite element model.

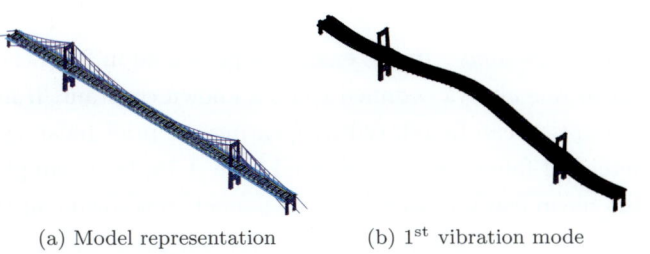

(a) Model representation        (b) 1st vibration mode

simulations is obtained by evaluating the finite element model for several sets of parameters that are generated from the prior $f(\boldsymbol{\theta})$. There are five model parameters that are treated as covariates of the GP regression model, $\mathbf{x}_i = [x_1 \ x_2 \ \cdots \ x_5]_i^{\mathsf{T}}$: the *stiffness* of the three expansion joints located on the bridge deck, along with the *initial strain* in the cables on the main and side spans. The locations of elements affected by these parameters are illustrated in figure 8.26. This problem employs the square-exponential correla-

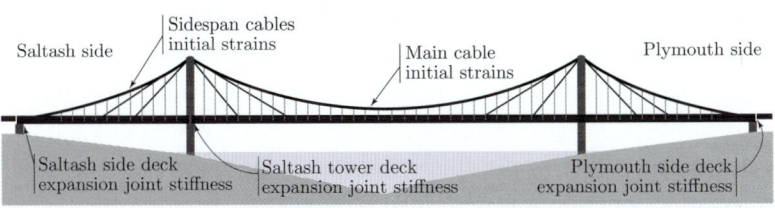

Figure 8.26: Tamar Bridge model parameter description.

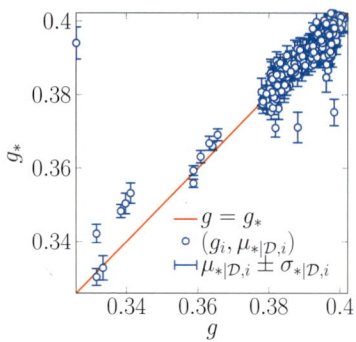

Figure 8.27: Comparison of the GPR model predictions with the finite-element model predictions using the leave-one-out cross-validation procedure.

tion function, so there is a total of six parameters to be learned,

$$\boldsymbol{\theta} = \{\ell_1, \ell_2, \ell_3, \ell_4, \ell_5, \sigma_G\}.$$

Figure 8.27 compares the GPR model predictions with the true finite element model outputs $g_i$. Note that in order to obtain a meaningful comparison between predicted and measured values, it is essential to test the model iteratively using a cross-validation procedure, where at each step, the observation corresponding to the prediction location in the validation set is removed from the training set.

## 8.2.7   Advanced Considerations

There are several aspects that should be considered in order to take advantage of the full potential of Gaussian process regression. Key aspects are related to (1) the definition of the prior mean functions, (2) the choice of covariance and correlation functions, and (3) the formulation required for modeling cases involving heteroscedastic errors.

*Prior mean functions*   In the examples presented in §8.2, the prior mean was always assumed to be a known constant. If needed, this assumption can be relaxed by treating the prior mean as a parameterized function to be inferred from data. In its simplest case, the mean can be assumed to be a single unknown constant. In more advanced cases we could describe our prior knowledge of the mean as a parameterized function that depends on covariates **x**. For example, if there is a single covariate, the mean could be modeled by the affine function $m(x) = ax + b$, where $\{a, b\}$ are parameters that need to be estimated from data. When using complex functions to describe the prior mean, it is essential to keep in mind that this procedure is sensitive to *overfitting* in a similar way as linear regression (see §8.1.2). Employing a highly flexible parameterization may lead to an increased log-likelihood in the training set, yet it may not materialize in the prediction performance evaluated on the test set. It is therefore essential to choose the best prior parameterization using methods such as *cross-validation* (see §8.1.2) or *Bayesian model selection* (see §6.8).

*Covariance functions*   In this section, only the square-exponential correlation function was described; nonetheless, several other choices exist. In his thesis, Duvenaud[8] presents several covariance models. A common one is the *Matern*-class correlation function, for

[8] Duvenaud, D. (2014). *Automatic model construction with Gaussian processes.* PhD thesis, University of Cambridge.

which a special case is defined as

$$\rho(x_i, x_j) = \exp\left(-\frac{|x_i - x_j|}{\ell}\right).$$

Figure 8.28 represents this correlation function for several length-scale values.

When using the square-exponential correlation function with covariate values that are close to each other, it is common to obtain correlation coefficients $\rho(\mathbf{x}_i, \mathbf{x}_j)$ close to one. This has the adverse effect of leading to singular or near-singular covariance matrices (see §2.4.2). Matern-class correlation functions can mitigate this problem because of its rapid reduction of the correlation coefficient, with an increase in the absolute distance between covariates.

Another correlation function particularly useful for engineering applications is the *periodic* one defined as

$$\rho(x_i, x_j) = \exp\left(-\frac{2}{\ell^2}\sin\left(\pi\frac{x_i - x_j}{p}\right)^2\right).$$

Figure 8.29 presents examples of the periodic correlation function for periods $p = \{5, 10\}$ and length-scale parameters $\ell = \{0.5, 1\}$. As its name indicates, this function is suited for modeling periodic phenomena, where events spaced by a distance close to the period $p$ will have a high correlation. The length-scale $\ell$ controls how fast the correlation decreases as the distance between covariates gets away from an integer that is a multiple of the period $p$.

Figure 8.30 presents an example of the periodic correlation function employed to model the historical daily average temperatures recorded at Montreal airport from 1953 to 1960. The blue circles represent the training data that were employed to learn the parameters: the prior mean value $\mu$, the Gaussian process prior standard deviation $\sigma_G$, the observation standard deviation $\sigma_V$, and

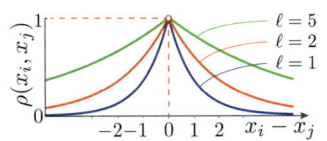

Figure 8.28: An example of Matern-class correlation function.

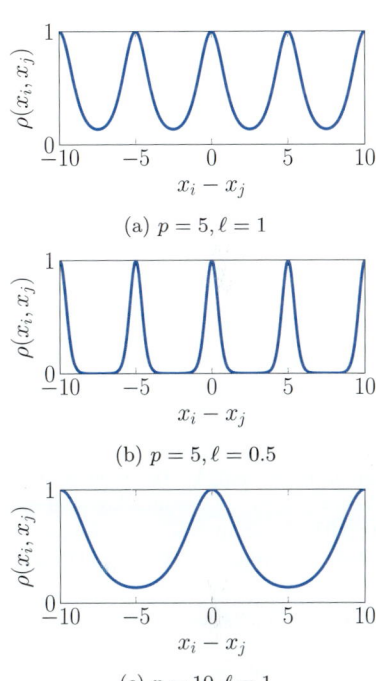

(a) $p = 5, \ell = 1$

(b) $p = 5, \ell = 0.5$

(c) $p = 10, \ell = 1$

Figure 8.29: Examples of periodic correlation functions.

Figure 8.30: Example of application of a periodic covariance function to model temperature data.

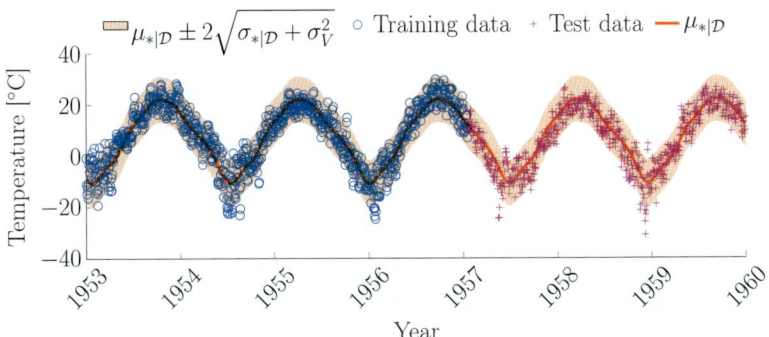

the period $p$. The pink crosses represent the test data that was not employed to perform predictions or to learn parameters. Most observations from the test set lie within the $\pm 2\sigma$ posterior interval. It shows how Gaussian process regression can, even with such a simple model with only four parameters, capture the temperature pattern as well as its uncertainty.

For more complex problems, highly complex covariance structures can be built by adding together simpler ones, such as those described in this section (see Rasmussen and Williams[9] for details). The choice regarding which correlation or covariance structure is most appropriate should, as for the prior mean functions, be based on methods such as *cross-validation* (see §8.1.2) or *Bayesian model selection* (see §6.8).

*Homo- and heteroscedasticity*   So far, this section has only treated cases associated with homoscedastic errors such as the generic case presented in figure 8.31a. In engineering applications, it is common to be confronted with cases where the variances $\sigma_V^2$ or $\sigma_G^2$ depend on the covariate values. A generic example of heteroscedastic cases is presented in figure 8.31b, which has been modeled by considering that the observation-error standard deviation $\sigma_V$ is itself modeled as a Gaussian process that depends on covariate values.[10] This method is implemented in the GPstuff[11] toolbox.

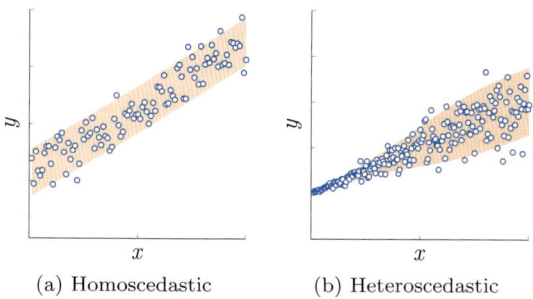

(a) Homoscedastic        (b) Heteroscedastic

[9] Rasmussen, C. E. and C. K. Williams (2006). *Gaussian processes for machine learning*. MIT Press.

[10] Tolvanen, V., P. Jylänki, and A. Vehtari (2014). Expectation propagation for nonstationary heteroscedastic Gaussian process regression. In *IEEE International Workshop on Machine Learning for Signal Processing (MLSP)*, 1–6. IEEE.

[11] Vanhatalo, J., J. Riihimäki, J. Hartikainen, P. Jylänki, V. Tolvanen, and A. Vehtari (2013). GPstuff: Bayesian modeling with Gaussian processes. *Journal of Machine Learning Research 14*, 1175–1179.

Figure 8.31: Examples of homo- and heteroscedastic cases where the shaded regions represent $\mu_Y \pm 2\sigma_Y$.

## 8.3   Neural Networks

The idea behind neural networks originates from the perceptron model[12] developed in the 1950s. Neural networks became extremely popular in the period from 1980 until the mid-1990s when they became overshadowed by other emerging methods. It was not until the early 2010s, with the appellation of *deep learning*, that they again took the leading role in the development of machine learning methods. In deep learning, the label *deep* refers to the great number

[12] Rosenblatt, F. (1958). The perceptron: A probabilistic model for information storage and organization in the brain. *Psychological Review 65*(6), 386.

of layers in the model. The rise of deep learning as the state-of-the-art method was caused by three main factors:[13] (1) the increasing size of the data sets to train on, (2) the increase in computational capacity brought by graphical processing units (GPUs), and (3) the improvements in the methods allowing to train deeper models. One limitation is that in order to achieve the performance it is renowned for, deep learning requires large data sets containing from thousands to millions of labeled examples for covariates $\mathbf{x}_i$ and system responses $y_i$. In the field of computer vision, speech recognition, and genomics, it is common to deal with data sets of such a large size; in the civil engineering context, it is not yet the case.

[13] Goodfellow, I., Y. Bengio, and A. Courville (2016). *Deep learning*. MIT Press.

This section presents the basic formulation for the simplest neural network architecture: the *feedforward neural network*. The reader interested in a detailed review of neural networks and their numerous architectures—such as *convolutional neural networks* (CNN), *generative adversarial networks* (GAN), and *recurrent neural networks* (RNN, LSTM)—should refer to specialized textbooks such as the one by Goodfellow, Bengio, and Courville (2016).

### 8.3.1 Feedforward Neural Networks

Feedforward is the simplest architecture for neural networks, where information is transferred from the input layer to the output layer by propagating it into layers of hidden variables. The idea is to approximate complex functions by a succession of simple combinations of hidden variables organized in layers.

*Linear regression*    The linear regression method presented in §8.1 is employed to introduce the concepts surrounding feedforward neural networks. Figure 8.32 presents a regression problem where three covariates $\mathbf{x} = [x_1 \ x_2 \ x_3]^\intercal$ are associated with a single observed system response $y = z + v$, $v : V \sim \mathcal{N}(v; 0, \sigma_V^2)$, where $v$ is a realization of a zero-mean normal observation error. $z$ represents a hidden variable, which is defined by the linear combination of covariate values using three weight parameters and one bias parameter,

$$z = w_1 x_1 + w_2 x_2 + w_3 x_3 + b.$$

In figure 8.32, the leftmost nodes represent covariates, the white node is a hidden variable, and the rightmost node is the observations $y_i : Y \sim \mathcal{N}(y; z, \sigma_V^2)$. Arrows represent the dependencies defined by the weight and bias parameters $\{w_1, w_2, w_3, b\}$. Weight parameter $w_i$ describes the slope of a plane with respect to each

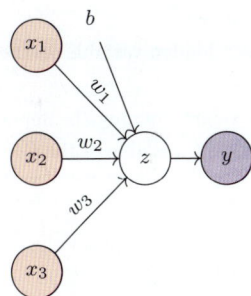

Figure 8.32: Representation of linear regression as a feedforward neural network. The leftmost nodes represent covariates, the white node is a hidden variable, and the rightmost node represents the observation.

variable, and $b$ the plane's intercept (e.g., see figure 2.2). With the current configuration, this model is limited because it is only a linear function, that is, the hidden variable $z$ is a linear function with respect to covariates $\mathbf{x}$.

*Linear regression with multiple layers*    Now we explore the effect of modifying the model as presented in figure 8.33 by adding multiple hidden variables, where each one is modeled as a linear combination of either the covariates or the hidden variables on the preceding layer. First, let us define the notation employed to describe this network: $z_i^{(l)}$ represents the $i^{\text{th}}$ hidden variable on the $l^{\text{th}}$ layer. In the first layer, the two hidden variables are defined by

$$
\begin{aligned}
z_1^{(1)} &= w_{1,1}^{(0)}x_1 + w_{1,2}^{(0)}x_2 + w_{1,3}^{(0)}x_3 + b_1^{(0)} \\
z_2^{(1)} &= w_{2,1}^{(0)}x_1 + w_{2,2}^{(0)}x_2 + w_{2,3}^{(0)}x_3 + b_2^{(0)},
\end{aligned}
$$

where $w_{i,j}^{(0)}$ is the weight defining the dependence between the $j^{\text{th}}$ variable of the $0^{\text{th}}$ layer (i.e., the first covariate itself) and the $i^{\text{th}}$ hidden variable of the $1^{\text{st}}$ layer. Following the same notation, $b_i^{(0)}$ describes the plane's intercept for the $i^{\text{th}}$ hidden variable of the $1^{\text{st}}$ layer. The only hidden variable on the second layer is defined as a linear combination of the two hidden variables on layer 1, so

$$
z_1^{(2)} = w_{1,1}^{(1)}z_1^{(1)} + w_{1,2}^{(1)}z_2^{(1)} + b^{(1)}.
$$

The observation model is now defined by $y = z_1^{(2)} + v$, $v : V \sim \mathcal{N}(v; 0, \sigma_V^2)$, so each observation of the system response is defined as $y_i : Y \sim \mathcal{N}(y; z_1^{(2)}, \sigma_V^2)$. Note that $z_1^{(2)}$ implicitly depends on the covariates $\mathbf{x}$, so that there are covariates implicitly associated with each observation $y_i$. The complete data set is $\mathcal{D} = \{\mathcal{D}_x, \mathcal{D}_y\} = \{(\mathbf{x}_i, y_i), \forall i \in \{1 : \mathtt{D}\}\}$.

In this second case, adding hidden variables did not help us to generalize our model to nonlinear system responses because linear functions of linear functions are themselves still linear. Therefore, no matter how many layers of hidden variables we add, the final model remains a hyperplane. In order to describe nonlinear functions, we need to introduce nonlinearities using *activation functions*.

*Activation functions*    An *activation function* describes a nonlinear transformation of a hidden variable $z$. Common activation functions are the *logistic*, the *hyperbolic tangent* (tanh), and the *rectified*

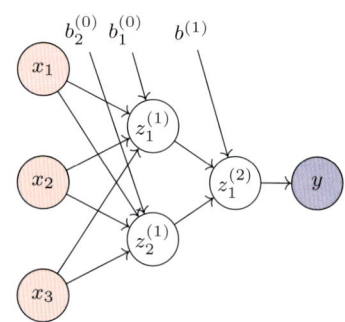

Figure 8.33: Representation of linear regression using multiple layers. The leftmost nodes represent covariates, the white nodes the hidden variables, and the rightmost node the observation.

**Nomenclature**

$z_i^{(l)}$ : the $i^{\text{th}}$ hidden variable on the $l^{\text{th}}$ layer

$w_{i,j}^{(l)}$ : the weight defining the dependence between the $i^{\text{th}}$ variable of the $(l+1)^{\text{th}}$ layer and the $j^{\text{th}}$ hidden variable of the $(l)^{\text{th}}$ layer

$b_i^{(l)}$: the plane's intercept for the $i^{\text{th}}$ hidden variable of the $(l+1)^{\text{th}}$ layer

$x_j$: $j^{\text{th}}$ covariate

*linear unit* (ReLU),

$$\sigma^{\mathbf{s}}(z) = \frac{1}{1+\exp(-z)} = \frac{\exp(z)}{\exp(z)+1} \quad \text{(logistic)}$$

$$\sigma^{\mathbf{t}}(z) = \frac{e^z - e^{-z}}{e^z + e^{-z}} \quad \text{(tanh)}$$

$$\sigma^{\mathbf{r}}(z) = \max(0, z) \quad \text{(ReLU)}.$$

These three activation functions are presented in figure 8.34. The ReLU activation function is currently widely employed because it facilitates the training of models by mitigating the problems associated with vanishing gradients (see §8.3.2).

*Feedforward neural networks: A simple case*   In a feedforward network, a hidden variable $z_i^{(j)}$ is passed through an activation function,

$$a_i^{(j)} = \sigma\left(z_i^{(j)}\right),$$

before getting fed to the next level. The output of the activation function is called the *activation unit*. Figure 8.35 presents the expanded and compact forms for a simple feedforward network with three covariates in its *input layer*, one *hidden layer* containing two activation units (green nodes), and one *output layer* containing a single activation unit (blue node). Note that in the compact forms (i.e., (b) and (c)), hidden variables $z_i^{(j)}$ are not explicitly represented anymore. The activation units on the first layer are

$$a_1^{(1)} = \sigma\left(z_1^{(1)}\right)$$
$$a_2^{(1)} = \sigma\left(z_2^{(1)}\right).$$

The output unit $a^{(0)}$ is equal to the hidden variable $z^{(0)}$, which is itself defined as a linear combination of the activation units on the hidden layer so that

$$a^{(0)} = z^{(0)} = w_1^{(1)}a_1^{(1)} + w_2^{(1)}a_2^{(1)} + b^{(1)}.$$

The observation model is now defined by $y = a^{(0)} + v$, $v : V \sim \mathcal{N}(v; 0, \sigma_V^2)$, so each observation of the system response is defined as $y_i : Y \sim \mathcal{N}(y; a^{(0)}, \sigma_V^2)$.

*Fully connected feedforward neural network*   We now describe formulation for the *fully connected feedforward neural network* (FCNN) with L hidden layers, each having A activation units. The example presented includes a single output observation. In practice, FCNN can be employed for any number of output variables.

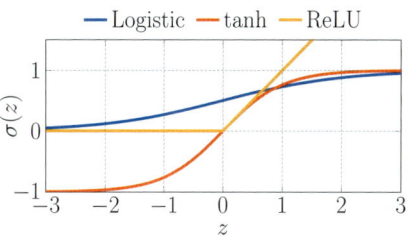

Figure 8.34: Comparison of the *logistic*, *tanh*, and *ReLU* activation functions.

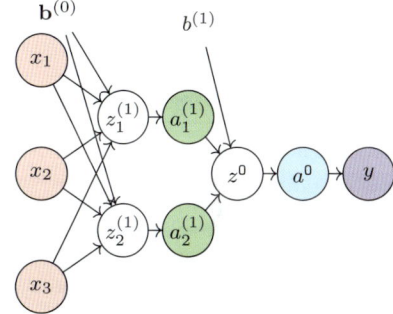

(a) Expanded-form variable nomenclature

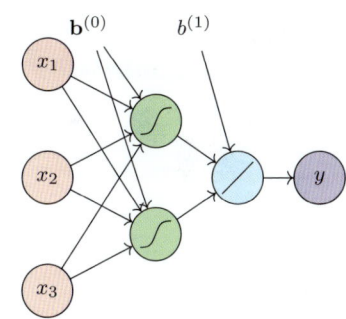

(b) Compact-form representation of activation functions

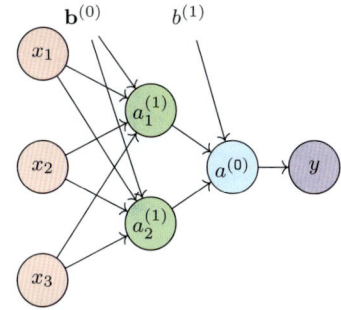

(c) Compact-form variable nomenclature

Figure 8.35: A simple feedforward neural network.

Figure 8.36 presents a graphical representation for an FCNN where (a) presents the connectivity with the activation functions and (b) the nomenclature employed for weights $w_{i,j}^{(l)}$ and hidden units $a_i^{(l)}$. The activation unit $i$ in the first layer is described by a linear combination of covariates,

$$a_i^{(1)} = \sigma\left(z_i^{(1)}\right) = \sigma\left(w_{i,1}^{(0)}x_1 + w_{i,2}^{(0)}x_2 + \cdots + w_{i,\mathbf{X}}^{(0)}x_\mathbf{X} + b_i^{(0)}\right).$$

In subsequent layers, the activation unit $i$ in layer $l+1$ is a linear combination of all the activation units in layer $l$ as described by

Figure 8.36: A fully connected neural network having L hidden layers, each having A activation units. The leftmost nodes describe the *input layer* containing X covariates. The green inner nodes describe the unobserved *hidden layers* where an activation unit $i$ in layer $l$ is $a_i^{(l)} = \sigma(z_i^{(l)})$. The blue node describes the *output layer* where $a^{(0)} = \sigma^0(z^{(0)}) = z^{(0)}$ for regression problems. The rightmost node describes the *observed system responses*, which are realizations $y_i : Y \sim \mathcal{N}(y; a^{(0)}, \sigma_V^2)$.

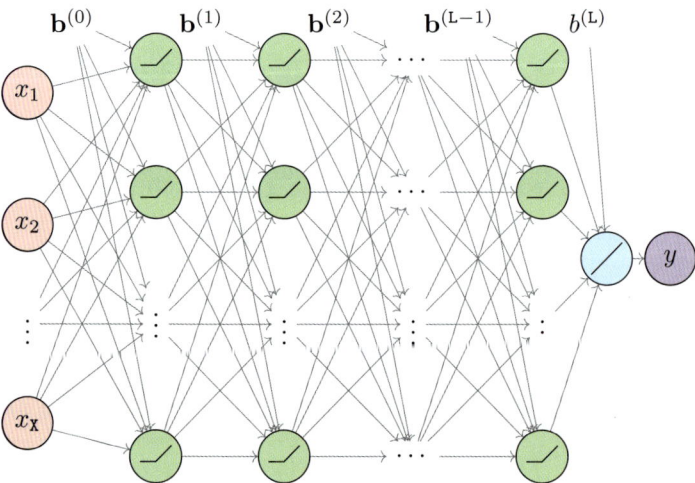

(a) Representation of the network structure with its activation functions

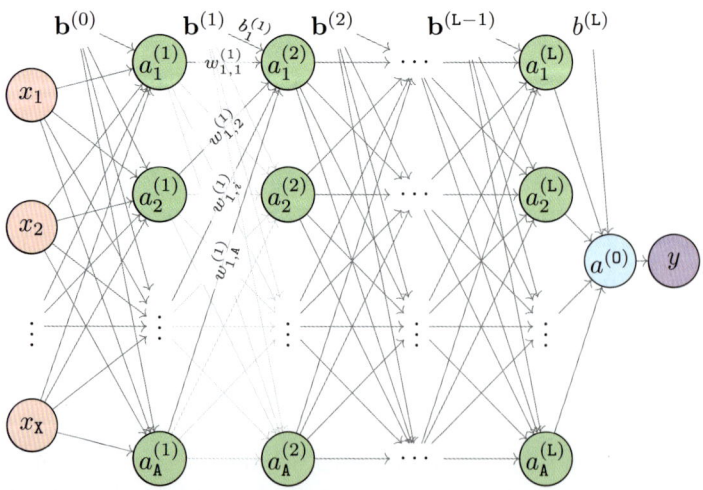

(b) Variable nomenclature: The *activation unit* 1 in layer 2 is
$$a_1^{(2)} = \sigma\left(z_1^{(2)}\right) = \sigma\left(w_{1,1}^{(1)}a_1^{(1)} + w_{1,2}^{(1)}a_2^{(1)} + \cdots + w_{1,\mathbf{A}}^{(1)}a_\mathbf{A}^{(1)} + b_1^{(1)}\right)$$

Activation unit in layer $l+1$

Hidden variable in layer $l+1$

Weights

Bias

$$a_i^{(l+1)} = \sigma\left(z_i^{(l+1)}\right) = \sigma\left(w_{i,1}^{(l)}a_1^{(l)} + w_{i,2}^{(l)}a_2^{(l)} + \cdots + w_{i,\mathtt{A}}^{(l)}a_{\mathtt{A}}^{(l)} + b_i^{(l)}\right).$$

Activation function    Activation units from layer $l$

The feedforward propagation for all the units contained in a layer can be expressed with a matrix notation as $\mathbf{a}^{(l+1)} = \sigma(\mathbf{z}^{(l+1)})$, that is,

$$\underbrace{\begin{bmatrix} a_1^{(l+1)} \\ a_2^{(l+1)} \\ \vdots \\ a_{\mathtt{A}}^{(l+1)} \end{bmatrix}_{\mathtt{A}\times 1}}_{\mathbf{a}^{(l+1)}} = \sigma\left(\underbrace{\begin{bmatrix} z_1^{(l+1)} \\ z_2^{(l+1)} \\ \vdots \\ z_{\mathtt{A}}^{(l+1)} \end{bmatrix}_{\mathtt{A}\times 1}}_{\mathbf{z}^{(l+1)}}\right),$$

where the hidden variables are given by $\mathbf{z}^{(l+1)} = \mathbf{W}^{(l)}\mathbf{a}^{(l)} + \mathbf{b}^{(l)}$, that is,

$$\underbrace{\begin{bmatrix} z_1^{(l+1)} \\ z_2^{(l+1)} \\ \vdots \\ z_{\mathtt{A}}^{(l+1)} \end{bmatrix}_{\mathtt{A}\times 1}}_{\mathbf{z}^{(l+1)}} = \underbrace{\begin{bmatrix} w_{1,1}^{(l)} & w_{1,2}^{(l)} & \cdots & w_{1,\mathtt{A}}^{(l)} \\ w_{2,1}^{(l)} & w_{2,2}^{(l)} & \cdots & w_{2,\mathtt{A}}^{(l)} \\ \vdots & \vdots & \ddots & \vdots \\ w_{\mathtt{A},1}^{(l)} & w_{\mathtt{A},2}^{(l)} & \cdots & w_{\mathtt{A},\mathtt{A}}^{(l)} \end{bmatrix}_{\mathtt{A}\times\mathtt{A}}}_{\mathbf{W}^{(l)}} \times \underbrace{\begin{bmatrix} a_1^{(l)} \\ a_2^{(l)} \\ \vdots \\ a_{\mathtt{A}}^{(l)} \end{bmatrix}_{\mathtt{A}\times 1}}_{\mathbf{a}^{(l)}} + \underbrace{\begin{bmatrix} b_1^{(l)} \\ b_2^{(l)} \\ \vdots \\ b_{\mathtt{A}}^{(l)} \end{bmatrix}_{\mathtt{A}\times 1}}_{\mathbf{b}^{(l)}}.$$

The variables $z^{(0)}$ and $a^{(0)}$ on the output layer are obtained from a linear combination of the activation units on the last hidden layer L,

$$a^{(0)} = \sigma^0(z^{(0)}) = z^{(0)} = w_1^{(\mathtt{L})}a_1^{(\mathtt{L})} + w_2^{(\mathtt{L})}a_2^{(\mathtt{L})} + \cdots + w_{\mathtt{A}}^{(\mathtt{L})}a_{\mathtt{A}}^{(\mathtt{L})} + b^{(\mathtt{L})}.$$

Because we are dealing with a regression problem, we can consider that the output activation function $\sigma^0(\cdot)$ is linear. The observation model is defined by $y = a^{(0)} + v \equiv a^{(0)}(\mathbf{x}) + v$, $v : V \sim \mathcal{N}(v; 0, \sigma_V^2)$, so each observation of the system response is defined as $y_i : Y \sim \mathcal{N}(y; a^{(0)}(\mathbf{x}), \sigma_V^2)$.

Neural networks are characterized by their large number of parameters to be estimated. For example, a fully connected network with L layers, each having A activation units, has $\mathtt{A} \times \mathtt{A} \times (\mathtt{L}-1)$ weight parameters and $\mathtt{A} \times \mathtt{L}$ bias parameters for its hidden layers. For $\mathtt{L} = \mathtt{A} = 10$, it leads to more than 1000 parameters to be learned from the data set $\mathcal{D}$. In comparison, the GPR model presented in §8.2 had fewer than 10 parameters. Moreover, it is not uncommon for modern deep neural networks to be parameterized by millions of parameters. As a consequence, learning these parameters (i.e., training the neural network) requires large data sets and is a computationally demanding task.

### 8.3.2    Parameter Estimation and Backpropagation

The task of learning model parameters is referred to as *training* a neural network. In general, there are so many parameters to be learned that the utilization of second-order optimization methods such as Newton-Raphson (see §5.2) are computationally prohibitive. Instead, neural networks rely on gradient descent algorithms (see §5.1), where the gradient of the loss function with respect to each parameter is obtained using the *backpropagation* method.

*Log-likelihood*    As for other types of regression models, the performance of a neural network is measured using quantities derived from the likelihood of observations $f(\mathcal{D}_y|\mathcal{D}_x, \boldsymbol{\theta}_w, \boldsymbol{\theta}_b)$, where $\{\boldsymbol{\theta}_w, \boldsymbol{\theta}_b\}$ are respectively the weight and bias parameters. By assuming that observations are conditionally independent given $a^{(0)}$, the joint likelihood for a data set $\mathcal{D} = \{\mathcal{D}_x, \mathcal{D}_y\} = \{(\mathbf{x}_i, y_i), \forall i \in \{1:D\}\}$ is defined as

$$f(\mathcal{D}_y|\mathcal{D}_x, \boldsymbol{\theta}_w, \boldsymbol{\theta}_b) = \prod_{i=1}^{D} \mathcal{N}(y_i; a^{(0)}(\mathbf{x}_i; \boldsymbol{\theta}_w, \boldsymbol{\theta}_b), \sigma_V^2),$$

where $a^{(0)}(\mathbf{x}_i; \boldsymbol{\theta}_w, \boldsymbol{\theta}_b) \equiv a^{(0)}$ explicitly defines the dependence of the output layer on the bias and weight parameters from all layers. In practice, neural networks are trained by minimizing a loss function $J(\mathcal{D}; \boldsymbol{\theta}_w, \boldsymbol{\theta}_b)$, which is derived from the negative log-likelihood,

**Note:** The derivation of the loss function for a *neural network* is analogous to the derivation of the loss function presented in §8.1.1 for *linear regression*.

$$
\begin{aligned}
\ln f(\mathcal{D}_y|\mathcal{D}_x, \boldsymbol{\theta}_w, \boldsymbol{\theta}_b) &= \sum_{i=1}^{D} \ln \left( \mathcal{N}(y_i; a^{(0)}(\mathbf{x}_i; \boldsymbol{\theta}_w, \boldsymbol{\theta}_b), \sigma_V^2) \right) \\
&\propto -\frac{1}{D} \sum_{i=1}^{D} \frac{1}{2} \left( y_i - a^{(0)}(\mathbf{x}_i; \boldsymbol{\theta}_w, \boldsymbol{\theta}_b) \right)^2 \\
&= -\frac{1}{D} \sum_{i=1}^{D} J(y_i; \mathbf{x}_i, \boldsymbol{\theta}_w, \boldsymbol{\theta}_b) \\
&= -J(\mathcal{D}; \boldsymbol{\theta}_w, \boldsymbol{\theta}_b).
\end{aligned}
$$

Note that the loss function $J(\mathcal{D}; \boldsymbol{\theta}_w, \boldsymbol{\theta}_b)$ corresponds to the *mean-square error* (MSE). The optimal set of parameters $\boldsymbol{\theta}^* = \{\boldsymbol{\theta}_w^*, \boldsymbol{\theta}_b^*\}$ are thus those which minimize the MSE,

$$\boldsymbol{\theta}^* = \arg\min_{\boldsymbol{\theta}} J(\mathcal{D}; \boldsymbol{\theta}_w, \boldsymbol{\theta}_b).$$

*Gradient descent* The optimal values $\boldsymbol{\theta}^*$ are sought iteratively using a gradient-descent algorithm, where the updated parameter values are

$$\boldsymbol{\theta} \leftarrow \boldsymbol{\theta} - \lambda \nabla_{\boldsymbol{\theta}} J(\mathcal{D}; \boldsymbol{\theta}_w, \boldsymbol{\theta}_b)$$

and where $\lambda$ is the *learning rate* (see §5.1). The gradients are estimated using the *backpropagation* algorithm that will be presented next. The parameters $\boldsymbol{\theta}_0 = \{\boldsymbol{\theta}_w, \boldsymbol{\theta}_b\}$ are typically initialized randomly following $\theta_i \sim \mathcal{N}(\theta; 0, \epsilon^2)$, where $\epsilon \approx 10^{-2}$. The random initialization is intended to avoid having an initial symmetry in the model. The objective of having initial values close to zero is that multiplying covariates and hidden variables by weights $w_{i,j}^{(l)} = 0$ will initially limit the model capacity. Note that a common practice is to normalize covariates and observations in the data set.

*Backpropagation* The *backpropagation*[14] method employs the *chain rule of derivation* in order to backpropagate the model prediction errors through the network. For a single observation $(\mathbf{x}_i, y_i)$, the prediction error $\delta_i^{(0)}$ is defined as the discrepancy between the predicted value $a^{(0)}(\mathbf{x}_i; \boldsymbol{\theta}_w, \boldsymbol{\theta}_b)$ and the observation $y_i$, times the gradient of the output's activation function,

$$
\begin{aligned}
\delta_i^{(0)} &= \tfrac{\partial}{\partial z^{(0)}} J(y_i; \mathbf{x}_i, \boldsymbol{\theta}_w, \boldsymbol{\theta}_b) \\
&= \tfrac{\partial}{\partial a^{(0)}} J(y_i; \mathbf{x}_i, \boldsymbol{\theta}_w, \boldsymbol{\theta}_b) \cdot \tfrac{\partial a^{(0)}}{\partial z^{(0)}} \\
&= -\big(y_i - a^{(0)}(\mathbf{x}_i; \boldsymbol{\theta}_w, \boldsymbol{\theta}_b)\big) \cdot \tfrac{\partial}{\partial z^{(0)}} \sigma^0(z^{(0)}) \quad \text{(general case)} \\
&= a^{(0)}(\mathbf{x}_i; \boldsymbol{\theta}_w, \boldsymbol{\theta}_b) - y_i \quad \text{(regression)}.
\end{aligned}
$$

In the context of regression, the term $\tfrac{\partial}{\partial z^{(0)}} \sigma^0(z^{(0)})$ is equal to one because $a^{(0)} = \sigma^0(z^{(0)}) = z^{(0)}$. Note that in order to evaluate the prediction error for an observation $y_i$, we first need to evaluate the model prediction for the current parameter values $\{\boldsymbol{\theta}_w, \boldsymbol{\theta}_b\}$ and a vector of covariates $\mathbf{x}_i$. For the $i^{\text{th}}$ observation in the training set, the gradient vector of the loss function with respect to either weight or bias parameters on the $l^{\text{th}}$ layer is

$$
\begin{aligned}
\nabla_{\mathbf{W}_i^{(l)}} J(y_i; \mathbf{x}_i, \boldsymbol{\theta}_w, \boldsymbol{\theta}_b) &= \tfrac{\partial J}{\partial \mathbf{W}^{(l)}} = \delta_i^{(l+1)} (\mathbf{a}_i^{(l)})^{\intercal} \\
\nabla_{\mathbf{b}_i^{(l)}} J(y_i; \mathbf{x}_i, \boldsymbol{\theta}_w, \boldsymbol{\theta}_b) &= \tfrac{\partial J}{\partial \mathbf{b}^{(l)}} = \delta_i^{(l+1)}.
\end{aligned}
$$

For hidden layers, the responsibility of each activation unit for the prediction error is estimated using

$$\boldsymbol{\delta}_i^{(l+1)} = \left( (\mathbf{W}^{(l+1)})^{\intercal} \boldsymbol{\delta}_i^{(l+2)} \right) \odot \nabla \sigma(\mathbf{z}^{(l+1)}),$$

where $\odot$ is the Hadamar (element-wise) product, and $\nabla$ denotes the gradient vector.

[14] Rumelhart, D. E., G. E. Hinton, and R. J. Williams (1986). Learning representations by back-propagating errors. *Nature 323*, 533–536.

**Feedforward neural network**

$$
\begin{aligned}
J &= \tfrac{1}{2}(y_i - a^{(0)})^2 \\
\mathbf{a}^{(0)} &= \sigma\big( \underbrace{\mathbf{W}^{(L)} \mathbf{a}^{(L)} + \mathbf{b}^{(L)}}_{\mathbf{z}^{(0)}} \big) \\
\mathbf{a}^{(l+1)} &= \sigma\big( \underbrace{\mathbf{W}^{(l)} \mathbf{a}^{(l)} + \mathbf{b}^{(l)}}_{\mathbf{z}^{(l+1)}} \big)
\end{aligned}
$$

**Chain rule of derivation**

$$\frac{\partial J}{\partial \mathbf{W}^{(L)}} = \underbrace{\frac{\partial J}{\partial \mathbf{a}^{(0)}} \frac{\partial \mathbf{a}^{(0)}}{\partial \mathbf{z}^{(0)}}}_{\delta_i^{(0)}} \underbrace{\frac{\partial \mathbf{z}^{(0)}}{\partial \mathbf{W}^{(L)}}}_{\mathbf{a}_i^{(L)}}$$

$$\frac{\partial J}{\partial \mathbf{b}^{(L)}} = \underbrace{\frac{\partial J}{\partial \mathbf{a}^{(0)}} \frac{\partial \mathbf{a}^{(0)}}{\partial \mathbf{z}^{(0)}}}_{} \underbrace{\frac{\partial \mathbf{z}^{(0)}}{\partial \mathbf{b}^{(L)}}}_{1}$$

$$\frac{\partial J}{\partial \mathbf{W}^{(l)}} = \underbrace{\frac{\partial J}{\partial \mathbf{z}^{(l+2)}} \frac{\partial \mathbf{z}^{(l+2)}}{\partial \mathbf{a}^{(l+1)}} \frac{\partial \mathbf{a}^{(l+1)}}{\partial \mathbf{z}^{(l+1)}}}_{\delta_i^{(l+1)}} \underbrace{\frac{\partial \mathbf{z}^{(l+1)}}{\partial \mathbf{W}^{(l)}}}_{\mathbf{a}_i^{(l)}}$$

$$\frac{\partial J}{\partial \mathbf{b}^{(l)}} = \underbrace{\underbrace{\frac{\partial J}{\partial \mathbf{z}^{(l+2)}}}_{\delta_i^{(l+2)}} \underbrace{\frac{\partial \mathbf{z}^{(l+2)}}{\partial \mathbf{a}^{(l+1)}}}_{\mathbf{W}^{(l+1)}} \underbrace{\frac{\partial \mathbf{a}^{(l+1)}}{\partial \mathbf{z}^{(l+1)}}}_{\nabla \sigma(\mathbf{z}^{(l+1)})}}_{} \underbrace{\frac{\partial \mathbf{z}^{(l+1)}}{\partial \mathbf{b}^{(l)}}}_{1}$$

Note: Several of the advances made in neural networks can be attributed to advancements in GPU calculations. Nowadays most nontrivial models are trained using GPUs.

The common procedure is to perform parameter updates not using each observation individually, but by using *mini-batches*. A mini-batch $\mathcal{D}^{\text{mb}} \subset \mathcal{D}$ is a subset of observations of size $n$ that is a power of 2 ranging from 32 to 256. The power of 2 constraint is there to maximize the computational efficiency when working with graphical processing units (GPUs). The gradient of the loss function is then the average of the gradients obtained for each observation in a mini-batch,

$$\nabla_{\mathbf{W}^{(l)}} J(\mathcal{D}^{\text{mb}}; \boldsymbol{\theta}_w, \boldsymbol{\theta}_b) = \frac{1}{n} \sum_{i:y_i \in \mathcal{D}^{\text{mb}}} \nabla_{\mathbf{W}_i^{(l)}} J(y_i; \mathbf{x}_i, \boldsymbol{\theta}_w, \boldsymbol{\theta}_b)$$

$$\nabla_{\mathbf{b}^{(l)}} J(\mathcal{D}^{\text{mb}}; \boldsymbol{\theta}_w, \boldsymbol{\theta}_b) = \frac{1}{n} \sum_{i:y_i \in \mathcal{D}^{\text{mb}}} \nabla_{\mathbf{b}_i^{(l)}} J(y_i; \mathbf{x}_i, \boldsymbol{\theta}_w, \boldsymbol{\theta}_b).$$

During training, an *epoch* corresponds to performing one pass of parameter update for all mini-batches contained in the training set. Neural networks are typically trained using *stochastic gradient descent*, which computes the gradient on the subset of data contained in mini-batches rather than on the full data set. The reader should refer to specialized literature[15] for descriptions of stochastic gradient descent optimization methods such as *Adagrad*, *RMSprop*, and *Adam*.

[15] Goodfellow, I., Y. Bengio, and A. Courville (2016). *Deep learning*. MIT Press.

### 8.3.3   Regularization

Because of the large number of parameters involved in the definition of a neural network, it is prone to overfitting, where the performance obtained during training will not generalize for the test set. Here, we present two *regularization* procedures to prevent overfitting: *weight decay* and *early stopping*.

*Weight decay*   The first regularization strategy is to employ maximum a-posteriori (MAP) estimation rather than a maximum likelihood estimate (MLE). The prior knowledge for model parameters in an MAP allows us to prevent overfitting by limiting the model complexity. For instance, we can define a prior for the weights in the vector $\boldsymbol{\theta}_w$ so that $f(w_{i,j}^{(l)}) = \mathcal{N}(w_{i,j}^{(l)}, 0, 1/\alpha)$. A large $\alpha$ value forces weight parameters to be close to zero, which in turn reduces the model capacity. If $\alpha$ is small, little penalization is applied on weights and the model is then free to be highly nonlinear and can more easily fit the data in the training set.

With MAP, the posterior is now approximated by the product of the likelihood and the prior, so

$$
\begin{aligned}
f(\boldsymbol{\theta}_w, \boldsymbol{\theta}_b | \mathcal{D}) \quad &\propto \quad f(\mathcal{D}_y | \mathcal{D}_x, \boldsymbol{\theta}_w, \boldsymbol{\theta}_b) \cdot f(\boldsymbol{\theta}_w, \boldsymbol{\theta}_b) \\
&= \quad \prod_{k=1}^{\mathrm{D}} \mathcal{N}(y_k; a^{(0)}(\mathbf{x}_k; \boldsymbol{\theta}_w, \boldsymbol{\theta}_b), \sigma_V^2) \cdot \prod_{i,j,l} \mathcal{N}(w_{i,j}^{(l)}; 0, 1/\alpha) \\
\ln f(\boldsymbol{\theta}_w, \boldsymbol{\theta}_b | \mathcal{D}) \quad &\propto \quad -\frac{1}{\mathrm{D}} \sum_{k=1}^{\mathrm{D}} J(y_k; \mathbf{x}_k, \boldsymbol{\theta}_w, \boldsymbol{\theta}_b) - \frac{\alpha}{2} \sum_{i,j,l} (w_{i,j}^{(l)})^2 \\
&\propto \quad -\left( J(\mathcal{D}; \boldsymbol{\theta}_w, \boldsymbol{\theta}_b) + \frac{\alpha}{2} \boldsymbol{\theta}_w^{\mathsf{T}} \boldsymbol{\theta}_w \right).
\end{aligned}
$$

We can then redefine the loss function as

$$
\tilde{J}(\mathcal{D}; \boldsymbol{\theta}_w, \boldsymbol{\theta}_b) = J(\mathcal{D}; \boldsymbol{\theta}_w, \boldsymbol{\theta}_b) + \alpha \Omega(\boldsymbol{\theta}_w),
$$

where $\Omega(\boldsymbol{\theta})$, the *regularization* or *weight decay* term, is

$$
\Omega(\boldsymbol{\theta}_w) = \frac{1}{2} ||\boldsymbol{\theta}_w||^2 = \frac{1}{2} \boldsymbol{\theta}_w^{\mathsf{T}} \boldsymbol{\theta}_w.
$$

Defining a Gaussian prior for weights corresponds to $L^2$-norm regularization. This choice for regularization is not the only one available, yet it is a common one. The gradient of the loss function with respect to weight parameters has to be modified to include the regularization term, so

$$
\nabla_{\mathbf{W}^{(l)}} \tilde{J}(\mathcal{D}^{\mathrm{mb}}; \boldsymbol{\theta}_w, \boldsymbol{\theta}_b) = J(\mathcal{D}^{\mathrm{mb}}; \boldsymbol{\theta}_w, \boldsymbol{\theta}_b) + \alpha \mathbf{W}^{(l)}.
$$

In short, weight decay allows controlling the model capacity with the hyperparameter $\alpha$. The hyperparameter $\alpha$ then needs to be learned using a method such as *cross-validation* (see §8.1.2).

*Early stopping*   Given that we initialize each parameter randomly, following $\theta_i \sim \mathcal{N}(\theta; 0, \epsilon^2)$, with $\epsilon \approx 10^{-2}$, the model will initially have a poor performance at fitting the observations in the training set. As the number of optimization epochs increases, the fitting ability of the model will improve and, as a result, the loss for the training set $J(\mathcal{D}^{\mathrm{train}}; \boldsymbol{\theta}_w, \boldsymbol{\theta}_b)$ will decrease. If we keep optimizing parameters, there is a point where the model will gain too much capacity, and it will start overfitting the data in the training set. Thus, a second option to prevent overfitting is to monitor the loss function evaluated at each epoch for an independent validation set $J(\mathcal{D}^{\mathrm{validation}}; \boldsymbol{\theta}_w, \boldsymbol{\theta}_b)$. Note that the data in the validation set must not be employed to train the model parameters. The optimal parameter values $\boldsymbol{\theta}^* = [\boldsymbol{\theta}_w^* \ \boldsymbol{\theta}_b^*]^{\mathsf{T}}$ are taken as those corresponding to the epoch for which the loss evaluated for the validation set is minimal. This procedure is equivalent to treating the number of

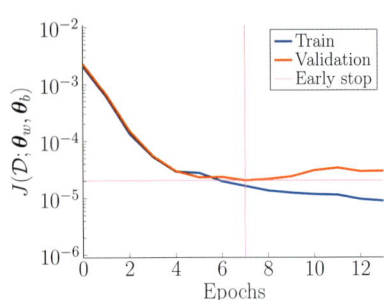

Figure 8.37: Example of early stopping when the loss function for the validation set stops decreasing.

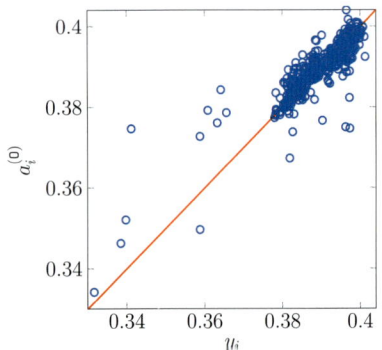

Figure 8.38: Comparison of the system responses with the model predictions obtained in a leave-one-out cross-validation setup.

training epochs as a hyperparameter to be optimized using the validation set.

Figure 8.37 presents an example of early stopping where we see that the loss for the training set keeps decreasing with the number of epochs. For the loss evaluated on the validation set, the minimum is reached at seven epochs. Beyond seven epochs, the fitting ability we gain on the training set does not generalize on the validation set.

### 8.3.4   Example: Metamodel

Here we revisit the example presented in §8.2.6, where we construct a metamodel for the responses of a structure using five covariates and a set of $D = 1000$ simulations. For this example, we employ a tanh activation function with $A = 10$ activation units on each of the $L = 2$ hidden layers. Figure 8.38 compares the system responses with the model predictions obtained in a leave-one-out cross-validation setup where for each fold, 90 percent of the data is employed for training the model and the remaining 10 percent to validate when the training should be stopped. The loss function $J(\mathcal{D}; \boldsymbol{\theta}_w, \boldsymbol{\theta}_b)$, evaluated using the leave-one-out cross-validation procedure, is equal to $1.32 \times 10^{-5}$. In comparison, the equivalent mean-square error obtained for the GPR model presented in §8.2.6 is $8.8 \times 10^{-6}$. It shows that neural networks are not necessarily outperforming other methods, because they are typically best suited for problems involving a large number of covariates and for which large data sets are available.

## Exercises

**P8.1**  Why does linear regression have an analytic formulation for identifying the MLE optimal parameters while other regression methods do not?

**P8.2**  How can linear regression be employed to model nonlinear relationships between $x$ and $y$?

**P8.3**  Explain the bias-variance tradeoff and its relationship with overfitting.

**P8.4**  Explain the approaches to prevent overfitting.

**P8.5**  Explain a correlation and covariance function in the context of a Gaussian process.

**P8.6**  For Gaussian process regression, what is the relationship between a length-scale $\ell$ and the importance of its associated covariate?

**P8.7**  Explain why methods such as linear regression and neural networks are more prone to overfitting than Gaussian process regression. In what case does the Gaussian process regression also become sensitive to overfitting?

**P8.8**  In the context of regression, formulate the likelihood for a set of observations $\mathcal{D} = \{\mathcal{D}_x, \mathcal{D}_y\}$, $\mathcal{D}_x = \{x_1, x_2, \cdots, x_D\}$, $\mathcal{D}_y = \{y_1, y_2, \cdots, y_D\}$ for the observation model

$$\mathbf{y} = \mathbf{g}(\mathbf{x}) + \mathbf{v}, \quad \begin{cases} \mathbf{g}(\mathbf{x}) : \mathbf{G}(\mathbf{x}) \sim \mathcal{N}(\mathbf{g}(\mathbf{x}); \mathbf{m_X}, \mathbf{\Sigma_X}) \\ \mathbf{v} : \mathbf{V} \sim \mathcal{N}(\mathbf{v}; \mathbf{0}, \sigma_V^2 \cdot \mathbf{I}), \ V_i \perp\!\!\!\perp V_j, \forall i \neq j. \end{cases}$$

**P8.9**  Given the data set $\mathcal{D} = \{(x_i, y_i), \forall i = \{1 : 10\}\}$ where $x$ describes the longitudinal position in kilometers along a pipeline and $y$ describes the temperature in degrees Celsius. Observations follow the equation $y_i = g(x_i) + v_i$, where $v_i$ is an observation error that is a realization of the random variable $v_i : V_i \sim \mathcal{N}(v_i; 0, \sigma_V^2)$.

$$\begin{aligned} \mathcal{D}_x &= \{81.47 \quad 90.58 \quad 12.70 \quad 91.34 \quad 63.24 \quad 9.75 \quad 27.85 \quad 54.69 \quad 95.75 \quad 96.49\} \\ \mathcal{D}_y &= \{-1.39 \quad 0.48 \quad 0.73 \quad 0.20 \quad -1.23 \quad 0.60 \quad 0.49 \quad -0.10 \quad 1.22 \quad 1.52\} \end{aligned}$$

a. Implement your own Gaussian process regression model for the data set, and learn the maximum likelihood estimate for hyperparameters $\mathcal{P} = \{\sigma_V, \sigma_g, \ell\}$. Use the hypothesis that the prior mean $\boldsymbol{\mu_g} = \mathbf{0}$.

b. Estimate the posterior covariance matrix for hyperparameters obtained in (a) using the Laplace approximation.

c. Plot the expected values $\mu_*$ as well as the $\mu_* \pm \sigma_*$ confidence region for the model in the range $x_* \in \{0, 1, \cdots, 100\}$.

# 9
# Classification

Classification is similar to regression, where the task is to predict a system response $y$, given some covariates $\mathbf{x}$ describing the system properties. With regression, the system responses were continuous quantities; with *classification*, the system responses are now *discrete classes*. Figure 9.1 presents two examples of classification applied to the civil engineering context; in (a), the sigmoid-like curve describes the probability of having pathogens in water, as a function of the chlorine concentration. In this problem, observations are either $-1$ or $+1$, respectively corresponding to one of the two classes {pathogens, ¬pathogens}. The second example, (b), presents the probability of having structural damage in a building given the peak ground acceleration (PGA) associated with an earthquake. Observations are either $-1$ or $+1$, respectively corresponding to one of the two classes {damage, ¬damage}.

For classification, the data consists in $\mathtt{D}$ pairs of covariates $\mathbf{x}_i \in \mathbb{R}^{\mathtt{X}}$ and system responses $y_i \in \{-1, +1\}$, so $\mathcal{D} = \{(\mathbf{x}_i, y_i), \forall i \in \{1 : \mathtt{D}\}\}$. For problems involving multiple classes, the system response is $y_i \in \{1, 2, \cdots, \mathtt{C}\}$. The typical goal of a classification method is to obtain a mathematical model for the probability of a system response given a covariate, for example, $\Pr(Y = +1|\mathbf{x})$. This problem can be modeled with either a *generative* or a *discriminative* approach. A generative classifier models the probability density function (PDF) of the covariates $\mathbf{x}$, for each class $y$, and then use these PDFs to compute the probability of classes given the covariate values. A discriminative classifier directly builds a model of $\Pr(Y = +1|\mathbf{x})$ without modeling the probability density function (PDF) of covariates $\mathbf{x}$, for each class $y$. In practice, discriminative approaches are most common because they typically perform better on medium and large data sets,[1] and are often easier to use. This chapter presents the generic formulation for probabilistic generative classifiers and three discriminative classifiers: *logistic regression*,

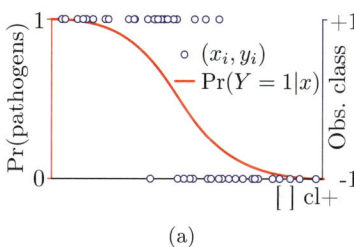

(a)

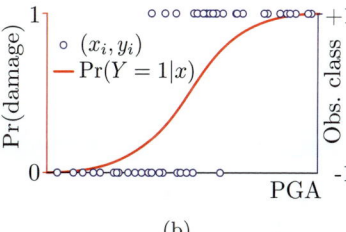

(b)

Figure 9.1: Examples of application of classification analysis in the context of civil engineering.

**Data**
$\mathcal{D} = \{(\mathbf{x}_i, y_i), \forall i \in \{1 : \mathtt{D}\}\}$

$\mathbf{x}_i \in \mathbb{R} : \begin{cases} \text{Covariate} \\ \text{Attribute} \\ \text{Regressor} \end{cases}$

$y_i \in \{-1, +1\}$ : Observation for binary classification

$y_i \in \{1, 2, \cdots, \mathtt{C}\}$ : Observation for multi-classes classification

[1] Ng, A. and M. Jordan (2002). On discriminative vs. generative classifiers: A comparison of logistic regression and naive Bayes. In *Advances in neural information processing systems*, Volume 14, pp. 841–848. MIT Press.

*Gaussian process classification*, and *neural networks*. We will see that *generative classifiers* are best suited for small data sets, *Gaussian process classification* is suited for medium-size ones, and *neural networks* for large ones. In order to simplify the explanations, this chapter presents each method while focusing on the context of *binary* classification, that is, $y \in \{-1, +1\}$.

## 9.1   Generative Classifiers

The basic idea of a *generative classifier* is to employ a set of observations $\mathcal{D}$ to model the conditional PDF of covariates $\mathbf{x} \in \mathbb{R}^{\mathtt{X}}$ given each class $y \in \{-1, +1\}$. Then these PDFs are employed in combination with the prior probability of each class in order to obtain the posterior probability of a class given a covariate $\mathbf{x}$. Generative classifiers are most appropriate when working with problems involving a small number of covariates and observations because they allow us to explicitly include domain-specific knowledge regarding $f(\mathbf{x}|y)$ or tailored features such as censored data.

### 9.1.1   Formulation

The posterior probability of a class $y$ given covariates $\mathbf{x}$ can be expressed using Bayes rule,

$$\underbrace{p(y|\mathbf{x})}_{\text{posterior}} = \frac{\overbrace{f(\mathbf{x}|y)}^{\text{likelihood}} \cdot \overbrace{p(y)}^{\text{prior}}}{\underbrace{f(\mathbf{x})}_{\text{evidence}}}.$$

The posterior probability of a class $y$ conditional on a covariate value $x$ is defined according to

$$p(y|\mathbf{x}) = \begin{cases} \Pr(Y = -1|\mathbf{x}) & \text{if } y = -1 \\ \Pr(Y = +1|\mathbf{x}) = 1 - \Pr(Y = -1|\mathbf{x}) & \text{if } y = +1. \end{cases}$$

**Note:** the prior knowledge can either be based on expert knowledge, i.e., $p(y)$, or be empirically estimated using observations, i.e.,

$$p(y|\mathcal{D}) = \frac{p(\mathcal{D}|y) \cdot p(y)}{p(\mathcal{D})}$$

The prior probability of a class is described by the probabilities

$$p(y) = \begin{cases} \Pr(Y = -1) & \text{if } y = -1 \\ \Pr(Y = +1) & \text{if } y = +1. \end{cases}$$

The prior and posterior are described by probability mass functions because $y \in \{-1, +1\}$ is a discrete variable. The likelihood of covariates $\mathbf{x}$ conditional on a class $y$ is a multivariate conditional probability density function that follows

$$f(\mathbf{x}|y) = \begin{cases} f(\mathbf{x}|y = -1) & \text{if } y = -1 \\ f(\mathbf{x}|y = +1) & \text{if } y = +1. \end{cases}$$

The normalization constant (i.e., the evidence) is obtained by summing the product of the likelihood and prior over all classes,

$$f(\mathbf{x}) = \sum_{y \in \{-1,1\}} f(\mathbf{x}|y) \cdot p(y).$$

A key step with a generative classifier is to pick a distribution type for each $f(\mathbf{x}|y_j; \boldsymbol{\theta}_j), \forall j$ and then estimate the parameters $\boldsymbol{\theta}_j$ for these PDFs using the data set $\mathcal{D}$, that is, $f(\boldsymbol{\theta}_j|\mathcal{D})$. For a class $y = j$, the parameters of $f(\mathbf{x}|y_j) \equiv f(\mathbf{x}|y_j; \boldsymbol{\theta}_j)$ are estimated using only observations $\mathcal{D}_j : \{(\mathbf{x}, y)_i, \forall i : y_i = j\}$, that is, the covariates $\mathbf{x}_i$ such that the associated system response is $y_i = j$. The estimation of parameters $\boldsymbol{\theta}$ can be performed using either a Bayesian approach or a deterministic one as detailed in §6.3. Like for the likelihood, the prior probability of each class $p(y|\mathcal{D})$ can be estimated using either a Bayesian approach or a deterministic one using the relative frequency of observations where $y_i = j$. With a Bayesian approach, the posterior predictive probability mass function (PMF) is obtained by marginalizing the uncertainty about the posterior parameter estimate,

$$\underbrace{p(y|\mathbf{x}, \mathcal{D})}_{\text{posterior predictive}} = \int \frac{\overbrace{f(\mathbf{x}|y; \boldsymbol{\theta})}^{\text{likelihood}} \cdot \overbrace{p(y; \boldsymbol{\theta})}^{\text{prior}}}{\underbrace{f(\mathbf{x}; \boldsymbol{\theta})}_{\text{normalization constant}}} \cdot \underbrace{f(\boldsymbol{\theta}|\mathcal{D})}_{\text{posterior } \boldsymbol{\theta}} \, d\boldsymbol{\theta}.$$

When either a maximum a-posteriori (MAP) or a maximum likelihood estimate (MLE) is employed instead of a Bayesian estimation approach, the posterior predictive is replaced by the approximation,

$$\hat{p}(y|\mathbf{x}, \mathcal{D}) = \frac{f(\mathbf{x}|y; \boldsymbol{\theta}^*) \cdot p(y; \boldsymbol{\theta}^*)}{f(\mathbf{x}; \boldsymbol{\theta}^*)},$$

where $\boldsymbol{\theta}^*$ are the MLE or MAP estimates for parameters.

*Maximum likelihood estimate* In the special case where $f(x|y = j; \boldsymbol{\theta}^*) = \mathcal{N}\big(x; \mu_{x|y_j}^*, \sigma_{x|y_j}^{2*}\big)$ and the number of available data $\mathsf{D}$ is large, one can employ the MLE approximation for the parameters $\mu_{x|y_j}$ and $\sigma_{x|y_j}^2$ which is defined by

$$\left. \begin{array}{rcl} \mu_{x|y_j}^* & = & \frac{1}{\mathsf{D}_j} \sum_i x_i \\ \sigma_{x|y_i}^{2*} & = & \frac{1}{\mathsf{D}_j} \sum_i (x_i - \mu_{x|y_j}^*)^2 \end{array} \right\}, \ \forall \{i : y_i = j\},$$

where $\mathsf{D}_j = \#\{i : y_i = j\}$ denotes the number of observations in a data set $\mathcal{D}$ that belongs to the $j^{\text{th}}$ class. The MLE approximation of $p(y = j; \boldsymbol{\theta}^*)$ is given by

$$p(y = j; \boldsymbol{\theta}^*) = \frac{\mathsf{D}_j}{\mathsf{D}}.$$

**Note:** The notation $f(\mathbf{x}|y; \boldsymbol{\theta}) \equiv f(\mathbf{x}|y, \boldsymbol{\theta})$ indicates that the conditional probability $f(\mathbf{x}|y)$ is parameterized by $\boldsymbol{\theta}$.

**MLE:** $\boldsymbol{\theta}^* = \arg\max_{\boldsymbol{\theta}} f(\mathcal{D}; \boldsymbol{\theta})$

$\boxed{\boldsymbol{\theta} = \{\mu_x, \sigma_x\}}$

$$\mathcal{D} = \{x_i \in \mathbb{R}, \forall i \in \{1 : \mathsf{D}\}\}$$
$$f(x; \boldsymbol{\theta}) = \mathcal{N}(x; \mu_x, \sigma_x^2)$$
$$f(\mathcal{D}; \boldsymbol{\theta}) = \prod_{i=1}^{\mathsf{D}} \mathcal{N}(x_i; \mu_x, \sigma_x^2)$$
$$\ln f(\mathcal{D}; \boldsymbol{\theta}) \propto \sum_{i=1}^{\mathsf{D}} \left( -\frac{1}{2} \frac{(x_i - \mu_x)^2}{\sigma_x^2} + \ln \frac{1}{\sigma_x} \right)$$
$$\frac{\partial \ln f(\mathcal{D}; \boldsymbol{\theta})}{\partial \mu_x} = \sum_{i=1}^{\mathsf{D}} x_i - \mathsf{D}\mu_x = 0$$
$$\rightarrow \mu_x^* = \frac{1}{\mathsf{D}} \sum_{i=1}^{\mathsf{D}} x_i$$
$$\frac{\partial \ln f(\mathcal{D}; \boldsymbol{\theta})}{\partial \sigma_x^2} = \sum_{i=1}^{\mathsf{D}} \frac{(x_i - \mu_x)^2}{(\sigma_x^2)^2} - \frac{\mathsf{D}}{\sigma_x^2} = 0$$
$$\rightarrow \sigma_x^{2*} = \frac{1}{\mathsf{D}} \sum_{i=1}^{\mathsf{D}} (x_i - \mu_x)^2$$

$\boxed{\boldsymbol{\theta} = p_j}$

$$\mathcal{D} = \{y_i \in \{1 : \mathsf{C}\}, \forall i \in \{1 : \mathsf{D}\}\}$$
$$\rightarrow \mathsf{D}_j = \#\{i : y_i = j\} \leq \mathsf{D}$$
$$f(y; \boldsymbol{\theta}) = \begin{cases} p_j & \text{for } y = j \\ 1 - p_j & \text{for } y \neq j \end{cases}$$
$$f(\mathcal{D}; \boldsymbol{\theta}) \propto p_j^{\mathsf{D}_j} (1 - p_j)^{(\mathsf{D} - \mathsf{D}_j)}$$
$$\ln f(\mathcal{D}; \boldsymbol{\theta}) \propto \mathsf{D}_j \ln p_j + (\mathsf{D} - \mathsf{D}_j) \ln(1 - p_j)$$
$$\frac{\partial \ln f(\mathcal{D}; \boldsymbol{\theta})}{\partial p_j} = \frac{\mathsf{D}_j}{p_j} - \frac{(\mathsf{D} - \mathsf{D}_j)}{1 - p_j} = 0$$
$$\rightarrow p_j^* = \frac{\mathsf{D}_j}{\mathsf{D}}$$

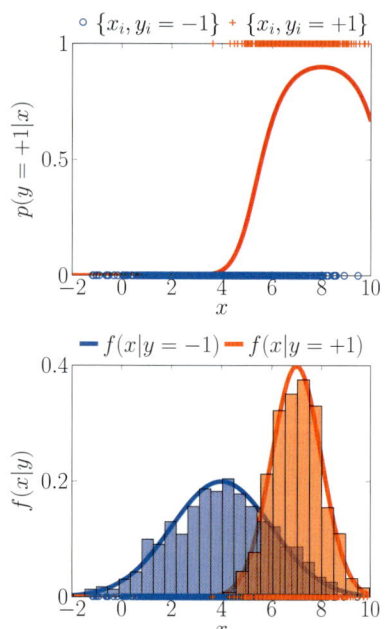

Figure 9.2: Example of generative classifier using marginal PDFs.

| Method | $f(\mathbf{x}|y=j)$ |
|---|---|
| Naive Bayes | $X_i \perp\!\!\!\perp X_k|y, \forall i \neq k$<br>e.g. $\mathcal{B}(x_1; \alpha, \beta) \cdot \mathcal{N}(x_2, \mu, \sigma^2)$ |
| LDA | $\mathcal{N}(\mathbf{x}; \boldsymbol{\mu}_{\mathbf{x},j}, \boldsymbol{\Sigma}_{\mathbf{x}})$ |
| QDA | $\mathcal{N}(\mathbf{x}; \boldsymbol{\mu}_{\mathbf{x},j}, \boldsymbol{\Sigma}_{\mathbf{x},j})$ |

(a) Naive Bayes

(b) LDA, $\boldsymbol{\Sigma}_{\mathbf{x},-1} = \boldsymbol{\Sigma}_{\mathbf{x},+1}$

(c) QDA, $\boldsymbol{\Sigma}_{\mathbf{x},-1} \neq \boldsymbol{\Sigma}_{\mathbf{x},+1}$

Figure 9.3: Examples of PDF $f(\mathbf{x}|y=j)$ for naive Bayes, LDA, and QDA.

Figure 9.2 presents an example of application of a generative classifier where the parameters for the normal PDFs are estimated using the MLE approximation. The bottom graph overlays the empirical histograms for the data associated with each class along with the PDFs $f(x|y=j, \boldsymbol{\theta}^*)$. The top graph presents the classifier $p(y=+1|x, \boldsymbol{\theta}^*)$, which is the probability of the class $y=+1$ given the covariate value $x$. If we assume that the prior probability of each class is equal, this conditional probability is computed as

$$p(y=+1|x) = \frac{f(x|y=+1) \cdot p(y=+1)}{f(x|y=-1) \cdot p(y=-1) + f(x|y=+1) \cdot p(y=+1)}$$

$$= \frac{f(x|y=+1)}{f(x|y=-1)+f(x|y=+1)}, \text{ for } p(y=-1) = p(y=+1).$$

*Multiple covariates*   When there is more than one covariate $\mathbf{x} = [x_1 \; x_2 \; \cdots \; x_\mathtt{X}]^\intercal$ describing a system, we must define the joint PDF $f(\mathbf{x}|y)$. A common approach to define this joint PDF is called *naive Bayes* (NB). It assumes that covariates $X_k$ are independent of each other so their joint PDF is obtained from the product of the marginals,

$$f(\mathbf{x}|y=j) = \prod_{k=1}^{\mathtt{X}} f(x_k|y=j).$$

The method employing the special case where the joint PDF of covariates is described by a multivariate normal $f(\mathbf{x}|y) = \mathcal{N}(\mathbf{x}; \boldsymbol{\mu}, \boldsymbol{\Sigma})$ is called *quadratic discriminant analysis* (QDA). The label *quadratic* refers to the shape of the boundary in the covariate domain where the probability of both classes are equal, $\Pr(y=+1|\mathbf{x}) = \Pr(y=-1|\mathbf{x})$. In the case where the covariance matrix $\boldsymbol{\Sigma}$ is the same for each class, the boundary becomes linear, and this special case is named *linear discriminant analysis* (LDA). Both the naive Bayes, QDA, and LDA methods typically employ an MLE approach to estimate the parameters of $f(\mathbf{x}|y=j)$. Figure 9.3 presents examples of $f(\mathbf{x}|y=j)$ for the three approaches: NB, LDA, and QDA. Figure 9.4 presents an example of an application of quadratic discriminant analysis where the joint probabilities $f(\mathbf{x}|y)$ are multivariate Normal PDFs as depicted by the contour plots. The mean vector and covariance matrices describing these PDFs are estimated using an MLE approximation. Because there are two covariates, the classifier $f(y=+1|\mathbf{x})$ is now a 2-D surface obtained following

$$p(y=+1|\mathbf{x}) = \frac{f(\mathbf{x}|y=+1) \cdot p(y=+1)}{f(\mathbf{x}|y=-1) \cdot p(y=-1) + f(\mathbf{x}|y=+1) \cdot p(y=+1)}.$$

Note that for many problems, generative methods such as naive Bayes and quadratic discriminant analysis are outperformed by discriminative approaches such as those presented in §9.2–§9.4.

### 9.1.2  Example: Post-Earthquake Structural Safety Assessment

We present here an example of a data-driven post-earthquake structural safety assessment. The goal is to assess right after an earthquake whether or not buildings are safe for occupation, that is, $y \in \{\text{safe}, \neg\text{safe}\}$. We want to guide this assessment based on empirical data where the observed covariate $x \in (0, 1)$ describes the ratio of a building's first natural frequency $f$ [Hz] after and before the earthquake,

$$x = \frac{f_{\text{post-earthquake}}}{f_{\text{pre-earthquake}}}.$$

Figure 9.5 presents a set of 45 empirical observations[2] collected in different countries and describing the frequency ratio $x$ as a function of the damage index $d \in \{0, 1, 2, 3, 4, 5\}$, which ranges from undamaged ($d = 0$) up to collapse ($d = 5$). Note that observations marked by an arrow are censored data, where $x$ is an upper-bound censored observation (see §6.3) for the frequency ratio. The first step is to employ this data set $\mathcal{D} = \{(\underbrace{x_i}_{\in(0,1)}, \underbrace{d_i}_{\in\{0:5\}}), \forall i \in \{1 : 45\}\}$ in order to learn the conditional PDF of the frequency ratio $X$, given a damage index $d$. Because $x \in (0, 1)$, we choose to describe its conditional PDF using a Beta distribution,

$$f(x; \boldsymbol{\theta}(d)) = \mathcal{B}(x; \underbrace{\mu(d), \sigma(d)}_{\boldsymbol{\theta}}).$$

Note that the Beta distribution is parameterized here by its mean and standard deviation and not by $\alpha$ and $\beta$ as described in §4.3. The justification for this choice of parameterization will become clear when we present the constraints that are linking the prior knowledge for different damage indexes $d$. For a given $d$, the posterior PDF for the parameters $\boldsymbol{\theta}(d) = \{\mu(d), \sigma(d)\}$ is obtained using Bayes,

$$\overbrace{f(\boldsymbol{\theta}(d)|\mathcal{D})}^{\text{posterior}} = \frac{\overbrace{f(\mathcal{D}|\boldsymbol{\theta}(d))}^{\text{likelihood}} \cdot \overbrace{f(\boldsymbol{\theta}(d))}^{\text{prior}}}{\underbrace{f(\mathcal{D})}_{\text{constant}}},$$

and the posterior predictive PDF of $X$ is obtained by marginalizing the uncertainty associated with the estimation of $\boldsymbol{\theta}(d)$,

$$f(x|d, \mathcal{D}) = \int f(x; \boldsymbol{\theta}(d)) \cdot f(\boldsymbol{\theta}(d)|\mathcal{D}) d\boldsymbol{\theta}.$$

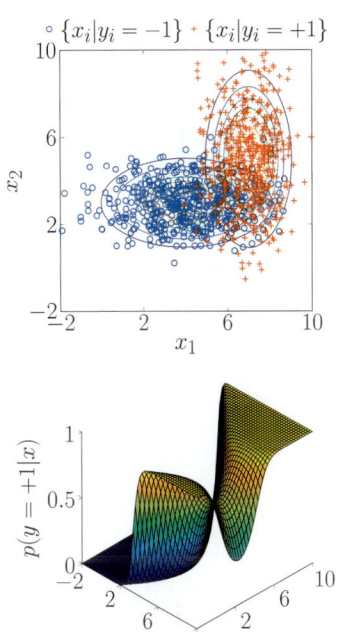

Figure 9.4: Example of quadratic discriminant analysis where the joint probabilities $f(\mathbf{x}|y_i)$ depicted by contour plots are described by multivariate Normal PDFs. The parameters of each PDF are estimated using an MLE approach.

[2] Goulet, J.-A., C. Michel, and A. Der Kiureghian (2015). Data-driven post-earthquake rapid structural safety assessment. *Earthquake Engineering & Structural Dynamics* 44(4), 549–562.

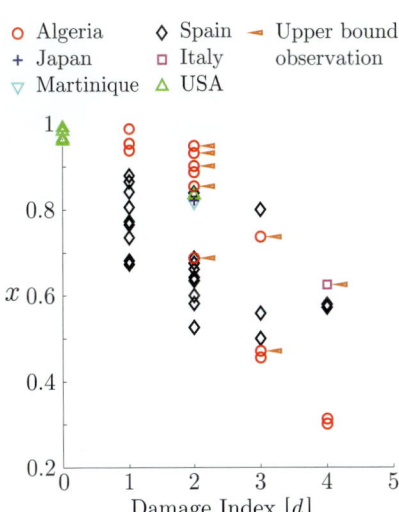

Figure 9.5: Empirical observations of the ratio between fundamental frequency of buildings after and before an earthquake.

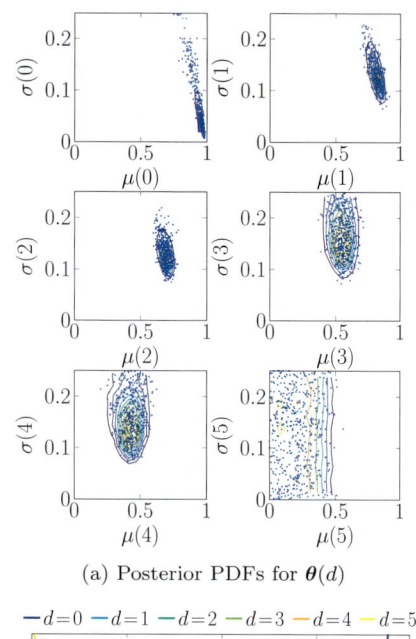

(a) Posterior PDFs for $\boldsymbol{\theta}(d)$

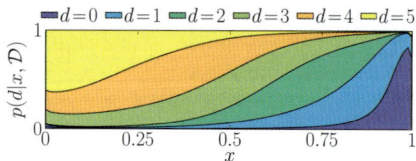

(b) Posterior predictive PDF

Figure 9.6: Posterior PDFs of means and standard deviations and posterior predictive PDFs of frequency ratios $x$ for each damage index $d$.

Figure 9.7: Posterior predictive probability of a damage index $d$ given a frequency ratio $x$.

By assuming that observations are conditionally independent given $x$, the joint likelihood of observations is

$$f(\mathcal{D}|\boldsymbol{\theta}(d)) = \prod_{\{i:d_i=d\}} \mathcal{L}(x_i|\mu(d),\sigma(d)),$$

where the marginal likelihood for each observation (see §6.3) is

$$\mathcal{L}(x_i|\mu(d),\sigma(d)) = \begin{cases} f(x_i|\mu(d),\sigma(d)), & x_i: \text{direct observation} \\ F(x_i|\mu(d),\sigma(d)), & x_i: \text{censored observation} \end{cases}$$

and where $f(\cdot)$ denotes the PDF and $F(\cdot)$ denotes the cumulative distribution function (CDF). The prior is assumed to be uniform in the range $(0,1)$ for $\mu(d)$ and uniform in the range $(0,0.25)$ for $\sigma(d)$. For the mean parameters $\mu(d)$, the prior is also constrained following $\mu(0) > \mu(1) > \cdots > \mu(5)$. These constraints reflect our domain-specific knowledge where expect the mean frequency ratio to decrease with increasing values of $d$.

The Metropolis algorithm presented in §7.1 is employed to generate $\mathtt{S} = 35{,}000$ samples ($\hat{R} \leq 1.005$ for $\mathtt{C} = 3$ chains) from the posterior for the parameters $\boldsymbol{\theta}(d)$. Figure 9.6a presents the contours of the posterior PDF for each pair of parameters $\{\mu(d),\sigma(s)\}$ along with a subset of samples. The predictive PDFs $f(x|d,\mathcal{D})$ of a frequency ratio $x$ conditional on $d$ are presented in figure 9.6b. The predictive probability of each damage index $d$ given a frequency ratio $x$ is obtained following

$$p(d|x,\mathcal{D}) = \int \frac{f(x;\boldsymbol{\theta}(d)) \cdot p(d)}{\sum_{d'=0}^{5} f(x;\boldsymbol{\theta}(d')) \cdot p(d')} \cdot f(\boldsymbol{\theta}(d)|\mathcal{D})d\boldsymbol{\theta}. \qquad (9.1)$$

The integral in equation 9.1 can be approximated using the Markov chain Monte Carlo (MCMC) samples as described in §7.5. Figure 9.7 presents the posterior predictive probability $p(d|x,\mathcal{D})$, which is computed while assuming that there is an equal prior probability $p(d)$ for each damage index. This classifier can be employed to assist inspectors during post-earthquake safety assessment of structures.

## 9.2    Logistic Regression

*Logistic regression* is not a state-of-the-art approach, yet its prevalence, historical importance, and simple formulation make it a good choice to introduce *discriminative* regression methods. Logistic regression is the extension of linear regression (see §8.1) for classification problems. In the context of regression, a linear function $\mathbb{R}^{\mathtt{X}} \to \mathbb{R}$ is defined so that it transforms an $\mathtt{X}$-dimensions covariate domain into a single output $g(\mathbf{x}) = \mathbf{Xb} \in \mathbb{R}$. In the context of classification, system responses are discrete, for example, $y \in \{-1, +1\}$. The idea with classification is to transform the output of a linear function $g(\mathbf{x})$ into the $(0, 1)$ interval describing the probability $\Pr(y = +1 | x)$. We can transform the output of a linear model $\mathbf{Xb}$ in the interval $(0, 1)$ by passing it through a logistic *sigmoid* function $\sigma(z)$,

$$\sigma(z) = \frac{1}{1 + \exp(-z)} = \frac{\exp(z)}{\exp(z) + 1},$$

as plotted in figure 9.8. The sigmoid is a transformation $\mathbb{R} \to (0, 1)$ such that an input defined in the real space is squashed in the interval $(0, 1)$. In short, logistic regression maps the covariates $\mathbf{x}$ to the probability of a class,

$$\underbrace{\mathbf{x} \in \mathbb{R}^{\mathtt{X}}}_{\text{covariates}} \to \underbrace{g(\mathbf{x}) = \mathbf{Xb} \in \mathbb{R}}_{\text{hidden/latent variable}} \to \underbrace{\sigma(g(\mathbf{x})) \in (0, 1)}_{\Pr(Y=y|\mathbf{x})}.$$

Note that $g(\mathbf{x})$ is considered as a *latent* or *hidden* variable because it is not directly observed; only the class $y$ and its associated covariate $\mathbf{x}$ are.

Figure 9.9 presents the example of a function $g(x) = 0.08x - 4$ (i.e., the red line on the horizontal plane), which is passed through a logistic sigmoid function. The blue curve on the leftmost vertical plane is the logistic sigmoid function, and the orange curve on the right should be interpreted as the probability the class $y = +1$, given a covariate $x$, $\Pr(y = +1 | x)$. For four different covariates $x_i$, simulated observations $y_i \in \{-1, +1\}$ are depicted by crosses.

In practice, we have to follow the reverse path, where observations $\mathcal{D} = \{(x_i, y_i), \forall i \in \{1 : \mathtt{D}\}\}$ are available and we need to estimate the parameters $\mathbf{b}$ and the basis functions $\phi(x)$ defining the model matrix $\mathbf{X}$. For a given data set and a choice of basis functions, we separate the observations $x_i$ in order to build two model matrices: one for $x_i : y_i = -1$, $\mathbf{X}_{(-1)}$, and another for $x_i : y_i = +1$, $\mathbf{X}_{(+1)}$. The likelihood of an observation $y = +1$ is given by $\Pr(y = +1 | x) = \sigma(\mathbf{Xb})$, and for an observation $y = -1$ the likelihood is $\Pr(y = -1 | x) = 1 - \sigma(\mathbf{Xb})$. With the hypothesis that observations are conditionally independent, the joint likelihood is obtained from the product of the marginal likelihood or, equivalently, the sum of the marginal log-likelihood of observations given parameters $\mathbf{b}$,

$$\ln p(\mathcal{D}|\mathbf{b}) = \sum \ln(\sigma(\mathbf{X}_{(+1)}\mathbf{b})) + \sum \ln(1 - \sigma(\mathbf{X}_{(-1)}\mathbf{b})).$$

Optimal parameters $\mathbf{b}^*$ can be inferred using an MLE procedure that maximizes $\ln p(\mathcal{D}|\mathbf{b})$. Unfortunately, contrarily to the optimiza-

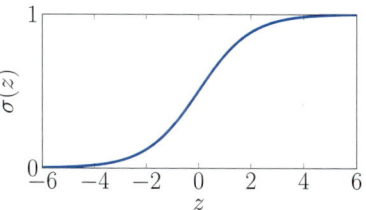

Figure 9.8: The logistic sigmoid function $\sigma(z) = (1 + \exp(-z))^{-1}$.

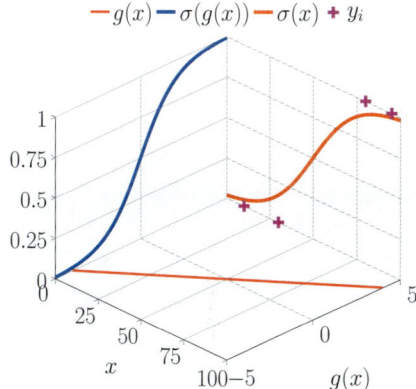

Figure 9.9: Example of a function $g(x) = 0.08x - 4$ passed through the logistic sigmoid function.

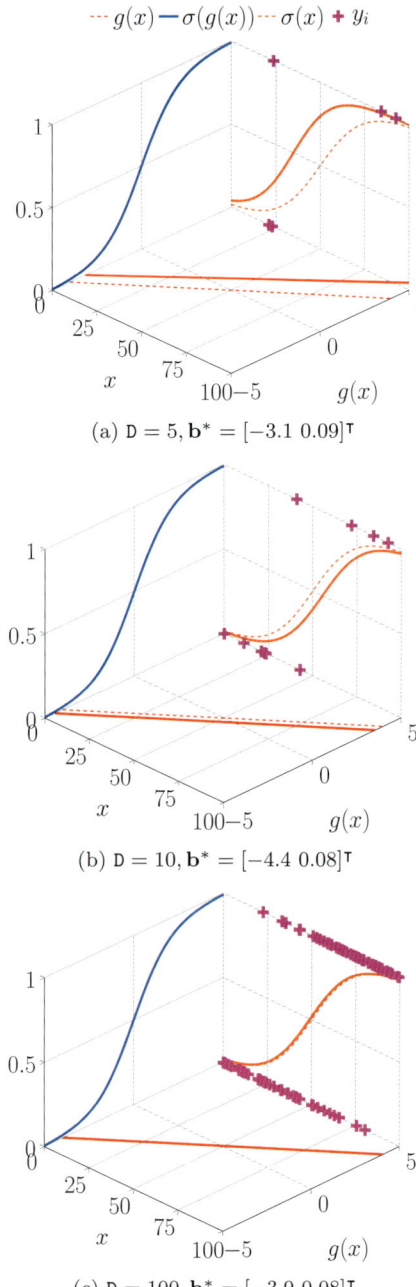

(a) $D = 5, \mathbf{b}^* = [-3.1\ 0.09]^{\intercal}$

(b) $D = 10, \mathbf{b}^* = [-4.4\ 0.08]^{\intercal}$

(c) $D = 100, \mathbf{b}^* = [-3.9\ 0.08]^{\intercal}$

Figure 9.10: Example of application of logistic regression.

[3] Ben-Akiva, M. E. and S. R. Lerman (1985). *Discrete choice analysis: Theory and application to travel demand.* MIT Press; and McFadden, D. (2001). Economic choices. *American Economic Review 91*(3), 351–378.

tion of parameters in the case of the linear regression presented in §8.1, no close-form analytic solution exists here to identify $\mathbf{b}^*$. With linear regression, the derivative of the log-likelihood (i.e., the loss function) was linear, leading to an analytic solution for $\mathbf{b}^*$ : $\frac{\partial J(\mathbf{b})}{\partial b} = 0$. With logistic regression, the derivative of the log-likelihood is a nonlinear function, so we have to resort to an optimization algorithm (see chapter 5) to identify $\mathbf{b}^*$.

*Example: Logistic regression*   Figure 9.10 presents three examples involving a single covariate $x$ and a linear model $g(x) = b_1 x + b_0$, and where parameters $\mathbf{b} = [b_0\ b_1]^{\intercal}$ are estimated with respectively 5, 10, and 100 observations. The true parameter values employed to generate simulated observations are $\check{\mathbf{b}} = [-4\ 0.08]^{\intercal}$. The corresponding functions $g(\mathbf{X}\check{\mathbf{b}})$ and $\sigma(g(\mathbf{X}\check{\mathbf{b}}))$ are represented by dashed lines, and those estimated using MLE parameters $g(\mathbf{X}\mathbf{b}^*)$ and $\sigma(g(\mathbf{X}\mathbf{b}^*))$ are represented by solid lines. We can observe that as the number of observations increases, the classifier converges toward the one employed to generate the data.

This case is a trivial one because, in this closed-loop simulation, the model structure (i.e., as defined by the basis functions in model matrix $\mathbf{X}$) was a perfect fit for the problem. In practical cases, we have to select an appropriate set of basis functions $\phi_j(x_i)$ to suit the problem at hand. Like for the linear regression, the selection of basis functions is prone to overfitting, so we have to employ either the Bayesian model selection (see §6.8) or cross-validation (see §8.1.2) for that purpose.

*Civil engineering perspectives*   In the field of transportation engineering, logistic regression has been extensively employed for *discrete choice modeling*[3] because of the interpretability of the model parameters $\mathbf{b}$ in the context of behavioral economics. However, for most benchmark problems, the predictive capacity of logistic regression is outperformed by more modern techniques such as *Gaussian process classification* and *neural networks*, which are presented in the next two sections.

## 9.3   Gaussian Process Classification

*Gaussian process classification* (GPC) is the extension of Gaussian process regression (GPR; see §8.2) to classification problems. In the context of GPR, a function $\mathbb{R}^{\mathtt{X}} \to \mathbb{R}$ is defined so that it transforms an $\mathtt{X}$-dimensions covariate domain into a single output $g(\mathbf{x}) \in \mathbb{R}$. In the context of classification, the system response is $y \in \{-1, +1\}$.

Again, the idea with classification is to transform $g(\mathbf{x})$'s outputs into the $(0, 1)$ interval describing the probability $\Pr(Y = +1|x)$. For GPC, the transformation in the interval $(0, 1)$ is done using the standard Normal CDF presented in figure 9.11, where $\Phi(z)$ denotes the CDF of $Z \sim \mathcal{N}(z; 0, 1)$.

Like the logistic sigmoid function presented in §9.2, $\Phi(z)$ is a transformation $\mathbb{R} \rightarrow (0, 1)$, such that an input defined in the real space is mapped in the interval $(0, 1)$. Note that choosing the transformation function $\Phi(z)$ instead of the logistic function is not arbitrary; we will see later that it allows maintaining the analytic tractability of GPC. We saw in §8.2 that the function $g(\mathbf{x}_*)$, describing the system responses at prediction locations $\mathbf{x}_*$, is modeled by a Gaussian process,

$$f_{\mathbf{G}_*|\mathbf{x}_*,\mathcal{D}}(\mathbf{g}_*|\mathbf{x}_*, \mathcal{D}) \equiv f(\mathbf{g}_*|\mathbf{x}_*, \mathcal{D}) = \mathcal{N}(\mathbf{g}_*; \boldsymbol{\mu}_{*|\mathcal{D}}, \boldsymbol{\Sigma}_{*|\mathcal{D}}).$$

In order to compute $\Pr(Y = +1|x_*, \mathcal{D})$, that is, the probability that $Y = +1$ for a covariate $x_*$ and a set of observations $\mathcal{D}$, we need to transform $g_*$ in the interval $(0, 1)$ and then marginalize the uncertainty associated with $f(g_*|x_*, \mathcal{D})$ so that

$$\Pr(Y = +1|x_*, \mathcal{D}) = \int \Phi(g_*) \cdot f(g_*|x_*, \mathcal{D}) dg_*. \tag{9.2}$$

If we choose the standard normal CDF, with $\Phi(z)$ as a transformation function, the integral in equation 9.2 follows the closed-form solution

$$\Pr(Y = +1|x_*, \mathcal{D}) = \Phi\left(\frac{\mathbb{E}[G_*|x_*, \mathcal{D}]}{\sqrt{1 + \text{var}[G_*|x_*, \mathcal{D}]}}\right), \tag{9.3}$$

where for the $i^{\text{th}}$ prediction locations, $\mathbb{E}[G_*|x_*, \mathcal{D}] \equiv [\boldsymbol{\mu}_{*|\mathcal{D}}]_i$ and $\text{var}[G_*|x_*, \mathcal{D}] \equiv [\boldsymbol{\Sigma}_{*|\mathcal{D}}]_{ii}$. Figure 9.12 illustrates how the uncertainty related to a Gaussian process evaluated through $\Phi(G(x))$ is marginalized in order to describe $\Pr(Y = +1|x, \mathcal{D})$.

So far, we have seen how to employ the standard normal CDF to transform a Gaussian process $g_* \in \mathbb{R}$, with PDF $f(g_*|x_*, \mathcal{D})$ obtained from a set of observations $\mathcal{D} = \{\mathcal{D}_x, \mathcal{D}_y\} = \{(\mathbf{x}_i, y_i \in \mathbb{R}), \forall i \in \{1 : \mathsf{D}\}\}$, into a space $\Phi(g_*) \in (0, 1)$. The issue is that in a classification setup, $g(x)$ is not directly observable; only $y_i \in \{-1, +1\}$ is. This requires inferring for each covariate $x \in \mathcal{D}_x$ the mean and standard deviations for *latent* variables $G(x)$. For that, we need to rewrite equation 9.3 in order to explicitly include the

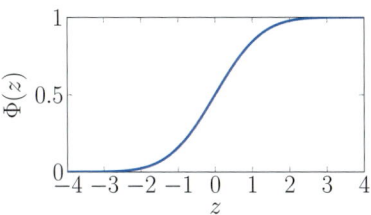

Figure 9.11: The standard Normal CDF, where $\Phi(z)$ denotes the CDF of $Z \sim \mathcal{N}(z; 0, 1)$.

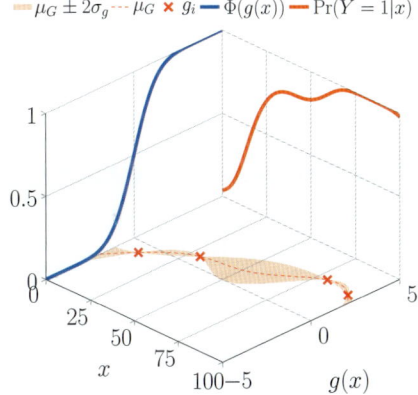

Figure 9.12: Example of Gaussian process $G(x)$ evaluated through $\Phi(G(x))$ and whose uncertainty is marginalized in order to describe $\Pr(Y = +1|x)$.

**Note:** The qualification *latent* refers to variables that are not directly observed and that need to be inferred from observations of the classes $y \in \{-1, +1\}$.

conditional dependence over **g** so that

$$
\begin{aligned}
\Pr(Y = +1|x_*, \mathcal{D}) &= \int \Phi(g_*) \cdot f(g_*|x_*, \mathbf{g}, \mathcal{D}) dg_* \\
&= \Phi \left( \frac{\mathbb{E}[G_*|x_*, \mathbf{g}, \mathcal{D}]}{\sqrt{1 + \mathrm{var}[G_*|x_*, \mathbf{g}, \mathcal{D}]}} \right).
\end{aligned}
$$

For the $i^{\text{th}}$ prediction locations, $\mathbb{E}[G_*|x_*, \mathbf{g}, \mathcal{D}] \equiv [\boldsymbol{\mu}_{*|\mathbf{g}, \mathcal{D}}]_i$ and $\mathrm{var}[G_*|x_*, \mathbf{g}, \mathcal{D}] \equiv [\boldsymbol{\Sigma}_{*|\mathbf{g}, \mathcal{D}}]_{ii}$, where

$$
\begin{aligned}
f(\mathbf{g}_*|\mathbf{g}) &\equiv f(\mathbf{g}_*|\mathbf{x}_*, \mathbf{g}, \mathcal{D}) \\
&= \mathcal{N}(\mathbf{g}_*; \boldsymbol{\mu}_{*|\mathbf{g}, \mathcal{D}}, \boldsymbol{\Sigma}_{*|\mathbf{g}, \mathcal{D}}).
\end{aligned}
\tag{9.4}
$$

Equation 9.4 presents the posterior probability of the Gaussian process outcomes $\mathbf{g}_*$ at prediction location $\mathbf{x}_*$, given the data set $\mathcal{D} = \{\mathcal{D}_x, \mathcal{D}_y\} = \{(\mathbf{x}_i, y_i \in \{-1, 1\}), \forall i \in \{1 : \mathtt{D}\}\}$, and the inferred values for $\mathbf{g}$ at observed locations $\mathbf{x}$. The mean and covariance matrix for the PDF in equation 9.4 are, respectively,

$$
\begin{aligned}
\boldsymbol{\mu}_{*|\mathbf{g}, \mathcal{D}} &= \overbrace{\boldsymbol{\mu}_*}^{=0} + \boldsymbol{\Sigma}_{\mathbf{G}*}^{\mathsf{T}} \boldsymbol{\Sigma}_{\mathbf{G}}^{-1} (\boldsymbol{\mu}_{\mathbf{G}|\mathcal{D}} - \overbrace{\boldsymbol{\mu}_{\mathbf{G}}}^{=0}) \\
\boldsymbol{\Sigma}_{*|\mathbf{g}, \mathcal{D}} &= \boldsymbol{\Sigma}_* - \boldsymbol{\Sigma}_{\mathbf{G}*}^{\mathsf{T}} \boldsymbol{\Sigma}_{\mathbf{G}|\mathcal{D}}^{-1} \boldsymbol{\Sigma}_{\mathbf{G}*},
\end{aligned}
$$

where $\boldsymbol{\mu}_{\mathbf{G}|\mathcal{D}}$ and $\boldsymbol{\Sigma}_{\mathbf{G}|\mathcal{D}}$ are the mean vector and covariance matrix for the inferred latent observations of $G(\mathbf{x}) \sim f(\mathbf{g}|\mathcal{D}) = \mathcal{N}(\mathbf{g}; \boldsymbol{\mu}_{\mathbf{G}|\mathcal{D}}, \boldsymbol{\Sigma}_{\mathbf{G}|\mathcal{D}})$. The posterior PDF for inferred latent variables $G(\mathbf{x})$ is

$$
\begin{aligned}
f(\mathbf{g}|\mathcal{D}) &= \frac{p(\mathcal{D}_y|\mathbf{g}) \cdot f(\mathbf{g}|\mathcal{D}_x)}{p(\mathcal{D}_y|\mathcal{D}_x)} \\
&\propto f(\mathcal{D}_y|\mathbf{g}) \cdot f(\mathbf{g}|\mathcal{D}_x).
\end{aligned}
\tag{9.5}
$$

The first term corresponds to the joint likelihood of observations $\mathcal{D}_y$ given the associated set of inferred latent variables $\mathbf{g}$. With the assumption that observations are conditionally independent given $g_i$, the joint likelihood is obtained from the product of the marginals so that

$$
\begin{aligned}
p(\mathcal{D}_y|\mathbf{g}) &= \prod_{i=1}^{\mathtt{D}} p(y_i|g_i) \\
&= \prod_{i=1}^{\mathtt{D}} \Phi(y_i \cdot g_i).
\end{aligned}
$$

The second term in equation 9.5, $f(\mathbf{g}|\mathcal{D}_x) = \mathcal{N}(\mathbf{g}; \boldsymbol{\mu}_{\mathbf{G}} = \mathbf{0}, \boldsymbol{\Sigma}_{\mathbf{G}})$ is the prior knowledge for the inferred latent variables $G(\mathbf{x})$. The posterior mean of $f(\mathbf{g}|\mathcal{D})$, that is, $\boldsymbol{\mu}_{\mathbf{G}|\mathcal{D}}$, is obtained by maximizing the logarithm of $f(\mathbf{g}|\mathcal{D})$ so that

$$
\begin{aligned}
\boldsymbol{\mu}_{\mathbf{G}|\mathcal{D}} &= \mathbf{g}^* \\
&= \underset{\mathbf{g}}{\arg\max} \ln f(\mathbf{g}|\mathcal{D}) \\
&= \underset{\mathbf{g}}{\arg\max} \ln f(\mathcal{D}_y|\mathbf{g}) - \tfrac{1}{2}\mathbf{g}^{\mathsf{T}}\boldsymbol{\Sigma}_{\mathbf{G}}^{-1}\mathbf{g} - \tfrac{1}{2}\ln|\boldsymbol{\Sigma}_{\mathbf{G}}| - \tfrac{\mathtt{D}}{2}\ln 2\pi.
\end{aligned}
$$

The maximum $\mathbf{g}^*$ corresponds to the location where the derivative equals zero so that

$$\frac{\partial \ln f(\mathbf{g}|\mathcal{D})}{\partial \mathbf{g}} = \nabla_\mathbf{g} \ln f(\mathbf{g}|\mathcal{D}) = \nabla \ln f(\mathcal{D}_y|\mathbf{g}) - \boldsymbol{\Sigma}_\mathbf{G}^{-1}\mathbf{g} = 0. \quad (9.6)$$

By isolating $\mathbf{g}$ in equation 9.6, we can obtain optimal values $\mathbf{g}^*$ iteratively using

$$\mathbf{g} \leftarrow \boldsymbol{\Sigma}_\mathbf{G} \nabla \ln f(\mathcal{D}_y|\mathbf{g}), \quad (9.7)$$

where the initial starting location can be taken as $\mathbf{g} = \mathbf{0}$. The uncertainty associated with the optimal set of inferred latent variables $\boldsymbol{\mu}_{\mathbf{G}|\mathcal{D}} = \mathbf{g}^*$ is estimated using the Laplace approximation (see §6.7.2). The Laplace approximation for the covariance matrix corresponds to the inverse of the Hessian for the negative log-likelihood evaluated at $\boldsymbol{\mu}_{\mathbf{G}|\mathcal{D}}$, so that

$$\begin{aligned} \boldsymbol{\Sigma}_{\mathbf{G}|\mathcal{D}} &= \mathbf{H}[-\ln f(\mathbf{g}|\mathcal{D})]^{-1} \\ &= \left( \boldsymbol{\Sigma}_\mathbf{G}^{-1} - \mathrm{diag}(\nabla\nabla \ln f(\mathcal{D}_y|\boldsymbol{\mu}_{\mathbf{G}|\mathcal{D}})) \right)^{-1}. \end{aligned}$$

Note that we can improve the efficiency for computing the MLE in equation 9.8 by using the information contained in the Hessian as we did with the Newton-Raphson method in §5.2. In that case, optimal values $\mathbf{g}^*$ are obtained iteratively using

$$\mathbf{g} \leftarrow \mathbf{g} - \mathbf{H}[-\ln f(\mathbf{g}|\mathcal{D})]^{-1} \cdot \nabla_\mathbf{g} \ln f(\mathbf{g}|\mathcal{D}). \quad (9.8)$$

Figure 9.13 presents an application of GPC using the Gaussian process for machine learning (GPML) package with a different number of observations, $\mathtt{D} \in \{25, 50, 100\}$. In this figure, the green solid lines on the horizontal planes describe the inferred confidence intervals for the latent variables, $\mu_{G|\mathcal{D}} \pm 2\sigma_{G|\mathcal{D}}$. We can see that as the number of observations increases, the inferred function $\Pr(Y = +1|x_*, \mathcal{D})$ (orange solid line) tends toward the true function (orange dashed line).

Figure 9.14 presents the application of Gaussian process classification to the post-earthquake structural safety assessment example introduced in §9.1.2. Here, the multiclass problem is transformed into a binary one by modeling the probability that the damage index is either 0 or 1. The main advantage of GPC is that the problem setup is trivial when using existing, pre-implemented packages. With the generative approaches presented in §9.1, the formulation has to be specifically tailored for the problem. Nevertheless, note that a generative approach allows including censored data, whereas a discriminative approach such as the GPC presented here cannot.

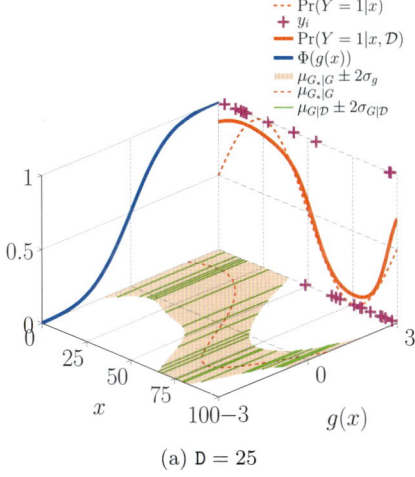

(a) D = 25

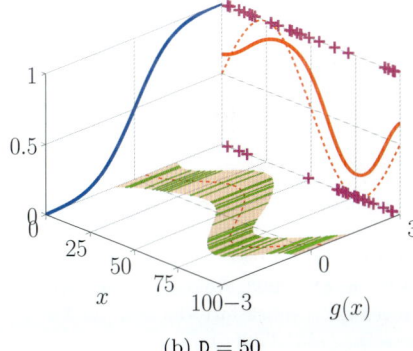

(b) D = 50

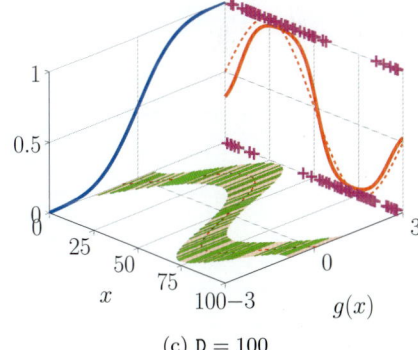

(c) D = 100

Figure 9.13: Example of application of Gaussian process classification using the package GPML with a different number of observations $\mathtt{D} \in \{25, 50, 100\}$.

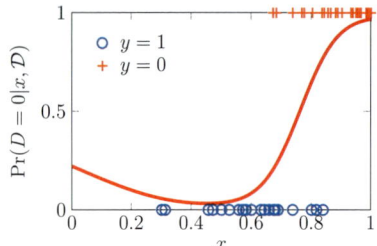

Figure 9.14: Example of application of Gaussian process classification to the post-earthquake structural safety evaluation data set.

[4] Rasmussen, C. E. and C. K. Williams (2006). *Gaussian processes for machine learning*. MIT Press.

[5] Ebden, M. (2008, August). Gaussian processes: A quick introduction. *arXiv preprint (1505.02965)*.

In practice, all the operations required to infer the latent variables $\mathbf{G}(\mathbf{x})$ as well as the parameter $\boldsymbol{\theta} = \{\sigma_g, \boldsymbol{\ell}\}$ in GPC are already implemented in the same open-source packages dealing with GPR, for example, GPML, GPstuff, and pyGPs (see §8.2).

*Strengths and limitations*    With the availability of Gaussian process classification implemented in a variety of open-source packages, the setup of a GPC problem is trivial, even for problems involving several covariates $\mathbf{x}$. When the number of observations is small, the task of inferring latent variables $\mathbf{g}$ becomes increasingly difficult and the performance decreases. Note also that the formulation presented assumes that we only have error-free direct observations. Moreover, because of the computationally demanding procedure of inferring latent variables $\mathbf{g}$, the performance is limited in the case of large data sets, for example, $\mathtt{D} > 10^5$. Given the limitations for both small and large data sets, the GPC presented here is thus best suited for medium-size data sets.

For more details about the Gaussian process classification and its extension to multiple classes, the reader should refer to dedicated publications such as the book by Rasmussen and Williams[4] or the tutorial by Ebden.[5]

## 9.4    Neural Networks

The formulation for the feedforward neural network presented in §8.3 can be adapted for binary classification problems by replacing the linear output activation function by a *sigmoid* function as illustrated in figure 9.15a. For an observation $y \in \{-1, +1\}$, the log of the probability for the outcome $y_i = +1$ is modeled as being *proportional* to the hidden variable on the output layer $z^{(0)}$, and the log of the probability for the outcome $y_i = -1$ is assumed to be proportional to 0,

$$
\begin{aligned}
\ln \Pr(Y = +1|\mathbf{x}, \boldsymbol{\theta}) \quad &\propto \quad z^{(0)} \\
&= \quad \mathbf{W}^{(L)}\mathbf{a}^{(L)} + \mathbf{b}^{(L)} \\
\ln \Pr(Y = -1|\mathbf{x}, \boldsymbol{\theta}) \quad &\propto \quad 0.
\end{aligned}
$$

These unnormalized log-probabilities are transformed into a probability for each possible outcome by taking the exponential of each of them and then normalizing by following

$$
\Pr(Y = +1|\mathbf{x}, \boldsymbol{\theta}) \quad = \quad \frac{\exp(z^{(0)})}{\exp(z^{(0)}) + \underbrace{\exp(0)}_{=1}} \quad = \quad \sigma(z^{(0)})
$$

$$
\Pr(Y = -1|\mathbf{x}, \boldsymbol{\theta}) \quad = \quad \frac{\exp(0)}{\exp(z^{(0)}) + \exp(0)} \quad = \quad \sigma(-z^{(0)}).
$$

As presented in §8.3.1, this procedure for normalizing the log-probabilities of $y$ is equivalent to evaluating the hidden variable $z^{(0)}$ (or its negative) in the logistic sigmoid function. Thus, the likelihood of an observation $y \in \{-1, +1\}$, given its associated covariates $\mathbf{x}$ and a set of parameters $\boldsymbol{\theta} = \{\boldsymbol{\theta}_w, \boldsymbol{\theta}_b\}$, is given by

$$p(y|\mathbf{x}, \boldsymbol{\theta}) = \sigma(y \cdot z^{(0)}).$$

With neural networks, it is common to minimize a loss function $J(\mathcal{D}; \boldsymbol{\theta}_w, \boldsymbol{\theta}_b)$ that is defined as the negative joint log-likelihood for a set of D observations,

$$
\begin{aligned}
J(\mathcal{D}; \boldsymbol{\theta}_w, \boldsymbol{\theta}_b) &= -\ln p(\mathcal{D}_y | \mathcal{D}_x, \boldsymbol{\theta}) \\
&= -\sum_{i=1}^{D} \ln \sigma(y_i \cdot z_i^{(0)}).
\end{aligned}
$$

In the case where the observed system responses can belong to multiple classes, $y \in \{1, 2, \cdots, \mathsf{C}\}$, the output layer needs to be modified to include as many hidden states as there are classes, $\mathbf{z}^{(0)} = [z_1^{(0)} \ z_2^{(0)} \ \cdots \ z_\mathsf{C}^{(0)}]^\mathsf{T}$, as presented in figure 9.15b. The log-probability of an observation $y = k$ is assumed to be proportional to the $k^{\text{th}}$ hidden state $z_k$,

$$\ln \Pr(Y = k|\mathbf{x}, \boldsymbol{\theta}) \propto z_k^{(0)}.$$

The normalization of the log-probabilities is done using the *softmax* activation function, where the probability of an observation $y = k$ is given by

$$
\begin{aligned}
p(y = k|\mathbf{x}, \boldsymbol{\theta}) &= \text{softmax}(\mathbf{z}^{(0)}, k) \\
&= \frac{\exp(z_k^{(0)})}{\sum_{j=1}^{\mathsf{C}} \exp(z_j^{(0)})}.
\end{aligned}
$$

By assuming that observations are conditionally independent from each other given $\mathbf{z}^{(0)}$, the loss function for the entire data set is obtained as in §8.3.2 by summing the log-probabilities (i.e., the marginal log-likelihoods) for each observation,

$$J(\mathcal{D}; \boldsymbol{\theta}_w, \boldsymbol{\theta}_b) = -\sum_{i=1}^{D} \ln p(y_i|\mathbf{x}_i, \boldsymbol{\theta}). \tag{9.9}$$

Minimizing the loss function $J(\mathcal{D}; \boldsymbol{\theta}_w, \boldsymbol{\theta}_b)$ defined in equation 9.9 corresponds to minimizing the *cross-entropy*. Cross-entropy is a concept from the field of *information theory*[6] that measures the similarity between two probability distributions or mass functions.

Note that in the context of classification, the gradient $\nabla \sigma^0(z^{(0)})$ in the *backpropagation* algorithm (see §8.3.2) is no longer equal to

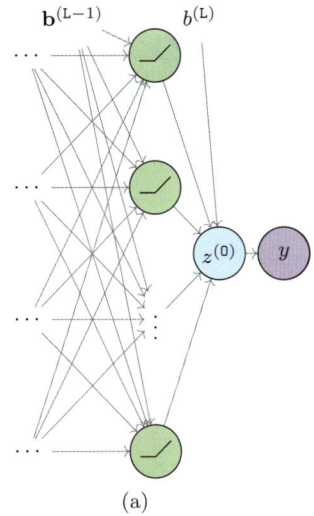

(a)

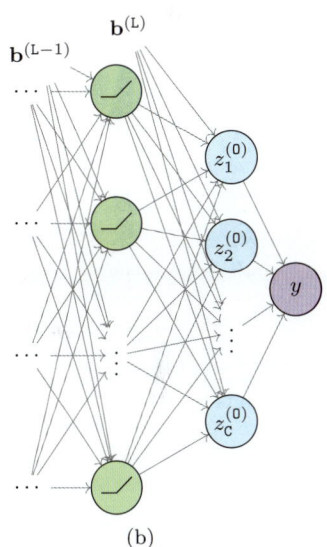

(b)

Figure 9.15: Nomenclature for a feed-forward neural network employed for classification: (a) represents the case for binary classification $y \in \{-1, +1\}$; (b) the case for C classes $y \in \{1, 2, \cdots, \mathsf{C}\}$.

**Note:** Given two PMFs, $p(x)$ and $q(x)$, the cross-entropy is defined as

$$H(p, q) = -\sum_x p(x) \ln q(x).$$

[6] MacKay, D. J. C. (2003). *Information theory, inference, and learning algorithms.* Cambridge University Press.

one because of the sigmoid function. Moreover, like for the regression setup, a neural network classifier is best suited for problems with a large number of covariates and for which large data sets are available.

## 9.5  *Regression versus Classification*

Note that when the data and context allow for it, it can be preferable to formulate a problem using a *regression* approach rather than using *classification*. Take, for example, the soil contamination example presented in §8.2.5. The data set available describes the contaminant concentration $c_i$ measured at coordinates $\mathbf{l}_i = \{x_i, y_i, z_i\}$. Using regression to model this problem results in a function describing the PDF of contaminant concentration as a function of coordinates $\mathbf{l}$, $f(\mathbf{c}|\mathbf{l}, \mathcal{D})$. A classification setup would instead model the probability that the soil contamination exceeds an admissible value $c_{\mathrm{adm.}}$, $\mathrm{Pr}(\mathbf{C} \geq c_{\mathrm{adm.}}|\mathbf{l}, \mathcal{D})$. Here, the drawback of classification is that information is lost when transforming a continuous system response $\mathbf{C}$ into a categorical event $\{\mathbf{C} \geq c_{\mathrm{adm.}}\}$. For the classification setup, the information available to build the model is whether the contaminant concentration is above $(c_i = +1)$ or below $(c_i = -1)$ the admissible concentration $c_{\mathrm{adm.}}$ for multiple locations $\mathbf{l}_i$. The issue with a classification approach is that because we work with categories, the information about how far or how close we are from the threshold $c_{\mathrm{adm.}}$ is lost.

*Exercises*

P9.1  Explain the differences between generative and discriminative classification approaches.

P9.2  Explain why it is easier for generative approaches to consider censored observations.

P9.3  Explain the differences between naive Bayes, LDA, and QDA.

P9.4  Explain the relation between linear regression and logistic regression.

P9.5  Given a logistic regression model $\sigma(b_0 + b_1 x)$, write the likelihood formulation for $\mathcal{D} = \{(x = 3, y = +1), (x = 8, y = -1)\}$.

P9.6  What are the parameters to infer in the context of logistic regression?

P9.7  What are the parameters and hidden variables to infer in the context of Gaussian process classification?

P9.8  Implement a generative classifier using Gaussian marginal PDFs and apply it to the data set $\mathcal{D} = \{\mathcal{D}_{+1}, \mathcal{D}_{-1}\}$,

$$\mathcal{D}_{+1} = \{\mathbf{y} = +1, \mathbf{x} = [1.4\ 2.2\ 2.0\ 3.5\ 1.7\ 0.4\ 2.2\ 1.5\ 1.3\ 2.3\ 0.5\ 2.3]^\mathsf{T}\}$$
$$\mathcal{D}_{-1} = \{\mathbf{y} = -1, \mathbf{x} = [6.1\ 2.9\ 3.9\ 5.5\ 3.4\ 4.4\ 2.3\ 5.9]^\mathsf{T}\}.$$

P9.9  Implement the logistic regression method and reproduce the results presented in figure 9.10b using the data set $\mathcal{D} = \{\mathcal{D}_{+1}, \mathcal{D}_{-1}\}$,

$$\mathcal{D}_{+1} = \{\mathbf{y} = +1, \mathbf{x} = [92.6\ 84.4\ 41.0\ 72.1]^\mathsf{T}\}$$
$$\mathcal{D}_{-1} = \{\mathbf{y} = -1, \mathbf{x} = [23.6\ 22.9\ 21.0\ 42.9\ 0.2\ 11.2]^\mathsf{T}\}.$$

P9.10  Implement the Gaussian process classification method and reproduce the results presented in figure 9.13a using the data set $\mathcal{D} = \{\mathcal{D}_{+1}, \mathcal{D}_{-1}\}$,

$$\mathcal{D}_{+1} = \{\mathbf{y} = +1, \mathbf{x} = [12.7\ \ 9.8\ 27.8\ 54.7\ 95.8\ 96.5\ 15.8\ 95.7\ 14.2\ 42.2\ 96.0\ \ 3.6]^\mathsf{T}\}$$
$$\mathcal{D}_{-1} = \{\mathbf{y} = -1, \mathbf{x} = [90.6\ 91.3\ 63.2\ 97.0\ 48.5\ 80.0\ 91.6\ 79.2\ 65.6\ 84.9\ 93.4\ 67.9\ 75.8]^\mathsf{T}\}.$$

# Part IV

# Unsupervised Learning

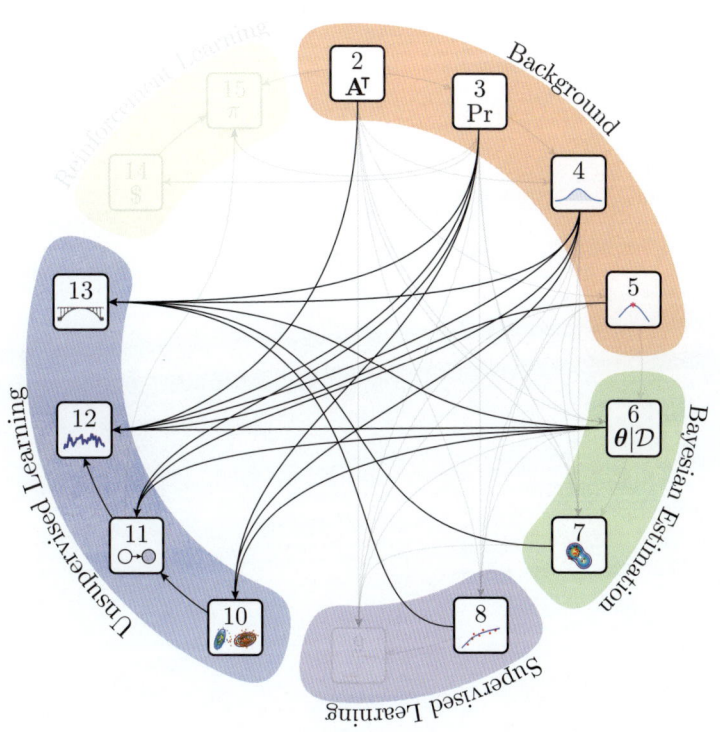

# 10
# Clustering and Dimension Reduction

This chapter covers two key problems associated with *unsupervised learning*: *clustering* and *dimension reduction*. Clustering consists in discovering patterns or subgroups among a set of covariates. Clustering is considered *unsupervised learning* because the cluster indexes are not available to train a model; cluster indexes have to be learned from the observed covariates themselves. With dimension reduction, the goal is to identify a lower-dimensional subspace for a set of covariates, in which the data can be represented with a negligible loss of information. In this chapter, we will introduce two clustering methods, Gaussian mixture models and K-means, and a dimension reduction technique, that is, principal component analysis.

## 10.1 Clustering

Figure 10.1 presents the Old Faithful geyser data set[1] where the covariates $\mathbf{x}_i = [x_1 \ x_2]_i^{\mathsf{T}}$ describe the eruption time and the time to the next eruption. Figure 10.1 displays two distinct behaviors: short eruptions with short times to the next ones and long eruptions with a long time to the next ones. The goal with clustering is to discover the existence of such an underlying structure in data.

Note that when working with data sets $\mathcal{D} = \mathcal{D}_x = \{\mathbf{x}_i, \forall i \in \{1 : \mathtt{D}\}\}, \mathbf{x}_i \in \mathbb{R}^{\mathtt{X}}$, with covariates defined in three or fewer dimensions ($\mathtt{X} \leq 3$), it is then trivial to visually identify clusters. The interest of having clustering methods is in identifying patterns in cases where $\mathtt{X} > 3$, such that the dimensionality of the data does not allow for a visual separation of clusters.

### 10.1.1 Gaussian Mixture Models

Clustering can be formulated as the task of fitting a *Gaussian*

(a) Old Faithful geyser

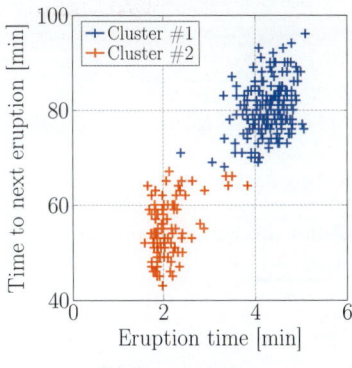

(b) Eruption data

Figure 10.1: Eruption durations and time between eruptions can be clustered into two groups.

[1] Adelchi, A. and A. W. Bowman (1990). A look at some data on the Old Faithful geyser. *Journal of the Royal Statistical Society. Series C (Applied Statistics) 39*(3), 357–365.

**Data**
$\mathcal{D} = \mathcal{D}_x = \{\mathbf{x}_i, \forall i \in \{1 : \mathtt{D}\}\}$

$\mathbf{x}_i \in \mathbb{R}^{\mathtt{X}} : \begin{cases} \text{Covariate} \\ \text{Attribute} \end{cases}$

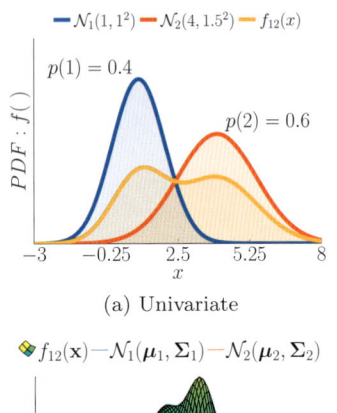

(a) Univariate

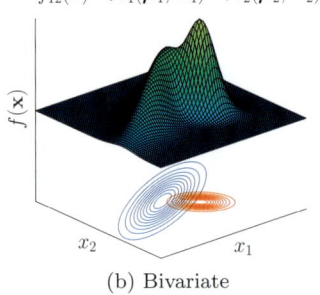

(b) Bivariate

Figure 10.2: Examples of Gaussian mixture PDFs.

**Gaussian mixture parameters**
$\mathcal{P} = \{(\boldsymbol{\mu}_k, \boldsymbol{\Sigma}_k), p(k), \forall k \in \{1 : \mathtt{K}\}\}$
$\underbrace{\qquad\qquad}_{\boldsymbol{\theta}}$

[2] Dempster, A. P., N. M. Laird, and D. B. Rubin (1977). Maximum likelihood from incomplete data via the EM algorithm. *Journal of the Royal Statistical Society. Series B (methodological)* 39(1), 1–38.

*mixture* model (GMM), on a set of observed covariates. A Gaussian mixture distribution is defined as the sum of $\mathtt{K}$ Normal (i.e., Gaussian) probability density functions (PDFs), each weighted by a probability $p(k), \forall k \in \{1 : \mathtt{K}\}$. The joint PDF for a Gaussian mixture is defined by

$$
\begin{aligned}
f(\mathbf{x}) &= \sum_{k=1}^{\mathtt{K}} f(\mathbf{x}, k) \\
&= \sum_{k=1}^{\mathtt{K}} f(\mathbf{x}|k) \cdot p(k) \\
&= \sum_{k=1}^{\mathtt{K}} \mathcal{N}(\mathbf{x}; \boldsymbol{\mu}_k, \boldsymbol{\Sigma}_k) \cdot p(k),
\end{aligned}
\tag{10.1}
$$

where $\boldsymbol{\theta} = \{(\boldsymbol{\mu}_k, \boldsymbol{\Sigma}_k), \forall k \in \{1 : \mathtt{K}\}\}$ describes the PDF parameters. Figure 10.2 presents examples of 1-D and 2-D Gaussian mixture PDFs. The issue is that in practice, we do not know $\boldsymbol{\theta}$ or $p(k)$, so we have to learn $\mathcal{P} = \{\boldsymbol{\theta}, p(k)\}$ from $\mathcal{D}_x$. Given the hypothesis that $\mathcal{D}_x$ contains observations that are conditionally independent given $\mathcal{P}$, the joint likelihood defined from equation 10.1 is

$$
f(\mathcal{D}_x|\boldsymbol{\theta}, p(k)) = \prod_{i=1}^{\mathtt{D}} \sum_{k=1}^{\mathtt{K}} \mathcal{N}(\mathbf{x}_i; \boldsymbol{\mu}_k, \boldsymbol{\Sigma}_k) \cdot p(k),
$$

and equivalently, the log-likelihood is

$$
\ln f(\mathcal{D}_x|\boldsymbol{\theta}, p(k)) = \sum_{i=1}^{\mathtt{D}} \ln \left( \sum_{k=1}^{\mathtt{K}} \mathcal{N}(\mathbf{x}_i; \boldsymbol{\mu}_k, \boldsymbol{\Sigma}_k) \cdot p(k) \right).
\tag{10.2}
$$

If we choose a maximum likelihood estimate (MLE) approach to estimate the parameters, their optimal values are defined as

$$
\mathcal{P}^* = \arg \max_{\mathcal{P}} \ln f(\mathcal{D}_x|\boldsymbol{\theta}, p(k)).
\tag{10.3}
$$

When the optimization in equation 10.3 employs methods such as those presented in chapter 5, it is referred to as *hard clustering*. Hard clustering is a difficult task because of several issues such as the constraints on $\boldsymbol{\Sigma}_k$ that require it to be positive semi-definite (see §3.3.5), and because of non-identifiability (see 6.3.4). *Soft clustering* solves these issues by employing the *expectation maximization* method to estimate $\mathcal{P}^*$.

*Expectation maximization*[2] (EM) is an iterative method that allows identifying optimal parameters $\mathcal{P}^*$ for problems involving *latent* (i.e., unobserved) variables. In the context of clustering, the latent variable is the unobserved cluster index $k$. If we could observe $\mathcal{D} = \{\mathcal{D}_x, \mathcal{D}_k\} = \{(\mathbf{x}_i, k_i), \forall i \in \{1 : \mathsf{D}\}\}$, the *complete data* log-likelihood for one observation $(\mathbf{x}_i, k_i)$ would be

$$
\begin{aligned}
\ln f(\mathbf{x}_i, k_i | \mathcal{P}) &= \ln f(\mathbf{x}_i, k_i | \mathcal{P}) \\
&= \ln \left( \mathcal{N}(\mathbf{x}_i; \boldsymbol{\mu}_{k_i}, \boldsymbol{\Sigma}_{k_i}) \cdot p(k_i) \right).
\end{aligned}
$$

Because complete data is not available, EM identifies optimal parameters by maximizing the *expected complete data log-likelihood,* $Q(\mathcal{P}^{(t)}, \mathcal{P}^{(t-1)})$, which depends on the sets of parameters defined for two successive iterations, $t - 1$ and $t$. For a single observation $\mathbf{x}_i$,

$$
Q(\mathcal{P}^{(t)}, \mathcal{P}^{(t-1)}) = \sum_{k=1}^{\mathsf{K}} \ln \big( \underbrace{\mathcal{N}(\mathbf{x}_i; \boldsymbol{\mu}_k^{(t)}, \boldsymbol{\Sigma}_k^{(t)}) \cdot p^{(t)}(k)}_{\text{fct.}(\mathcal{P}^{(t)})} \big) \cdot \underbrace{p(k|\mathbf{x}_i, \mathcal{P}^{(t-1)})}_{\text{fct.}(\mathcal{P}^{(t-1)})},
$$

where $p(k|\mathbf{x}_i, \mathcal{P}^{(t-1)})$ is the posterior probability mass function (PMF) for the latent cluster labels $k$ obtained at the previous iteration $t - 1$. This last quantity is referred to as the *responsibility* of a cluster $k$. For an entire data set $\mathcal{D} = \{\mathcal{D}_x\} = \{\mathbf{x}_i, \forall i \in \{1 : \mathsf{D}\}\}$, the expected complete data log-likelihood is

$$
Q(\mathcal{P}^{(t)}, \mathcal{P}^{(t-1)}) = \sum_{i=1}^{\mathsf{D}} \left( \sum_{k=1}^{\mathsf{K}} \ln \left( \mathcal{N}(\mathbf{x}_i; \boldsymbol{\mu}_k^{(t)}, \boldsymbol{\Sigma}_k^{(t)}) \cdot p^{(t)}(k) \right) \cdot p(k|\mathbf{x}_i, \mathcal{P}^{(t-1)}) \right).
$$

The EM method identifies the optimal parameter values $\mathcal{P}^*$ from $Q(\mathcal{P}^{(t)}, \mathcal{P}^{(t-1)})$ by iteratively performing the *expectation* and *maximization* steps detailed below.

*Expectation step*   For an iteration $t$, the expectation step computes the probability $p_{ik}^{(t-1)}$ that an observed covariate $\mathbf{x}_i$ is associated with the cluster $k$, given the parameters $\mathcal{P}^{(t-1)}$ so that

$$
\begin{aligned}
p_{ik}^{(t-1)} &= p(k|\mathbf{x}_i, \mathcal{P}^{(t-1)}) \\
&= \frac{\mathcal{N}(\mathbf{x}_i; \boldsymbol{\mu}_k^{(t-1)}, \boldsymbol{\Sigma}_k^{(t-1)}) \cdot p^{(t-1)}(k)}{\sum_{k=1}^{\mathsf{K}} \mathcal{N}(\mathbf{x}_i; \boldsymbol{\mu}_k^{(t-1)}, \boldsymbol{\Sigma}_k^{(t-1)}) \cdot p^{(t-1)}(k)}.
\end{aligned} \tag{10.4}
$$

The particularity of the expectation step is that the estimation of the responsibility $p_{ik}^{(t-1)}$ in equation 10.4 employs the PDF parameters $\boldsymbol{\theta}^{(t-1)} = \{(\boldsymbol{\mu}_k^{(t-1)}, \boldsymbol{\Sigma}_k^{(t-1)}), \forall k \in \{1 : \mathsf{K}\}\}$ and the marginal PMF $p^{(t-1)}(k)$ estimated at the previous iteration $t - 1$.

*Maximization step*   In the maximization step, the goal is to identify

$$
\mathcal{P}^{(t)*} = \{\boldsymbol{\theta}^{(t)*}, p^{(t)*}(k)\} = \arg \max_{\mathcal{P}^{(t)}} Q(\mathcal{P}^{(t)}, \mathcal{P}^{(t-1)}),
$$

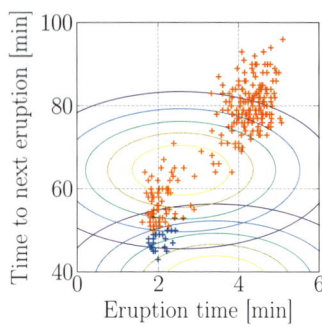

(a) Expectation step, $t = 1$

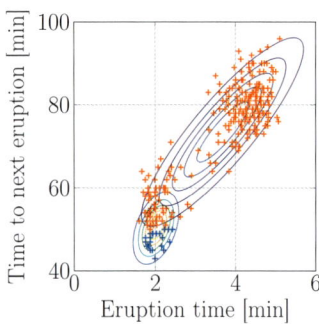

(b) Maximization step, $t = 1$

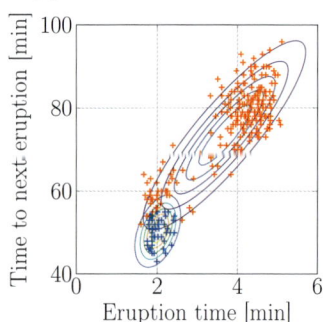

(c) Expectation step, $t = 2$

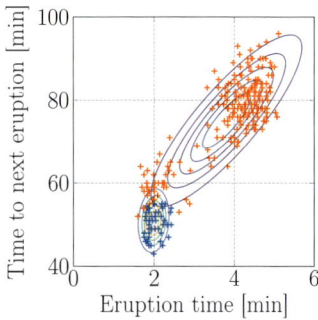

(d) Maximization step, $t = 2$

Figure 10.3: The first two iterations of the EM algorithm for a Gaussian mixture model applied on the Old Faithful data.

where the optimization is broken into two parts. The PDF's new set of optimal parameters $\boldsymbol{\theta}^{(t)}$ is obtained through the maximization problem

$$\boldsymbol{\theta}^{(t)*} = \arg\max_{\boldsymbol{\theta}^{(t)}} Q(\mathcal{P}^{(t)}, \mathcal{P}^{(t-1)}),$$

where the maximum corresponds to the value $\boldsymbol{\theta}$ such that the derivative of the function is zero,

$$
\begin{aligned}
0 &= \frac{\partial \ln Q(\mathcal{P}^{(t)}, \mathcal{P}^{(t-1)})}{\partial \boldsymbol{\theta}} \\
&= \frac{\partial \sum_{i=1}^{D} \left( \sum_{k=1}^{K} \ln\left(\mathcal{N}(\mathbf{x}_i; \boldsymbol{\mu}_k^{(t)}, \boldsymbol{\Sigma}_k^{(t)}) \cdot p^{(t)}(k)\right) \cdot p_{ik}^{(t-1)} \right)}{\partial \boldsymbol{\theta}} \\
&= \frac{\partial \sum_{i=1}^{D} \sum_{k=1}^{K} p_{ik}^{(t-1)} \cdot \ln \mathcal{N}(\mathbf{x}_i; \boldsymbol{\mu}_k^{(t)}, \boldsymbol{\Sigma}_k^{(t)})}{\partial \boldsymbol{\theta}} + \overbrace{\frac{\sum_{i=1}^{D} \sum_{k=1}^{K} p_{ik}^{(t-1)} \cdot \ln p^{(t)}(k)}{\partial \boldsymbol{\theta}}}^{=0} \\
&= \frac{\partial \sum_{i=1}^{D} p_{ik}^{(t-1)} \left[ \ln \det \boldsymbol{\Sigma}_k^{(t)} + (\mathbf{x}_i - \boldsymbol{\mu}_k^{(t)})^\mathsf{T} (\boldsymbol{\Sigma}_k^{(t)})^{-1} (\mathbf{x}_i - \boldsymbol{\mu}_k^{(t)}) \right]}{\partial \boldsymbol{\theta}}.
\end{aligned}
$$

By isolating the mean vector $\boldsymbol{\mu}_k^{(t)}$ and the covariance matrix $\boldsymbol{\Sigma}_k^{(t)}$, the new estimates are

$$
\left.
\begin{aligned}
\boldsymbol{\mu}_k^{(t)} &= \frac{\sum_{i=1}^{D} \mathbf{x}_i \cdot p_{ik}^{(t-1)}}{p_k^{(t-1)}} \\
\boldsymbol{\Sigma}_k^{(t)} &= \frac{\sum_{i=1}^{D} (\mathbf{x}_i - \boldsymbol{\mu}_k^{(t)})(\mathbf{x}_i - \boldsymbol{\mu}_k^{(t)})^\mathsf{T} \cdot p_{ik}^{(t-1)}}{p_k^{(t-1)}}
\end{aligned}
\right\}
p_k^{(t-1)} = \sum_{i=1}^{D} p_{ik}^{(t-1)}.
$$

By taking the derivative of $\ln Q(\mathcal{P}^{(t)}, \mathcal{P}^{(t-1)})$ with respect to $p(k)$ and repeating the same procedure, the new marginal PMF for each cluster $k$ is given by

$$p^{(t)}(k) = \frac{p_k^{(t-1)}}{D}.$$

In practice, the values $\boldsymbol{\mu}_k^{(0)}$ are initialized randomly, $\boldsymbol{\Sigma}_k^{(0)} = \sigma^2 \mathbf{I}$ is taken as a diagonal matrix, and $p^{(0)}(k) = 1/K$ is initially assumed to be equal for each cluster. Algorithm 8 summarizes the steps of the expectation maximization algorithm for estimating the parameters of a Gaussian mixture model.

Figure 10.3 presents the first two steps of the EM algorithm for a Gaussian mixture model applied to the Old Faithful example introduced in figure 10.1. At the first iteration $t = 1$, the two cluster centers $\boldsymbol{\mu}_k^{(0)}$ as well as their covariances $\boldsymbol{\Sigma}_k^{(0)}$ are initialized randomly. In the expectation step, the cluster responsibility is estimated for each data point $\mathbf{x}_i$. As depicted in figure 10.3a, a data

---

**Algorithm 8:** EM for Gaussian mixture models

1  define $\mathcal{D}_x$, $\epsilon$, $\ln f(\mathcal{D}_x|\mathcal{P}^{(0)}) = -\infty$, $\ln f(\mathcal{D}_x|\mathcal{P}^{(1)}) = 0$, $t = 1$;

2  initialize $\boldsymbol{\theta}^{(0)} = \{\boldsymbol{\mu}_k^{(0)}, \boldsymbol{\Sigma}_k^{(0)}\}$, $p^{(0)}(k) = 1/\texttt{K}$;

3  **while** $|\ln f(\mathcal{D}_x|\mathcal{P}^{(t)}) - \ln f(\mathcal{D}_x|\mathcal{P}^{(t-1)})| > \epsilon$ **do**

4  $\quad$ **for** $i \in \{1, 2, \cdots, \texttt{D}\}$ *(Expectation step)* **do**

5  $\quad\quad$ $p_i^{(t-1)} = \sum_{k=1}^{\texttt{K}} \mathcal{N}(\mathbf{x}_i; \boldsymbol{\mu}_k^{(t-1)}, \boldsymbol{\Sigma}_k^{(t-1)}) \cdot p^{(t-1)}(k)$

6  $\quad\quad$ **for** $k \in \{1, 2, \cdots, \texttt{K}\}$ **do**

7  $\quad\quad\quad$ $p_{ik}^{(t-1)} = \dfrac{\mathcal{N}(\mathbf{x}_i; \boldsymbol{\mu}_k^{(t-1)}, \boldsymbol{\Sigma}_k^{(t-1)}) \cdot p^{(t-1)}(k)}{p_i^{(t-1)}}$

8  $\quad$ **for** $k \in \{1, 2, \cdots, \texttt{K}\}$ *(Maximization step)* **do**

9  $\quad\quad$ $p_k^{(t-1)} = \sum_{i=1}^{\texttt{D}} p_{ik}^{(t-1)}$

10  $\quad\quad$ $\boldsymbol{\mu}_k^{(t)} = \dfrac{\sum_{i=1}^{\texttt{D}} \mathbf{x}_i \cdot p_{ik}^{(t-1)}}{p_k^{(t-1)}}$

11  $\quad\quad$ $\boldsymbol{\Sigma}_k^{(t)} = \dfrac{\sum_{i=1}^{\texttt{D}} (\mathbf{x}_i - \boldsymbol{\mu}_k^{(t)})(\mathbf{x}_i - \boldsymbol{\mu}_k^{(t)})^{\mathsf{T}} \cdot p_{ik}^{(t-1)}}{p_k^{(t-1)}}$

12  $\quad\quad$ $p^{(t)}(k) = \dfrac{p_k^{(t-1)}}{\texttt{D}}$

13  $\quad$ $\ln f(\mathcal{D}_x|\mathcal{P}^{(t)}) = \sum_{i=1}^{\texttt{D}} \ln \left[ \sum_{k=1}^{\texttt{K}} \mathcal{N}(\mathbf{x}_i; \boldsymbol{\mu}_k^{(t)}, \boldsymbol{\Sigma}_k^{(t)}) \cdot p^{(t)}(k) \right]$

14  $\quad$ $t = t + 1$

---

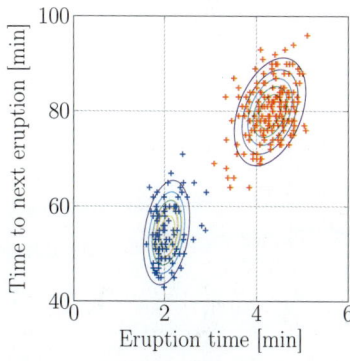

Figure 10.4: Example of an application of a Gaussian mixture model to the Old Faithful geyser data set. Contours represent the clusters identified using the EM algorithm after $t = 12$ iterations.

point belongs to the cluster associated with the highest responsibility. At this point, due to the crude initialization of $\{\boldsymbol{\mu}_k^{(0)}, \boldsymbol{\Sigma}_k^{(0)}\}$, the quality of the cluster membership assignment is poor. The maximization step for $t = 1$ updates the initial cluster centers and covariances. The updated contours for the clusters using $\boldsymbol{\mu}_k^{(1)}$ and $\boldsymbol{\Sigma}_k^{(1)}$ are depicted in figure 10.3b. Note that the responsibility for each data point $\mathbf{x}_i$ has not changed, so the cluster membership assignment remains the same. Figure 10.3c presents the expectation step for the second iteration, where we can notice the change in cluster membership assignment in comparison with the previous steps in (a) and (b). In (d), the contours of each cluster are presented for parameters $\{\boldsymbol{\mu}_k^{(2)}, \boldsymbol{\Sigma}_k^{(2)}\}$, which have been updated for the second time. Figure 10.4 presents the final clusters along with their contours obtained after $t = 12$ iterations.

Figure 10.5 presents a second example of a Gaussian mixture model for $\texttt{K} = 3$ generic clusters. Note that despite the overlap between clusters, the method performed well because, like in the

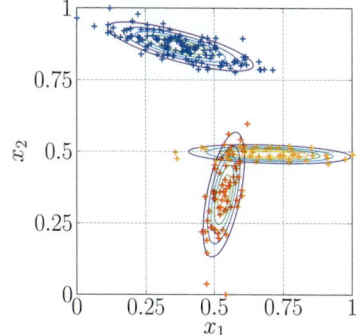

Figure 10.5: Example of an application of a Gaussian mixture model to overlapping clusters. Contours represent the clusters identified using the EM algorithm.

[3] Ng, A., M. Jordan, and Y. Weiss (2002). On spectral clustering: Analysis and an algorithm. In *Advances in neural information processing systems*, Volume 14, pp. 849–856. MIT Press.

[4] Murphy, K. P. (2012). *Machine learning: A probabilistic perspective*. MIT Press.

[5] Bishop, C. M. (2006). *Pattern recognition and machine learning*. Springer.

[6] Lloyd, S. (1982). Least squares quantization in PCM. *IEEE Transactions on Information Theory* 28(2), 129–137.

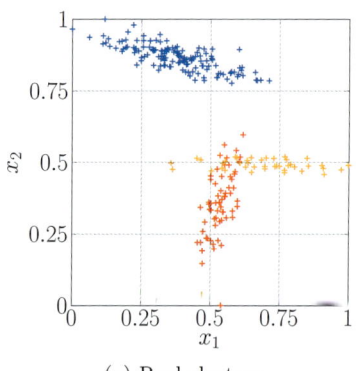

(a) Real clusters

(b) Clusters approximated with K-means, where crosses describe each cluster center.

Figure 10.6: Example of application of K-means. Notice how some samples were misclassified due to the hypothesis related to the circular shape of each cluster, that is, $\boldsymbol{\Sigma}_k = \sigma^2 \mathbf{I}$.

previous case, each cluster is well approximated by a multivariate Normal. Gaussian mixture models may display a poor performance when the clusters do not have ellipsoidal shapes. In such a case, we should resort to more advanced methods such as *spectral clustering*.[3] Further details regarding the derivation and theoretical justification of the EM methods are presented by Murphy[4] and Bishop.[5]

*Estimating the number of clusters*   In the previous examples, we have assumed that we knew the number of clusters K. In most practical applications, this is not the case and K must be estimated using methods such as cross-validation (see §8.1.2) or Bayesian model selection (see §6.8).

### 10.1.2   K-Means

*K-means*[6] is one of the most widespread clustering methods due to its simplicity. It can be seen as a special case of the Gaussian mixture model presented in §10.1.1, where the probability of each cluster is $p(k) = 1/$K, the covariance of each cluster is diagonal $\boldsymbol{\Sigma}_k = \sigma^2 \mathbf{I}$, and we seek to infer the cluster means $\boldsymbol{\mu}_k \in \mathbb{R}^{\mathtt{X}}$. With Gaussian mixture models, we were computing the probability that the covariates $\mathbf{x}_i$ belong to the cluster $k$; with K-means, the cluster membership assignment $k_i^*$ for a covariate $\mathbf{x}_i$ is deterministic so that it corresponds to

$$p(k|\mathbf{x}_i, \boldsymbol{\theta}) = \begin{cases} 1, & \text{if } k = k_i^* \\ 0, & \text{otherwise,} \end{cases}$$

where a covariate $\mathbf{x}_i$ is assigned to the cluster associated with the mean it is the closest to,

$$k_i^* = \arg \min_k ||\mathbf{x}_i - \boldsymbol{\mu}_k||.$$

The mean vector for the $k^{\text{th}}$ cluster is estimated as the average location of the covariates $\mathbf{x}_i$ associated with a cluster $k$ so that

$$\boldsymbol{\mu}_k = \frac{1}{\#\{i : k_i^* = k\}} \sum_{i:k_i^*=k} \mathbf{x}_i,$$

where $\#\{i : k_i^* = k\}$ is the number of observed covariates assigned to the cluster $k$.

The general procedure consists in initializing $\boldsymbol{\mu}_k^{(0)}$ randomly, iteratively assigning each observation to the cluster it is the closest to, and recomputing the mean of each cluster. Figure 10.6 presents in (a) the true cluster and in (b) the cluster membership assignment computed using K-means, where the mean of each cluster is presented by a cross. Note that because clusters are assumed to be spherical (i.e., $\boldsymbol{\Sigma}_k = \sigma^2 \mathbf{I}$), the cluster membership assignment is only based on the Euclidian distance between a covariate and the center of each cluster. This explains why in figure 10.6b several observed covariates are assigned to the wrong cluster. Algorithm 9 summarizes the steps for K-means clustering.

---

**Algorithm 9:** K-means

---

1  define $\mathbf{x}$, $\epsilon$, $t = 0$
2  initialize $\boldsymbol{\mu}_k^{(0)}$, and $\boldsymbol{\mu}_k^{(1)} = \boldsymbol{\mu}_k^{(0)} + \epsilon$
3  **while** $||\boldsymbol{\mu}_k^{(t)} - \boldsymbol{\mu}_k^{(t-1)}|| \geq \epsilon$ **do**
4       **for** $i \in \{1, 2, \cdots, \mathsf{D}\}$ **do**
5           $k_i^* = \arg\min_k ||\mathbf{x}_i - \boldsymbol{\mu}_k^{(t-1)}||^2$
6       **for** $k \in \{1, 2, \cdots, \mathsf{K}\}$ **do**
7           $\boldsymbol{\mu}_k^{(t)} = \frac{1}{\#\{i:k_i^*=k\}} \sum_{i:k_i^*=k} \mathbf{x}_i$
8       $t = t + 1$

---

## 10.2  Principal Component Analysis

*Principal component analysis* (PCA)[7] is a dimension reduction technique often employed in conjunction with clustering. The goal of PCA is to transform the covariate space in a new set of orthogonal axes for which the variance is maximized for the first dimension. Note that in order to perform PCA, we must work with normalized data with mean zero and with variance equal to one for each covariate.

[7] Hotelling, H. (1933). Analysis of a complex of statistical variables into principal components. *Journal of Educational Psychology* 24(6), 417.

Let us assume, our observations consist of a set of raw unnormalized covariates $\mathcal{D} = \{\mathbf{x}_i, \forall i \in \{1 : \mathsf{D}\}\}$, where $\mathbf{x}_i = [x_1 \ x_2 \ \cdots \ x_\mathsf{X}]_i^\mathsf{T}$. We normalize each value $x_{ij} = [\mathbf{x}_i]_j$ in this data set by subtracting from it the empirical average $\hat{\mu}_j = \frac{1}{\mathsf{D}} \sum_{i=1}^{\mathsf{D}} x_{ij}$ and dividing by the empirical standard deviation $\hat{\sigma}_j = \left(\frac{1}{\mathsf{D}-1} \sum_{i=1}^{\mathsf{D}} (x_{ij} - \hat{\mu}_j)^2\right)^{1/2}$ so that

$$\tilde{x}_{ij} = \frac{x_{ij} - \hat{\mu}_j}{\hat{\sigma}_j}.$$

From our set of normalized covariates $\tilde{\mathbf{x}}_i = [\tilde{x}_1 \ \tilde{x}_2 \ \cdots \ \tilde{x}_\mathsf{X}]_i^\mathsf{T}$, we seek orthogonal vectors $\boldsymbol{\nu}_i \in \mathbb{R}^\mathsf{X}$, $\forall i \in \{1 : \mathsf{X}\}$ so that our transformation matrix,

$$\mathbf{V} = [\boldsymbol{\nu}_1 \ \boldsymbol{\nu}_2 \ \cdots \ \boldsymbol{\nu}_\mathsf{X}], \tag{10.5}$$

is orthonormal, that is, $\mathbf{V}^\mathsf{T}\mathbf{V} = \mathbf{I}$. We can transform the covariates $\tilde{\mathbf{x}}_i$ into the new space using $\mathbf{z}_i = \mathbf{V}^\mathsf{T}\tilde{\mathbf{x}}_i$. The transformed data is then referred to as the *score*. Here, we want to find the orthogonal transformation matrix $\mathbf{V}$ that minimizes the reconstruction error

$$J(\mathbf{V}) = \frac{1}{D} \sum_{i=1}^{D} ||\tilde{\mathbf{x}}_i - \mathbf{V}\mathbf{z}_i||^2. \tag{10.6}$$

The optimal solution minimizing $J(\mathbf{V})$ is obtained by taking $\mathbf{V}$ as the matrix containing the eigenvectors (see §2.4.2) from the empirical covariance matrix estimated from the set of normalized covariates,

$$\hat{\mathbf{\Sigma}} = \frac{1}{D-1} \sum_{i=1}^{D} \tilde{\mathbf{x}}_i \tilde{\mathbf{x}}_i^{\mathsf{T}}.$$

With PCA, the eigenvectors $\boldsymbol{\nu}_i$ in $\mathbf{V}$ are sorted according to the decreasing order of its associated eigenvalues. PCA finds a new set of orthogonal reference axes so that the variance for the first principal component is maximized in the transformed space. This explains why we need to work with normalized data so that the variance is similar for each dimension in the original space. If it is not the case, PCA will identify a transformation that is biased by the difference in the scale or the units of covariates. Note that an alternative to performing principal component analysis using the eigen decomposition is to employ the *singular value decomposition*[8] procedure.

Figure 10.7 presents a PCA transformation applied to the Old Faithful geyser example. In (a), the data is normalized by subtracting the empirical average and dividing by the empirical standard deviation for each covariate. We then obtain the PCA transformation matrix by performing the eigen decomposition of the empirical covariance matrix estimated from normalized covariates,

$$\mathbf{V} = [\boldsymbol{\nu}_1 \ \boldsymbol{\nu}_2] = \begin{bmatrix} -0.71 & -0.71 \\ 0.71 & -0.71 \end{bmatrix}.$$

We can see in figure 10.7b that when transformed in the PCA space, the variance is maximized along the first principal component.

*Fraction of the variance explained*   In the PCA space, the eigenvalues describe the variance associated with each principal component. The fraction of the variance explained by each principal component $\mathcal{I}$ is quantified by the ratio of the eigenvalue associated with the $i^{\text{th}}$ component divided by the sum of eigenvalues,

$$\mathcal{I} = \frac{\lambda_i}{\sum_{i=1}^{X} \lambda_i}.$$

In the example presented in figure 10.7, the fraction of the variance explained by each principal component is $\mathcal{I} = \{0.9504, 0.0496\}$.

[8] Van Loan, C. F. and G. H. Golub (1983). *Matrix computations.* Johns Hopkins University Press.

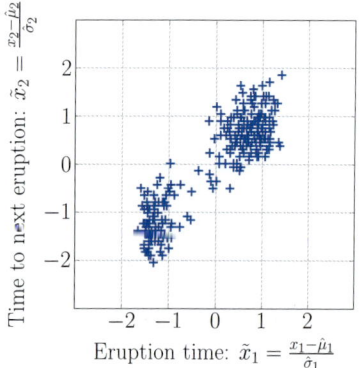

(a) Normalized and centered space

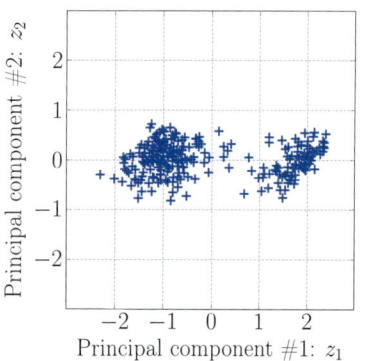

(b) Principal component space

Figure 10.7: Example of application of principal component analysis to the normalized and centered Old Faithful data set.

*Dimension reduction* PCA can be employed to reduce the number of dimensions describing the covariate space. This is achieved by removing the eigenvectors, that is, columns from the matrix $\mathbf{V}$, that are associated with a negligible fraction of the variance explained. Figure 10.8a presents the example of a data set with $\mathsf{K} = 3$ clusters and where covariates $\mathbf{x}_i \in \mathbb{R}^3$. In figure 10.8b, we see clusters identified using the K-means algorithm. By performing PCA on the normalized and centered data, we find the transformation matrix

$$\mathbf{V} = \begin{bmatrix} 0.7064 & -0.3103 & -0.6362 \\ 0.1438 & 0.9430 & -0.3002 \\ 0.6931 & 0.1206 & 0.7107 \end{bmatrix},$$

where the fraction of the variance explained by each principal component is $\mathcal{I} = \{85.5, 14.2, 0.3\}$. We can see here that the third principal component explains a negligible fraction of the variance. This indicates that the original data set defined in a 3-D space can be represented in a 2-D space with a negligible loss of information. Such a 2-D space is obtained by removing the last column from the original PCA matrix,

$$\mathbf{V}_2 = \begin{bmatrix} 0.7064 & -0.3103 \\ 0.1438 & 0.9430 \\ 0.6931 & 0.1206 \end{bmatrix},$$

and then performing the transformation $\mathbf{z}_i = \mathbf{V}_2^\mathsf{T} \tilde{\mathbf{x}}_i$. Figure 10.8c presents K-means clusters for the covariates transformed in the 2-D PCA space, $\mathbf{z}_i \in \mathbb{R}^2$.

Note that principal component analysis is limited to linear dimensional reduction. For nonlinear cases, we can opt for more advanced methods such as *t-distributed stochastic neighbor embedding* (t-SNE).[9] In addition to being employed in unsupervised learning, PCA and other dimensionality reduction techniques can be employed in the context of regression and classification (see chapters 8 and 9). In such a case, instead of modeling the relationship between the observed system responses and the covariates, the relationship is modeled between the observed system responses and the transformed covariates represented in a lower-dimensional space where inter-covariate dependence is removed using PCA or any other dimension-reduction technique.

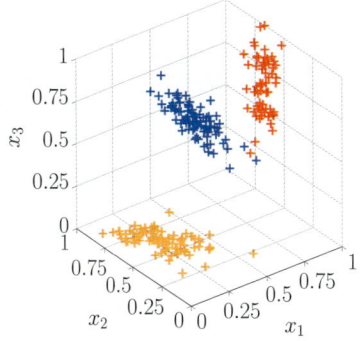

(a) Real clusters in a 3-D space

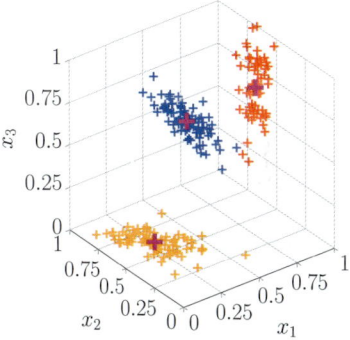

(b) K-means clusters in a 3-D space

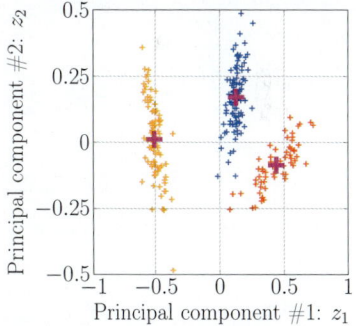

(c) K-means clusters in the 2-D principal component space

Figure 10.8: Example of application of the principal component analysis for dimensionality reduction.

[9] Van der Maaten, L. and G. Hinton (2008). Visualizing data using t-SNE. *Journal of Machine Learning Research 9*, 2579–2605.

*Exercises*

P10.1   Explain what makes clustering different from classification.

P10.2   Explain why the expectation maximization algorithm is needed for estimating the parameters of a Gaussian mixture model.

P10.3   Explain the relations between K-means and Gaussian mixture models.

P10.4   Implement the K-mean algorithm and apply it to the Old Faithful geyser data set. Compare your result with figure 10.1b.

P10.5   Implement a Gaussian mixture model with the EM algorithm and apply it to the Old Faithful geyser data set. Compare your result with figure 10.1b.

P10.6   Use principal component analysis to reproduce the results presented in figure 10.7b.

# 11

# *Bayesian Networks*

Bayesian networks were introduced by Pearl[1] and are also known as *belief networks* and *directed graphical models*. They are the result of the combination of *probability theory* covered in chapter 3 with *graph theory*, which employs graphs defined by links and nodes.

We saw in §3.2 that the *chain rule* allows formulating the joint probability for a set of random variables using conditional and marginal probabilities, for example, $p(x_1, x_2, x_3) = p(x_3|x_2, x_1)p(x_2|x_1)p(x_1)$. Bayesian networks (BNs) employ *nodes* to represent random variables and *directed links* to describe the dependencies between them. Bayesian networks are probabilistic models where the goal is to learn the joint probability defined over the entire network. The joint probability for the structure encoded by these nodes and links is formulated using the *chain rule*. The key with Bayesian networks is that they allow building sparse models for which efficient variable elimination algorithms exist in order to estimate any conditional probabilities from the joint probability.

BNs can be categorized as *unsupervised learning*[2] where the goal is to estimate the joint probability density function (PDF) for a set of observed variables. In its most general form we may seek to learn the structure of the BN itself. In this chapter, we restrict ourselves to the case where we know the graph structure, and the goal is to learn to predict unobserved quantities given some observed ones.

*Flu virus example*  Section 3.3.5 presented the distinction between *correlation* and *causality* using the flu virus example. A Bayesian network can be employed to model the dependence between the *temperature*, the presence of the flu *virus*, and being sick from the *flu*. We model our knowledge of these three quantities using discrete random variables that are represented by the nodes in figure 11.1. The arrows represent the dependencies between variables: The temperature $T$ affects the virus prevalence $V$, which in turns affects

[1] Pearl, J. (1988). *Probabilistic reasoning in intelligent systems: Networks of plausible inference.* Morgan Kaufmann.

[2] Ghahramani, Z. (2004). Unsupervised learning. In *Advanced lectures on machine learning*, Volume 3176, pp. 72–112. Springer.

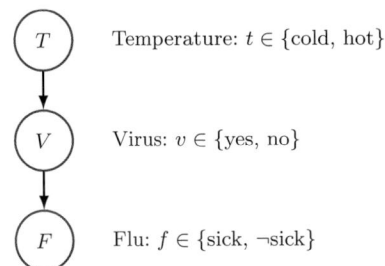

Temperature: $t \in \{\text{cold, hot}\}$

Virus: $v \in \{\text{yes, no}\}$

Flu: $f \in \{\text{sick}, \neg\text{sick}\}$

Figure 11.1: Example of Bayesian network for representing the relationships between temperature, the presence of the flu virus, and being sick from the flu virus.

**Virus example: Marginal and conditional probability tables**

$$p(t) = \{p(\text{cold}), p(\text{hot})\} = \{0.4, 0.6\}$$

$$p(v|t) = \left\{ \begin{array}{c|cc} & t = \text{cold} & t = \text{hot} \\ \hline v = \text{yes} & 0.8 & 0.1 \\ v = \text{no} & 0.2 & 0.9 \end{array} \right.$$

$$p(f|v) = \left\{ \begin{array}{c|cc} & v = \text{yes} & v = \text{no} \\ \hline f = \text{sick} & 0.7 & 0 \\ f = \neg\text{sick} & 0.3 & 1 \end{array} \right.$$

**Joint probability using chain rule**

$$p(v, t) = p(v|t) \cdot p(t)$$

$$= \left\{ \begin{array}{c|cc} & t = \text{cold} & t = \text{hot} \\ \hline v = \text{yes} & 0.8 \times 0.4 & 0.1 \times 0.6 \\ v = \text{no} & 0.2 \times 0.4 & 0.9 \times 0.6 \end{array} \right.$$

$$= \left\{ \begin{array}{c|cc} & t = \text{cold} & t = \text{hot} \\ \hline v = \text{yes} & 0.32 & 0.06 \\ v = \text{no} & 0.08 & 0.54 \end{array} \right.$$

$$p(f, v, t) = p(f|v) \cdot p(v, t)$$

$$= \left\{ \begin{array}{llll} f = \text{sick} & t = \text{cold} & t = \text{hot} \\ \hline v = \text{yes} & 0.32 \times 0.7 & 0.06 \times 0.7 \\ v = \text{no} & 0.08 \times 0 & 0.54 \times 0 \\ \\ f = \neg\text{sick} & t = \text{cold} & t = \text{hot} \\ \hline v = \text{yes} & 0.32 \times 0.3 & 0.06 \times 0.3 \\ v = \text{no} & 0.08 \times 1 & 0.54 \times 1 \end{array} \right.$$

$$= \left\{ \begin{array}{llll} f = \text{sick} & t = \text{cold} & t = \text{hot} \\ \hline v = \text{yes} & 0.224 & 0.042 \\ v = \text{no} & 0 & 0 \\ \\ f = \neg\text{sick} & t = \text{cold} & t = \text{hot} \\ \hline v = \text{yes} & 0.096 & 0.018 \\ v = \text{no} & 0.08 & 0.54 \end{array} \right.$$

**Variable elimination: Marginalization**

$$p(f, t) = \sum_v p(f, v, t)$$

$$= \left\{ \begin{array}{c|cc} & t = \text{cold} & t = \text{hot} \\ \hline f = \text{sick} & 0.224 & 0.042 \\ f = \neg\text{sick} & 0.176 & 0.558 \end{array} \right.$$

$$p(f) = \sum_t p(f, t)$$

$$= \left\{ \begin{array}{c|c} f = \text{sick} & 0.266 \\ f = \neg\text{sick} & 0.734 \end{array} \right.$$

$$p(t|f) = p(f, t)/p(f)$$

$$= \left\{ \begin{array}{c|cc} & t = \text{cold} & t = \text{hot} \\ \hline f = \text{sick} & 0.84 & 0.16 \\ f = \neg\text{sick} & 0.24 & 0.76 \end{array} \right.$$

[3] Nielsen, T. D. and F. V. Jensen (2007). *Bayesian networks and decision graphs.* Springer.

[4] Murphy, K. P. (2002). *Dynamic Bayesian networks: representation, inference and learning.* PhD thesis, University of California, Berkeley.

the probability of catching the virus and being sick $F$. The absence of a link between $T$ and $F$ indicates that the temperature and being sick from the flu are *conditionally independent* from each other. In the context of this example, conditional independence implies that $T$ and $F$ are independent when $V$ is known. The joint probability for $T$, $V$, and $F$,

$$p(f, v, t) = \overbrace{p(f|v)}^{=p(f|v,t)} \cdot \underbrace{p(v|t) \cdot p(t)}_{p(v,t)},$$

is obtained using the chain rule where, for each arrow, the conditional probabilities are described in a *conditional probability table*.

In the case where we observe $F = f$, we can employ the *marginalization* operation in order to obtain a conditional probability quantifying how the observation $f \in \{\text{sick}, \neg\text{sick}\}$ changes the probability for the temperature $T$,

$$\begin{aligned} p(t|f) &= \frac{p(f, t)}{p(f)} \\ &= \frac{\sum_v p(f, v, t)}{\sum_t \sum_v p(f, v, t)}. \end{aligned}$$

In minimalistic problems such as this one, it is trivial to calculate the joint probability using the chain rule and eliminating variables using marginalization. However, in practical cases involving dozens of variables with as many links between them, these calculations become computationally demanding. Moreover, in practice, we seldom know the marginal and conditional probabilities tables. A key interest of working with directed graphs is that efficient estimation methods are available to perform all those tasks.

Bayesian networks are applicable not only for discrete random variables but also for continuous ones, or a mix of both. In this chapter, we restrict ourselves to the study of BN for discrete random variables. Note that the state-space models presented in chapter 12 can be seen as a time-dependent Bayesian network using Gaussian random variables with linear dependence models. This chapter presents the nomenclature employed to define graphical models, the methods for performing inference, and the methods allowing us to learn the conditional probabilities defining the dependencies between random variables. In addition, we present an introduction to time-dependent Bayesian networks that are referred to as *dynamic Bayesian networks*. For advanced topics regarding Bayesian networks, the reader is invited to consult specialized textbooks such as the one by Nielsen and Jensen[3] or Murphy's PhD thesis.[4]

## 11.1 Graphical Models Nomenclature

Bayesian networks employ a special type of graph: *directed acyclic graph* (DAG). A DAG $\mathcal{G} = \{\mathcal{U}, \mathcal{E}\}$ is defined by a set of nodes $\mathcal{U}$ interconnected by a set of directed links $\mathcal{E}$. In order to be acyclic, the directed links between variables cannot be defined such that there are self-loops or cycles in the graph. For a set of random variables, there are many ways to define links between variables, each one leading to the same joint probability. Note that directed links between variables *are not required to describe causal relationships*. Despite causality not being a requirement, it is a key to efficiency; if the directed links in a graphical model are assigned following the causal relationships, it generally produces sparse models requiring the definition of a smaller number of conditional probabilities than noncausal counterparts.

Figure 11.2a presents a directed acyclic graph and (b) presents a directed graph containing a cycle so that it cannot be modeled as a Bayesian network. In figure 11.2a random variables are represented by nodes, where the observed variable $X_4$ depends on the hidden variables $X_2$ and $X_3$. The directions of links indicate that $X_2$ and $X_3$ are the *parent* of $X_4$, that is, parents$(X_4) = \{X_2, X_3\}$, and consequently, $X_4$ is the *child* of $X_2$ and $X_3$. Each child is associated with a *conditional probability table* (CPT) whose size depends on the number of parents, $p(x_i|\text{parents}(X_i))$. Nodes without parents are described by their marginal prior probabilities $p(x_i)$. The joint PDF $p(\mathcal{U})$ for the entire Bayesian network is formulated using the chain rule,

$$p(\mathcal{U}) = p(x_1, x_2, \cdots, x_{\mathtt{X}}) = \prod_{i=1}^{\mathtt{X}} p(x_i|\text{parents}(X_i)).$$

This application of the chain rule requires that given its parents, each node is independent of its other ancestors.

*Cyanobacteria example*   We explore the example illustrated in figure 11.3 of cyanobacteria blooms that can occur in lakes, rivers, and estuaries. Cyanobacteria blooms are typically caused by the use of fertilizers that wash into a water body, combined with warm temperatures that allow for bacteria reproduction. Cyanobacteria blooms can cause a change in the water color and can cause fish or marine life mortality. We employ a Bayesian network to describe the joint probability for a set of random variables consisting of the temperature $T$, the use of fertilizer $F$, the presence of cyanobacteria in water $C$, fish mortality $M$, and the water color $W$. In figure 11.4,

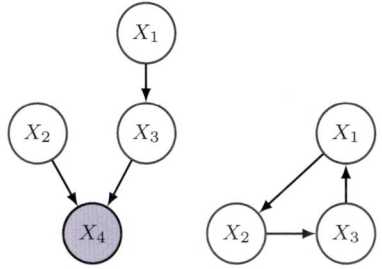

(a) Directed acyclic graph (DAG)   (b) Directed graph containing a cycle

Figure 11.2: Example of dependence represented by directed links between hidden (white) and an observed (shaded) nodes describing random variables.

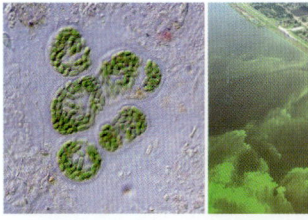

(a) Cyanobacteria seen under a microscope   (b) Cyanobacteria bloom in a rural environment

Figure 11.3: Example of cyanobacteria bloom. (Photo: NASA and USGS)

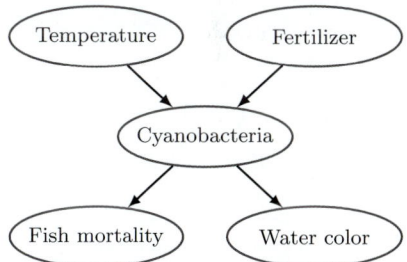

(a) Bayesian network semantic representation

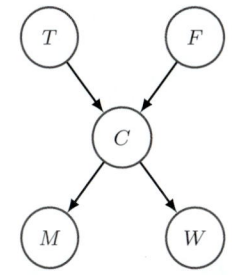

(b) Bayesian network represented with random variables

Figure 11.4: Bayesian network for the cyanobacteria example.

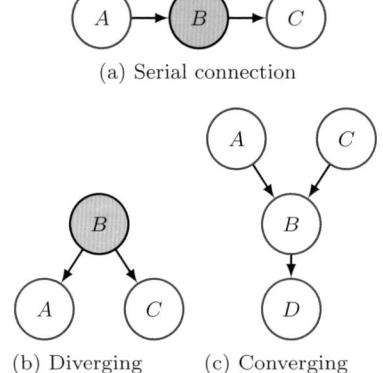

(a) Serial connection

(b) Diverging connection

(c) Converging connection

Figure 11.5: Cases where $A$ and $C$ are conditionally independent for the different types of connection in Bayesian networks.

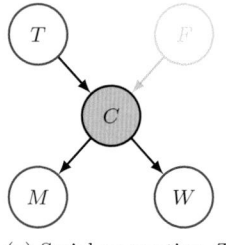

(a) Serial connection: $T \perp\!\!\!\perp M|c$ and $T \perp\!\!\!\perp W|c$

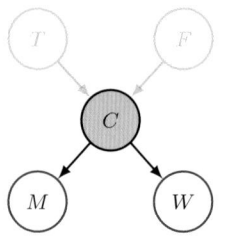

(b) Diverging connection: $M \perp\!\!\!\perp W|c$

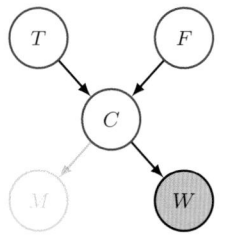

(c) Converging connection: $T \not\!\perp\!\!\!\perp F$

Figure 11.6: The concept of conditional independence illustrated using the cyanobacteria example.

the definition of the graph structure and its links follow the causal direction where the presence of cyanobacteria in water $C$ depends on the temperature $T$ and the presence of fertilizer $F$. Both the fish mortality $M$ and water color $W$ depend on the presence of cyanobacteria in water $C$.

## 11.2    Conditional Independence

As stated earlier, the absence of a link between two variables indicates that the pair is *conditionally independent*. In the case of a *serial connection*, as illustrated in figure 11.5a, the absence of a link between $A$ and $C$ indicates that these two are independent if $B$ is known, that is, $A \perp\!\!\!\perp C|\{B = b\}$. Therefore, as long as $B$ is *not* observed, $A$ and $C$ depend on each other through $B$. It is equivalent to say that given its parent (i.e., $B$), $C$ is independent of all its other non-descendants. In the case of a *diverging connection* (figure 11.5b), again the absence of link between $A$ and $C$ indicates that these two are independent if $B$ is observed. The case of a *converging connection* represented in figure 11.5c is different from the two others; the absence of link between $A$ and $C$ implies that both variables are independent unless $B$, or one of its descendants, is observed. When $B$, or one of its descendants is observed, the knowledge gained for $B$ also modifies the knowledge for $A$ and $C$. We say that $A$ and $C$ are *d-separated* in the case of a serial or diverging connection, where the intermediary variable $B$ is observed, or in the case of a converging connection, where neither the intermediary variable $B$ nor one of its descendants is observed. For any $d$-separated variables, a change in the knowledge for one variable does not affect the knowledge of the others.

*Cyanobacteria example*   In the example presented in figure 11.4b, the sets of variables $\{T, C, M\}$, $\{T, C, W\}$, $\{F, C, M\}$, and $\{F, C, W\}$ are all examples of serial connections. If as illustrated in figure 11.6a, we observe the presence of cyanobacteria, that is, $C = \text{yes}$, then the temperature $T$ becomes independent of fish mortality $M$ and water color $W$, that is, $T \perp\!\!\!\perp M|c$, $T \perp\!\!\!\perp W|c$. It implies that gaining knowledge about water temperature would not change our knowledge about either fish mortality or water color because these two quantities only depend on the presence of cyanobacteria, which is now a certainty given that $C$ was observed. For the diverging connection between variables $\{M, C, W\}$ illustrated in figure 11.6b, observing $C = c$ causes the fish mortality $M$ to be independent from the water color $W$, $M \perp\!\!\!\perp W|c$. Gaining knowledge about the occurrence of fish mortality would not change our knowledge about water color because these two quantities only depend on the presence of cyanobacteria, which is a certainty when $C$ is observed. For the converging connection between variables $\{T, C, F\}$ represented in figure 11.6c, despite the absence of a link between $T$ and $F$, the temperature is *not* independent from the use of fertilizer $F$ if we observe $C$ or one of its descendants, that is, $M$ or $W$. Without observing $C$, $M$, or $W$, the knowledge gained about the use of fertilizer has no impact on our knowledge of the temperature. On the other hand, if we observe the presence of cyanobacteria ($C = $ yes), then knowing that no fertilizer is present in the environment ($F = $ no) would increase the probability that the temperature is high because there is only a small probability of having cyanobacteria without fertilizer and with cold temperature.

## 11.3   Inference

The finality of Bayesian networks is not only to define the joint probability for random variables; it is to perform *inference*, that is, to compute the conditional probability for a set of unobserved random variables, given a set of observed ones. Let us consider $\mathcal{U} = \{X_1, X_2, \cdots, X_{\mathtt{X}}\}$, the set of random variables corresponding to the nodes defining a system, a subset of *observed variables* $\mathcal{D} \subset \mathcal{U}$, and another subset of query variables $\mathcal{Q} \subset \mathcal{U}$, such that $\mathcal{Q} \cap \mathcal{D} = \emptyset$. Following the definition of conditional probabilities presented in §3.2, the posterior probability for the variables in $\mathcal{Q}$ given observations $\mathcal{D}$ is

**Cyanobacteria example**

$$\mathcal{U} = \left\{ \begin{array}{ll} T: & t \in \{\text{cold, hot}\} \\ F: & f \in \{\text{yes, no}\} \\ C: & c \in \{\text{yes, no}\} \\ M: & m \in \{\text{yes, no}\} \\ W: & w \in \{\text{clear, green}\} \end{array} \right\}$$

$$
\begin{aligned}
p(\mathcal{Q}|\mathcal{D}) &= \frac{p(\mathcal{Q}, \mathcal{D})}{p(\mathcal{D})} \\
&= \frac{\sum_{\mathcal{U} \setminus \{\mathcal{Q}, \mathcal{D}\}} p(\mathcal{U})}{\sum_{\mathcal{U} \setminus \{\mathcal{D}\}} p(\mathcal{U})},
\end{aligned}
$$

where $\mathcal{U} \setminus \{\mathcal{Q}, \mathcal{D}\}$ describes the set of variables belonging to $\mathcal{U}$ while excluding those in $\{\mathcal{Q}, \mathcal{D}\}$. For the cyanobacteria example presented in figure 11.4b, let us consider we want the posterior probability $p(m|w = \text{green})$, that is, the probability of fish mortality $M$ given that we have observed colored water, $W = $ green. This joint probability is described by

$$p(m|w = \text{green}) = \frac{p(m, w = \text{green})}{p(w = \text{green})},$$

**Cyanobacteria example: CPT**

$$p(t) = \{p(\text{cold}), p(\text{hot})\} = \{0.4, 0.6\}$$

$$p(f) = \{p(\text{yes}), p(\text{no})\} = \{0.2, 0.8\}$$

$$p(c|t,f) = \begin{cases} \begin{array}{c|cc} c = \text{yes} & t = \text{cold} & t = \text{hot} \\ \hline f = \text{yes} & 0.5 & 0.95 \\ f = \text{no} & 0.05 & 0.8 \end{array} \\ \begin{array}{c|cc} c = \text{no} & t = \text{cold} & t = \text{hot} \\ \hline f = \text{yes} & 0.5 & 0.05 \\ f = \text{no} & 0.95 & 0.2 \end{array} \end{cases}$$

$$p(m|c) = \begin{cases} \begin{array}{c|cc} & c = \text{yes} & c = \text{no} \\ \hline m = \text{yes} & 0.6 & 0.1 \\ m = \text{no} & 0.4 & 0.9 \end{array} \end{cases}$$

$$p(w|c) = \begin{cases} \begin{array}{c|cc} & c = \text{yes} & c = \text{no} \\ \hline w = \text{clear} & 0.7 & 0.2 \\ w = \text{green} & 0.3 & 0.8 \end{array} \end{cases}$$

**Variable elimination**

$$p(c,f|t) = \begin{cases} \begin{array}{c|cc} c = \text{yes} & t = \text{cold} & t = \text{hot} \\ \hline f = \text{yes} & 0.5 \cdot 0.2 & 0.95 \cdot 0.2 \\ f = \text{no} & 0.05 \cdot 0.8 & 0.8 \cdot 0.8 \end{array} \\ \begin{array}{c|cc} c = \text{no} & t = \text{cold} & t = \text{hot} \\ \hline f = \text{yes} & 0.5 \cdot 0.2 & 0.05 \cdot 0.2 \\ f = \text{no} & 0.95 \cdot 0.8 & 0.2 \cdot 0.8 \end{array} \end{cases}$$

$$p(c|t) = \begin{cases} \begin{array}{c|cc} & t = \text{cold} & t = \text{hot} \\ \hline c = \text{yes} & 0.14 & 0.83 \\ c = \text{no} & 0.86 & 0.17 \end{array} \end{cases}$$

$$p(c,t) = \begin{cases} \begin{array}{c|cc} & t = \text{cold} & t = \text{hot} \\ \hline c = \text{yes} & 0.14 \cdot 0.4 & 0.83 \cdot 0.6 \\ c = \text{no} & 0.86 \cdot 0.4 & 0.17 \cdot 0.6 \end{array} \end{cases}$$

$$p(c) = \begin{cases} c = \text{yes} & 0.554 \\ c = \text{no} & 0.446 \end{cases}$$

$$p(m,c) = \begin{cases} \begin{array}{c|cc} & c = \text{yes} & c = \text{no} \\ \hline m = \text{yes} & 0.6 \cdot 0.554 & 0.1 \cdot 0.446 \\ m = \text{no} & 0.4 \cdot 0.554 & 0.9 \cdot 0.446 \end{array} \end{cases}$$

$$p(m, w = \text{green}, c) =$$

$$\begin{cases} \begin{array}{c|cc} & c = \text{yes} & c = \text{no} \\ \hline m = \text{yes} & 0.3324 \cdot 0.3 & 0.0446 \cdot 0.8 \\ m = \text{no} & 0.2216 \cdot 0.3 & 0.4014 \cdot 0.8 \end{array} \end{cases}$$

$$p(m, w = \text{green}) =$$

$$\begin{cases} \begin{array}{c|c} & w = \text{green} \\ \hline m = \text{yes} & 0.1354 \\ m = \text{no} & 0.3876 \end{array} \end{cases}$$

$$p(w = \text{green}) = 0.1354 + 0.3876 = 0.523$$

**Inference**

$$p(m|w = \text{green}) = \frac{p(m, w = \text{green})}{p(w = \text{green})}$$

$$= \begin{cases} \begin{array}{c|c} & w = \text{green} \\ \hline m = \text{yes} & 0.1354/0.523 \\ m = \text{no} & 0.3876/0.523 \end{array} \end{cases}$$

$$= \begin{cases} \begin{array}{c|c} & w = \text{green} \\ \hline m = \text{yes} & 0.26 \\ m = \text{no} & 0.74 \end{array} \end{cases}$$

where both terms on the right-hand side can be obtained through the marginalization of the joint probability mass function (PMF),

$$p(m, w = \text{green}) = \sum_t \sum_f \sum_c p(t, f, c, m, w = \text{green})$$

$$p(w = \text{green}) = \sum_m p(m, w = \text{green}).$$

This approach is theoretically correct; however, in practice, calculating the joint probability $p(\mathcal{U})$ quickly becomes computationally intractable with the increase in the number of variables in $\mathcal{U}$. The solution is to avoid computing the full joint probability table and instead proceed by eliminating variables sequentially.

*Variable elimination*    The goal of *variable elimination* is to avoid computing the full joint probability table $p(\mathcal{U})$ by working with its chain-rule factorization. For the cyanobacteria example, the joint probability is

$$p(\mathcal{U}) = p(t) \cdot p(f) \cdot p(c|t,f) \cdot p(w|c) \cdot p(m|c), \qquad (11.1)$$

where computing $p(m, w = \text{green})$ corresponds to

$$p(m, w = \text{green})$$

$$= \sum_t \sum_f \sum_c p(t) \cdot p(f) \cdot p(c|t,f) \cdot p(w = \text{green}|c) \cdot p(m|c)$$

$$= \sum_c \underbrace{p(w = \text{green}|c)}_{2 \times 1} \cdot \underbrace{p(m|c)}_{2 \times 2} \cdot \sum_t \underbrace{p(t)}_{2 \times 2} \cdot \sum_f \underbrace{p(f)}_{2 \times 2} \cdot \underbrace{p(c|t,f)}_{2 \times 2 \times 2}.$$

$$\underbrace{\qquad}_{p(c,f|t): \, 2 \times 2 \times 2}$$
$$\underbrace{\qquad}_{p(c|t): \, 2 \times 2}$$
$$\underbrace{\qquad}_{p(c,t): \, 2 \times 2}$$
$$\underbrace{\qquad}_{p(c): \, 2 \times 1}$$
$$\underbrace{\qquad}_{p(m,c): \, 2 \times 2}$$
$$\underbrace{\qquad}_{p(m,w=\text{green},c): \, 2 \times 2}$$
$$\underbrace{\qquad}_{p(m,w=\text{green}): \, 2 \times 1}$$

$$= \{0.1354, 0.3876\}.$$

Note how several terms in equation 11.1 are independent from the variables in the summation operators. It allows us to take out of the sums the terms that do not depend on them and then perform the marginalization sequentially. This procedure is more efficient than working with full probability tables. In the previous equation, braces refer to the ordering of operations and indicate the size of each probability table. The variable elimination procedure allows working with probability tables containing no more than $2^3 = 8$

entries. In comparison, the full joint probability table for $p(\mathcal{U})$ contains $2^5 = 32$ entries.

The efficiency of the variable elimination approach depends on the ordering of operations. The common method for ordering operations while performing variable elimination is the *junction tree* method. This method transforms the graph into a tree and then employs clustering to group nodes. The reader interested in the details of this inference method should refer to specialized textbooks. Note that the methods covered above are *exact inference methods*. In the case where the number of variables is so large that exact methods become computationally prohibitive, we can also resort to approximate methods based on Monte Carlo sampling.

## 11.4   Conditional Probability Estimation

In the previous sections, it was always assumed that the probabilities contained in conditional probability tables (CPTs) were known. In practice this is seldom the case; CPTs must be learned from data. There are two typical learning setups; in the first setup, the Bayesian network is *fully observed* so each observation $\mathcal{D}_i = \{\mathbf{x}_i\}$ consists in a joint realization $\mathbf{x}_i : \mathbf{X} \sim p(\mathcal{U})$. Figure 11.7a presents a table where each line is one observation for the fully observed BN employed in the cyanobacteria example. In the second learning setup, the BN is only *partially observed* so that some variables are not observed for the realization of $\mathbf{x}_i : \mathbf{X} \sim p(\mathcal{U})$. Figure 11.7b presents a data set for the cyanobacteria example where the BN is partially observed.

### 11.4.1   Fully Observed Bayesian Network

For a fully observed Bayesian network, the maximum likelihood estimate (MLE) for the conditional probability $\Pr(X_1 = x_1 | X_2 = x_2)$ can be estimated as the ratio between the number of realizations of a specific joint outcome $\{X_1 = x_1, X_2 = x_2\}$, divided by the total number of realizations of the outcome $\{X_2 = x_2\}$,

$$\hat{\Pr}(X_1 = x_1 | X_2 = x_2) = \hat{p}(x_1 | x_2) = \frac{\#\{X_1 = x_1, X_2 = x_2\}}{\#\{X_2 = x_2\}}. \quad (11.2)$$

We must be careful in the case where a specific outcome is not observed in the data set so $\#\{X_1 = x_1, X_2 = x_2\} = 0$. In such a case, the MLE will lead to a conditional probability equal to zero. This situation is problematic because we might not have observed this specific outcome yet, simply because the number of observations is too limited. A solution is to employ a maximum

| $\mathcal{U} = \mathbf{x}_i$ | T | F | C | M | W |
|---|---|---|---|---|---|
| 1 | cold | no | yes | yes | clear |
| 2 | hot | no | yes | yes | clear |
| 3 | hot | yes | no | yes | green |
| ⋮ | | | | | |
| D | hot | no | no | yes | green |

(a) Fully observed Bayesian network

| $\mathcal{U} = \mathbf{x}_i$ | T | F | C | M | W |
|---|---|---|---|---|---|
| 1 | cold | ? | yes | ? | clear |
| 2 | ? | ? | yes | yes | clear |
| 3 | ? | yes | no | yes | ? |
| ⋮ | | | | | |
| D | hot | no | no | ? | green |

(b) Partially observed Bayesian network

Figure 11.7: Example of data set for learning conditional probability tables.

**Reminder:** $\#\{\cdot\}$ denotes the number of elements in a set.

a-posteriori (MAP) estimation by adding one observation count to each possible outcome. Given that $x_1 \in \{1, 2, \cdots n\}$, equation 11.2 becomes

$$\hat{\Pr}(X_1 = x_1 | X_2 = x_2) = \frac{\#\{X_1 = x_1, X_2 = x_2\} + 1}{\#\{X_2 = x_2\} + n}. \qquad (11.3)$$

*Cyanobacteria example*   From the data set $\mathcal{D}$ presented in figure 11.7a, we can estimate each CPT involved in the definition of $p(\mathcal{U})$ in equation 11.1. For example, we might want to obtain the MLE estimate for the probability of having cyanobacteria given a hot temperature and the presence of fertilizer, $p(c = \text{yes} | t = \text{hot}, f = \text{yes})$. In such a case, we compute the number of realizations of $\{c = \text{yes}, t = \text{hot}, f = \text{yes}\}$ in $\mathcal{D}$, divided by the number of realizations of $\{t = \text{hot}, f = \text{yes}\}$. If we want to employ the MAP estimate instead of the MLE, we add one count to the numerator and two to the denominator so that

$$\hat{p}(c = \text{yes} | t = \text{hot}, f = \text{yes}) = \frac{\#\{c = \text{yes}, t = \text{hot}, f = \text{yes}\} + 1}{\#\{t = \text{hot}, f = \text{yes}\} + 2}. $$
$$(11.4)$$

The same method applies for any entry in a conditional probability table. In the case of the estimation of marginal probabilities such as $p(t = \text{hot})$, the calculation simplifies to the number of events where $\{t = \text{hot}\}$ divided by $\mathsf{D} + 2$, the number of joint observations in the data set plus the number of possible outcomes for $t$,

$$\hat{p}(t = \text{hot}) = \frac{\#\{t = \text{hot}\} + 1}{\mathsf{D} + 2}. $$

Before going further, we will explore the theoretical justification for the MLE and MAP in equations 11.2 and 11.3. Let us consider the simplified case where a network is made from $\mathsf{X}$ random variables $\mathcal{U} = \{\mathbf{X}\} = \{[X_1 \; X_2 \; \cdots \; X_{\mathsf{X}}]^{\mathsf{T}}\}$, where $x_k \in \{0, 1\}, \forall k \in \{1 : \mathsf{X}\}$. Here, the definition of a specific structure for the network is not necessary; we simply assume that the network is structured as a DAG, so that for each $X_k$, we have $p(x_k | \text{parents}(X_k))$. The quantity we seek to estimate from the data is the conditional probability $\theta_k = p(x_k = 1 | \text{parents}(X_k))$, where because $x_k$ is a binary variable, we also have $1 - \theta_k = p(x_k = 0 | \text{parents}(X_k))$. We describe the prior probability for $\theta_k$ by a Beta PDF (see §4.3),

$$\begin{aligned} f(\theta_k) &= \mathcal{B}(\theta_k; \alpha_0 + 1, \beta_0 + 1) \\ &= \frac{\theta_k^{\alpha_0}(1 - \theta_k)^{\beta_0}}{\mathrm{B}(\alpha_0 + 1, \beta_0 + 1)} \\ &\propto \theta_k^{\alpha_0}(1 - \theta_k)^{\beta_0}. \end{aligned}$$

**Note:** We add $+1$ to the parameters of the Beta prior because without them,

$$\mathcal{B}(\theta_k; \alpha_0, \beta_0) \propto \theta_k^{\alpha_0 - 1}(1 - \theta_k)^{\beta_0 - 1},$$

which would lead to an MAP equal to

$$\hat{\theta}_k = \frac{\alpha_0 + \alpha - 1}{\alpha_0 + \beta_0 + \alpha + \beta - 2}.$$

We have a data set

$$\begin{aligned} \mathcal{D} &= \{\mathcal{D}_1, \mathcal{D}_2, \cdots, \mathcal{D}_\mathsf{D}\} \\ &= \{\mathcal{U} = \mathbf{x}_1, \mathcal{U} = \mathbf{x}_2, \cdots, \mathcal{U} = \mathbf{x}_\mathsf{D}\} \end{aligned}$$

containing $\mathsf{D}$ realizations from the fully observed Bayesian network $\mathcal{U}$. Here, we are interested in the probability of the joint realization of $X_k = 1$, along with a specific combination of its parents, that is, parents$(X_k) \equiv \mathbf{x}^{\mathrm{pa}(k)}$. The number of such outcomes in the data set $\mathcal{D}$ is $\alpha = \#\{x_k = 1, \mathbf{x}^{\mathrm{pa}(k)}\} \in \{0, 1, \cdots, \mathsf{D}\}$. Analogously, the number of realizations in the data set of $X_k = 0$ along with the same specific combination of its parents is $\beta = \#\{x_k = 0, \mathbf{x}^{\mathrm{pa}(k)}\} \in \{0, 1, \cdots, \mathsf{D}\}$, and note that $\alpha + \beta = \#\{\mathbf{x}^{\mathrm{pa}(k)}\}$. The posterior PDF $f(\theta_k | \mathcal{D})$ is formulated following

$$\begin{aligned} f(\theta_k | \mathcal{D}) &\propto f(\mathcal{D} | \theta_k) \cdot f(\theta_k) \\ &\propto \left( \prod_{j=1}^{\mathsf{D}} p(\mathbf{x}_j | \theta_k) \right) \cdot \theta_k^{\alpha_0} (1 - \theta_k)^{\beta_0} \\ &\propto \left( \prod_{j=1}^{\mathsf{D}} \prod_{i=1}^{\mathsf{X}} p_{X_i} \left( [\mathbf{x}_j]_i | \mathbf{x}_j^{\mathrm{pa}(i)}, \theta_k \right) \right) \cdot \theta_k^{\alpha_0} (1 - \theta_k)^{\beta_0}. \end{aligned}$$

We saw in §6.3 that it is the same value $\hat{\theta}_k$ that maximizes either $f(\theta_k | \mathcal{D})$ or $\ln f(\theta_k | \mathcal{D})$. Therefore, the log-posterior is formulated following

$$\ln f(\theta_k | \mathcal{D}) \propto \ln \left( \theta_k^{\alpha_0} (1 - \theta_k)^{\beta_0} \right) + \sum_{j=1}^{\mathsf{D}} \sum_{i=1}^{\mathsf{X}} \ln p_{X_i} \left( [\mathbf{x}_j]_i | \mathbf{x}_j^{\mathrm{pa}(i)}, \theta_k \right),$$

where products were replaced by sums. We saw in §5.1 that the MAP estimator $\hat{\theta}_k$ corresponds to the location where the derivative of the log-posterior equals zero. The derivative of the log-posterior is given by

$$\begin{aligned} \frac{\partial \ln f(\theta_k | \mathcal{D})}{\partial \theta_k} &= \frac{\partial \ln(\theta_k^{\alpha_0}(1 - \theta_k)^{\beta_0})}{\partial \theta_k} + \frac{\partial \sum_{j=1}^{\mathsf{D}} \ln p_{X_k}\left([\mathbf{x}_j]_k | \mathbf{x}_j^{\mathrm{pa}(k)}, \theta_k\right)}{\partial \theta_k} \\ &= \frac{\partial \ln(\theta_k^{\alpha_0}(1 - \theta_k)^{\beta_0})}{\partial \theta_k} + \frac{\partial \ln(\theta_k^{\alpha}(1 - \theta_k)^{\beta})}{\partial \theta_k} \\ &= \frac{\partial \ln(\theta_k^{\alpha_0}(1 - \theta_k)^{\beta_0})}{\partial \theta_k} + \frac{\partial \left(\alpha \ln \theta_k + \beta \ln(1 - \theta_k)\right)}{\partial \theta_k} \\ &= \frac{(\alpha_0 + \beta_0)\theta_k - \alpha_0}{\theta_k(\theta_k - 1)} + \frac{(\alpha + \beta)\theta_k - \alpha}{\theta_k(\theta_k - 1)} \\ &= \frac{(\alpha_0 + \beta_0 + \alpha + \beta)\theta_k - (\alpha_0 + \alpha)}{\theta_k(\theta_k - 1)}. \end{aligned}$$

$$\alpha = \#\{x_k = 1, \mathbf{x}^{\mathrm{pa}(k)}\} \in \{0, 1, \cdots, \mathsf{D}\}$$
$$\beta = \#\{x_k = 0, \mathbf{x}^{\mathrm{pa}(k)}\} \in \{0, 1, \cdots, \mathsf{D}\}$$

Note that when taking the derivative of $\ln f(\theta_k | \mathcal{D})$ with respect to $\theta_k$, the sum with respect to $i$ simplifies to a single term involving the parameter $\theta_k$ we are trying to estimate, that is, the conditional log-probability $\ln p_{X_k}(x_k|\text{parents}(X_k))$. By setting the derivative $\frac{\partial \ln f(\theta_k|\mathcal{D})}{\partial \theta_k} = 0$, we obtain the MAP estimator

$$\frac{(\alpha_0 + \beta_0 + \alpha + \beta)\theta_k - (\alpha_0 + \alpha)}{\theta_k(\theta_k - 1)} = 0 \rightarrow \hat{\theta}_k = \frac{\alpha_0 + \alpha}{\alpha_0 + \beta_0 + \alpha + \beta}.$$

When we employ the prior parameters $\alpha_0 = \beta_0 = 1$, the MAP estimator simplifies to

$$\hat{\theta}_k = \frac{\alpha + 1}{\alpha + \beta + 2},$$

which is analogous to the formulation given in equation 11.3. If instead of having a binary state $x_k \in \{0, 1\}$ we have $x_k \in \{0, 1, \cdots, n\}$, then the prior PDF is a Dirichlet distribution, and the same principles apply for the derivation of the MAP estimator.

The Dirichlet distribution is a generalization of the Beta distribution for multivariate domains.

### 11.4.2   Partially Observed Bayesian Network

It is common in practice not to be able to observe every variable in a Bayesian network. In such a case, the method presented for a fully observed BN is not directly applicable, and we can resort to the *expectation maximization* (EM) method[5] (see §10.1.1). The EM method consists in repeating two steps recursively until convergence: in the *expectation step*, we employ the current values contained in the conditional probability tables, along with the inference procedure presented in §11.3 in order to replace observation counts by the expected number of observation counts. Take the cyanobacteria example, where we want to estimate $p(c = \text{yes}|t = \text{hot}, f = \text{yes})$ from a partially observed Bayesian network. We need to replace the explicit number of realizations $\#\{c = \text{yes}, t = \text{hot}, f = \text{yes}\}$ in equation 11.4 by the *expected number of counts*,

[5] **Note:** Here, we need to satisfy a key hypothesis; the underlying process controlling whether a data $x_i$ is available or missing must be independent of its actual value so that we can say that data is *missing at random*. For the cyanobacteria example, the hypothesis of *independence for the missing data* would not be satisfied if, for example, experimentalists were reluctant to go and collect samples when the water temperature is $T = \text{cold}$.

$$\mathbb{E}[\#\{c = \text{yes}, t = \text{hot}, f = \text{yes}\}] = \sum_{i=1}^{\text{D}} p(c = \text{yes}, t = \text{hot}, f = \text{yes}|\mathcal{D}_i).$$

For the first observation $\mathcal{D}_1 = \{\mathbf{x}_1\} = \{\text{cold}, \boxed{?}, \text{yes}, \boxed{?}, \text{clear}\}$ contained in the data set presented in figure 11.8, the conditional probability equals zero because the event where the temperature is $t = \text{hot}$ is impossible because we already know for this observation that the temperature is $t = \text{cold}$,

$$p(c = \text{yes}, t = \text{hot}, f = \text{yes}|\mathcal{D}_1) = 0.$$

| $\mathcal{U} = \mathbf{x}_i$ | T | F | C | M | W |
|---|---|---|---|---|---|
| 1 | cold | $\boxed{?}$ | yes | $\boxed{?}$ | clear |
| 2 | $\boxed{?}$ | $\boxed{?}$ | yes | yes | clear |
| 3 | $\boxed{?}$ | yes | no | yes | $\boxed{?}$ |
| 4 | hot | yes | yes | yes | green |
| $\vdots$ | | | | | |
| D | hot | no | no | $\boxed{?}$ | green |

Figure 11.8: Example of partially observed Bayesian network for the cyanobacteria data set.

For the second observation $\mathcal{D}_2 = \{\mathbf{x}_2\} = \{\boxed{?}, \boxed{?}, \text{yes, yes, clear}\}$, the joint probability to be inferred using the procedure presented in §11.3 and the current values contained in the CPTs is

$$p(c = \text{yes}, t = \text{hot}, f = \text{yes}|\mathcal{D}_2) = p(t = \text{hot}, f = \text{yes}|c = \text{yes}, m = \text{yes}, w = \text{clear}) \in (0, 1).$$

In the case where no variables are missing for an observation, as in the fourth observation $\mathcal{D}_4$, then

$$p(c = \text{yes}, t = \text{hot}, f = \text{yes}|\mathcal{D}_4) = 1.$$

Once the expectation procedure is completed for all observations in $\mathcal{D}$, the *maximization step* consists in computing either the MLE or MAP for updating the probabilities contained in CPTs. The maximization step employs the method presented in §11.4.1 for fully observed Bayesian networks, this time using the expected number of counts, for example,

$$\hat{p}(c = \text{yes}|t = \text{hot}, f = \text{yes}) = \frac{\mathbb{E}[\#\{c = \text{yes}, t = \text{hot}, f = \text{yes}\}] + 1}{\mathbb{E}[\#\{t = \text{hot}, f = \text{yes}\}] + 2}.$$

The expectation maximization method is intrinsically iterative, whereas the expectation step employs the CPT entries found at the last iteration during the maximization step. Both steps are repeated until a steady solution is reached.

## 11.5 Dynamic Bayesian Network

So far we have looked at Bayesian networks for modeling static systems that do not have a temporal component. When a BN is defined over time steps, we call it a *dynamic Bayesian network* (DBN). Just like for Bayesian networks, the dependence between variables is described by directed links and conditional probability tables; the same holds for the dependence between variables over successive time steps. Figure 11.9a presents the expanded representation of a dynamic Bayesian network, where pairs of hidden states $X_{t-1}$ and $X_t$ are linked by an arrow that points in the direction of time. Figure 11.9b represents the same network using a compact notation where the representation of the same state at different time steps is replaced by a single state $X_t$ with the double arrows indicating the links between different time steps. Figure 11.10 shows how the notion of time could be introduced in the cyanobacteria example where the double self-loop arrows indicate the presence of links between these variables over subsequent time steps.

For DBNs, we can apply the same *conditional probability estimation* methods we presented for Bayesian networks in §11.4. In the

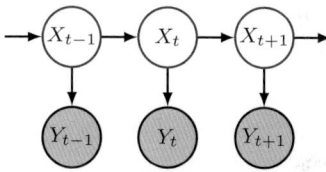

(a) Expanded DBN

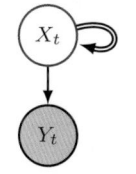

(b) Compact DBN

Figure 11.9: Equivalent expanded and compact representations of a dynamic Bayesian network. In (b), the double-lined arrow represents the conditional relationship between time steps.

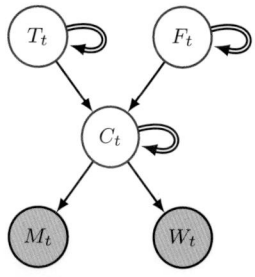

Figure 11.10: Dynamic Bayesian network for the cyanobacteria example.

[6] Murphy, K. P. (2002). *Dynamic Bayesian networks: representation, inference and learning*. PhD thesis, University of California, Berkeley.

[7] Diard, J., P. Bessiere, and E. Mazer (2003). A survey of probabilistic models using the Bayesian programming methodology as a unifying framework. In *International Conference on Computational Intelligence, Robotics and Autonomous Systems (IEEE-CIRAS)*, Singapore.

[8] Ghahramani, Z. (2001). An introduction to hidden Markov models and Bayesian networks. *International Journal of Pattern Recognition and Artificial Intelligence 15*(1), 9–42.

case of inference, specialized algorithms are available to perform variable elimination for DBN. The details for such algorithms can be found in dedicated literature.[6]

*Hidden Markov models*   Note that the dynamic Bayesian network presented in figure 11.9 represents a special case called a *hidden Markov model* (HMM). An HMM has only one *hidden-state variable* at each time $t$ along with one observed variable. The model is *Markovian* because the future $(X_{t+1})$ is independent of the past $(X_{t-1})$ given the present $(X_t)$, that is, $X_{t+1} \perp\!\!\!\perp X_{t-1}|x_t$. The HMM model is thus defined by an *observation model* $p(y_t|x_t)$ and a *transition model* $p(x_{t+1}|x_t)$. The fire alarm example presented in §6.2.2 can be described by a hidden Markov model. The reader interested in specialized inference algorithms for HMM should refer to dedicated literature. In this book we will instead focus our attention on the case of a dynamic Bayesian network using continuous state variables, that is, *state-space models*. These models are described at length in the next chapter. The reader interested about the relationships between Bayesian networks, HMM, state-space models, and the Markov decision process (see chapter 15) should consult the papers by Diard, Bessière and Mazer;[7] and Ghahramani.[8]

*Exercises*

P11.1  Explain the notion of conditional independence.

P11.2  Explain the notion of *d*-separation in the context of Bayesian networks.

P11.3  Explain what a directed acyclic graph is.

P11.4  Explain how the expectation maximization algorithm is employed to learn the conditional probability tables in partially observed Bayesian networks.

P11.5  For the cyanobacteria example presented in figure 11.4, use variable elimination to compute $p(w = \text{green}|m = \text{no})$, the conditional probability of having green water given that there is no fish mortality.

# 12
# State-Space Models

State-space models are suited to analyze time-series data. In chapter 8, we saw that it is possible to model time series using *regression* methods (e.g., see figure 8.30). Regression is suited for modeling time series displaying a *stationary* or *trend-stationary* baseline on which is added recurrent patterns, as displayed in figures 12.1a–b. In the case of a time series consisting in nonstationary processes such as the one in figure 12.1c, regression methods typically have a limited predictive capacity when extrapolating beyond the cases covered in the training set.

State-space models (SSM) analyze time series in a different way than regression models do. Regression is used in a supervised setup in order to describe the dependence between system responses and covariates, whereas in the case of time series, the covariates are related to time. On the other hand, *state-space models* bears that name because it uses a *dynamic model* to describe the dependence between the *hidden states* of a system at subsequent time steps. Take, for example, the task illustrated in figure 12.2 of predicting the landing location of a projectile; using kinematic equations, we can derive a linear model of the projectile position at time $t_0 + \Delta t$, from the initial position $x_0$, the initial velocity $\dot{x}_0$, and the exit angle $\theta_0$. This task is not well suited for regression methods, yet it can easily be formulated as an SSM because there is a kinematic model available for describing the dependence between the hidden states at a time $t$ and $t + \Delta t$. SSM is a probabilistic method where, given a model's structure, the task is to learn how to predict the hidden-state variables $\mathbf{x}_{t+1}$ given the current knowledge for $\mathbf{x}_t$. State-space models can be classified as *unsupervised learning* because it builds a joint probability density for a set of hidden-state variables while most of them are typically never observed.

In the context of civil engineering we can, for example, use SSMs for modeling the degradation of infrastructures over time.

**Note:** A time series or process is said to be *stationary* when its statistical properties, such as its mean and variance, are constant through time, as illustrated in figure 12.1d. A time series is said to be *trend-stationary* if the trend (linear or not) can be removed from it.

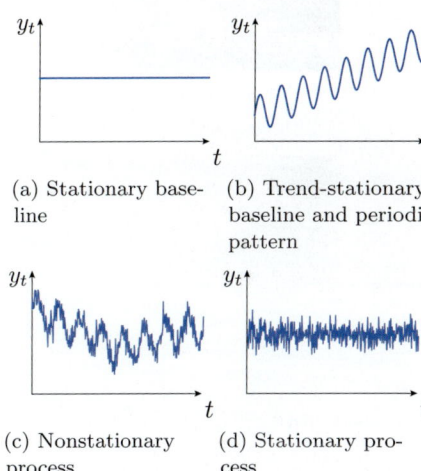

(a) Stationary baseline    (b) Trend-stationary baseline and periodic pattern

(c) Nonstationary process    (d) Stationary process

Figure 12.1: The concept of time-series stationarity.

Figure 12.2: A cannon-man trajectory can be described using the kinematics equations

$$
\begin{aligned}
x_{t+1} &= x_t + \dot{x}_t \Delta t + \tfrac{1}{2}\ddot{x}_t \Delta t^2 + w \\
\dot{x}_{t+1} &= \dot{x}_t + \ddot{x}_t \Delta t + \dot{w} \\
\ddot{x}_{t+1} &= \ddot{x}_t + \ddot{w},
\end{aligned}
$$

where $\mathbf{w} = [w \; \dot{w} \; \ddot{w}]^\intercal$ is a perturbation in the trajectory.

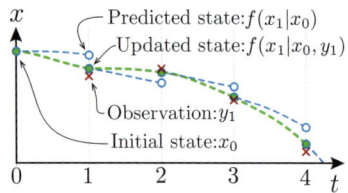

Figure 12.3: Example of state-space model where the state $x_t$ at a time $t$ is estimated from the combination of an observation $y_t$, a dynamic model describing the conditional probability $f(x_t|x_{t-1})$, and the prior knowledge for $x_{t-1}$.

Figure 12.4: Apollo mission service module. (Photo: NASA)

As illustrated in figure 12.3, we employ the state variable $x$ to describe the condition of a structure. Starting from a known initial condition $x_0$, we may develop a dynamic model $f(x_1|x_0)$, similar to the one in the projectile example, in order to describe the kinematic process behind degradation. Because our model will inevitably be an approximation of the real degradation kinematics, we will recursively update our predictions using observations $y_t$ as they become available. The theory presented in this chapter allows combining probabilistically the information contained in the prior knowledge $f(x_{t-1})$, the model $f(x_t|x_{t-1})$, and the observation $y_t$, in order to obtain a posterior estimate $f(x_t|y_1, \cdots, y_t)$. The procedure is employed recursively to go from a time step $t-1$ to $t$.

SSMs allow performing several tasks such as removing (i.e., filtering) the observation errors, forecasting future states, and estimating *hidden states* that are not directly observable. For example, in the infrastructure degradation example, even if only the condition $x_t$ is observed, an SSM allows estimating two additional *hidden states*: the degradation speed $\dot{x}_t$ and its acceleration $\ddot{x}_t$. State-space models take their origin from the field of *control theory*, where the breakthrough application of the method was the guiding system for the *Apollo mission* in the 1960s (see figure 12.4).

## 12.1   Linear Gaussian State-Space Models

We can describe the mathematical formulation for SSMs using the same theory we covered in §3.4.1, for the linear functions of random variables, and in §4.1.3, where we explored the analytic formulation of the multivariate Normal conditional probability density function (PDF). Let us consider a linear model $\mathbf{Y} = \mathbf{AX} + \mathbf{b}$, where $\mathbf{X} \sim \mathcal{N}(\mathbf{x}; \boldsymbol{\mu}_{\mathbf{X}}, \boldsymbol{\Sigma}_{\mathbf{X}})$, and a vector of observations $\mathbf{y}$. We saw in §3.4.1 that the output for linear functions of Normal random variables is also Normal,

$$
\left. \begin{aligned}
\boldsymbol{\mu}_{\mathbf{Y}} &= \mathbf{A}\boldsymbol{\mu}_{\mathbf{X}} + \mathbf{b} \\
\boldsymbol{\Sigma}_{\mathbf{Y}} &= \mathbf{A}\boldsymbol{\Sigma}_{\mathbf{X}}\mathbf{A}^{\mathsf{T}} \\
\boldsymbol{\Sigma}_{\mathbf{XY}} &= \boldsymbol{\Sigma}_{\mathbf{X}}\mathbf{A}^{\mathsf{T}}
\end{aligned} \right\}
\left[ \begin{array}{c} \mathbf{X} \\ \mathbf{Y} \end{array} \right], \left[ \begin{array}{c} \boldsymbol{\mu}_{\mathbf{X}} \\ \boldsymbol{\mu}_{\mathbf{Y}} \end{array} \right], \left[ \begin{array}{cc} \boldsymbol{\Sigma}_{\mathbf{X}} & \boldsymbol{\Sigma}_{\mathbf{XY}} \\ \boldsymbol{\Sigma}_{\mathbf{XY}}^{\mathsf{T}} & \boldsymbol{\Sigma}_{\mathbf{Y}} \end{array} \right].
$$

In §3.4.1, we saw that the conditional PDF of $\mathbf{X}$ given the observations $\mathbf{y}$ is obtained by dividing the joint PDF of $\mathbf{X}$ and $\mathbf{Y}$ by the PDF of $\mathbf{Y}$ evaluated for observations $\mathbf{y}$,

$$
\underbrace{f_{\mathbf{X}|\mathbf{y}}(\mathbf{x}|\underbrace{\mathbf{Y} = \mathbf{y}}_{\text{observations}})}_{\text{posterior knowledge}} = \frac{\overbrace{f_{\mathbf{XY}}(\mathbf{x}, \mathbf{y})}^{\text{prior knowledge}}}{f_{\mathbf{Y}}(\mathbf{y})} = \mathcal{N}(\mathbf{x}; \boldsymbol{\mu}_{\mathbf{X}|\mathbf{y}}, \boldsymbol{\Sigma}_{\mathbf{X}|\mathbf{y}}),
$$

where the conditional mean vector and covariance matrix are

$$\boldsymbol{\mu}_{\mathbf{X}|\mathbf{y}} = \boldsymbol{\mu}_{\mathbf{X}} + \boldsymbol{\Sigma}_{\mathbf{XY}}\boldsymbol{\Sigma}_{\mathbf{Y}}^{-1}(\mathbf{y} - \boldsymbol{\mu}_{\mathbf{Y}}), \ \boldsymbol{\Sigma}_{\mathbf{X}|\mathbf{y}} = \boldsymbol{\Sigma}_{\mathbf{X}} - \boldsymbol{\Sigma}_{\mathbf{XY}}\boldsymbol{\Sigma}_{\mathbf{Y}}^{-1}\boldsymbol{\Sigma}_{\mathbf{XY}}^{\mathsf{T}}.$$

### 12.1.1  Basic Problem Setup

We employ a basic setup involving a single hidden-state variable in order to illustrate how the notions of linear functions of random variables and the multivariate Normal conditional PDF are behind the state-space model theory. We use $t \in \{0, 1, \cdots, \mathsf{T}\}$ to describe discrete time stamps, $x_t$ to describe a hidden state of a system at a time $t$, and $y_t$ to describe an imperfect observation of $x_t$.

A hidden state $x_t$ is a deterministic, yet unknown, quantity.

*Observation and transition models*  The relation between an observation and a hidden state is described by the *observation model*,

$$\underbrace{y_t = x_t + v_t}_{\text{observation model}}, \ \overbrace{v_t : V \sim \mathcal{N}(v; 0, \sigma_V^2)}^{\text{observation error}}.$$

We call the observation $y_t$ imperfect because what we observe is the true state contaminated by zero-mean Normal-distributed observation errors. $x_t$ is referred to as a hidden state because it is not directly observed; only $y_t$ is. The dynamic model describing the possible transitions between successive time steps from a state $x_{t-1}$ to $x_t$ is described by the *transition model*,

$$\underbrace{x_t = x_{t-1} + w_t}_{\text{transition model}}, \ \overbrace{w_t : W \sim \mathcal{N}(w; 0, \sigma_W^2)}^{\text{model error}}, \qquad (12.1)$$

where $w_t$ is a zero-mean Normal-distributed model error describing the discrepancy between the reality and our idealized transition model. The transition model in equation 12.1 is *Markovian* (see chapter 7) because a state $x_t$ only depends on the state $x_{t-1}$ and is thus independent of all previous states. This aspect is key in the computational efficiency of the method.

*Notation*  In order to keep the notation simple, we opt for compact abbreviations; a set of observations $\{y_1, y_2, \cdots, y_t\}$ is described by $y_{1:t}$ and the posterior conditional expected value for $X_t$ given $y_{1:t}$, $\mathbb{E}[X_t|y_{1:t}]$, is summarized by $\mu_{t|t}$. For the subscript, $_{t|t}$, the first index refers to $X_t$ and the second to $y_{1:t}$. Following the same notation, $\mathbb{E}[X_t|y_{1:t-1}]$ is equivalent to $\mu_{t|t-1}$. The posterior conditional variance for $X_t$ given $y_{1:t}$, $\mathrm{var}[X_t|y_{1:t}]$, is summarized by $\sigma_{t|t}^2$.

**Notation**

$$
\begin{aligned}
y_{1:t} &\equiv \{y_1, y_2, \cdots, y_t\} \\
\mu_{t|t} &\equiv \mathbb{E}[X_t|y_{1:t}] \\
\mu_{t|t-1} &\equiv \mathbb{E}[X_t|y_{1:t-1}] \\
\sigma_{t|t}^2 &\equiv \mathrm{var}[X_t|y_{1:t}]
\end{aligned}
$$

**Note:** An absence of prior knowledge for $X_0$ can be described by $\text{var}[X_0] \to \infty$.

*Prediction step*   At time $t = 0$, only our prior knowledge $\mathbb{E}[X_0] = \mu_0$ and $\text{var}[X_0] = \sigma_0^2$ are available. From our prior knowledge at $t = 0$, the transition model $x_t = x_{t-1} + w_t$, and the observation model $y_t = x_t + v_t$, we can compute our joint prior knowledge at $t = 1$ for $X_1$ and $Y_1$, $f(x_1, y_1) = \mathcal{N}([x_1, y_1]^\mathsf{T}; \boldsymbol{\mu}, \boldsymbol{\Sigma})$, where the mean vector and covariance matrix are described by

$$\boldsymbol{\mu} = \begin{bmatrix} \mu_X \\ \mu_Y \end{bmatrix} = \begin{bmatrix} \mathbb{E}[X_1] \\ \mathbb{E}[Y_1] \end{bmatrix} \tag{12.2}$$

$$\boldsymbol{\Sigma} = \begin{bmatrix} \Sigma_X & \Sigma_{XY} \\ \Sigma_{YX} & \Sigma_Y \end{bmatrix} = \begin{bmatrix} \text{var}[X_1] & \text{cov}(X_1, Y_1) \\ \text{cov}(Y_1, X_1) & \text{var}[Y_1] \end{bmatrix}. \tag{12.3}$$

The expected value $\mathbb{E}[X_1]$ and variance $\text{var}[X_1]$ are obtained by propagating the uncertainty associated with the prior knowledge for $X_0$ through the transition model ($x_t = x_{t-1} + w_t$),

**Note:** We use the notation $\mu_1$ and $\sigma_1$ rather than $\mu_{1|0}$ and $\sigma_{1|0}$ because there is no observation at $t = 0$.

$$\begin{aligned} \mathbb{E}[X_1] &\equiv \mu_1 &= \mu_0 \\ \text{var}[X_1] &\equiv \sigma_1^2 &= \sigma_0^2 + \sigma_W^2. \end{aligned}$$

The expected value $\mathbb{E}[Y_1]$ and variance $\text{var}[Y_1]$ are obtained by propagating the uncertainty associated with the prior knowledge for $X_1$ through the observation model ($y_t = x_t + v_t$),

$$\begin{aligned} \mathbb{E}[Y_1] &= \mu_0 \\ \text{var}[Y_1] &= \underbrace{\sigma_0^2 + \sigma_W^2}_{\sigma_1^2} + \sigma_V^2. \end{aligned}$$

The covariance $\text{cov}(X_1, Y_1) = \sigma_1^2 = \sigma_0^2 + \sigma_W^2$ is equal to $\text{var}[X_1]$.

*Update step*   We can obtain the posterior knowledge at time $t = 1$ using

$$f(x_1|y_1) = \frac{f(x_1, y_1)}{f(y_1)} = \mathcal{N}(x_1; \mu_{1|1}, \sigma_{1|1}^2),$$

where the posterior mean and variance are

$$\begin{aligned} \mu_{1|1} &= \mu_X + \Sigma_{XY}\Sigma_Y^{-1}(y_1 - \mu_Y) \\ \sigma_{1|1}^2 &= \Sigma_X - \Sigma_{XY}\Sigma_Y^{-1}\Sigma_{XY}^\mathsf{T}. \end{aligned} \tag{12.4}$$

Note that the terms in equation 12.4 correspond to those defined in equations 12.2 and 12.3. The posterior $f(x_1|y_1)$ contains the information coming from the prior knowledge at $t = 0$, the system dynamics described by the transition model, and the observation at time $t = 1$.

The prediction and update steps are now generalized to transition from $t - 1 \to t$. From our posterior knowledge at $t - 1$, the transition model $x_t = x_{t-1} + w_t$, and the observation model $y_t = x_t + v_t$, we compute our joint prior knowledge at $t$, $f(x_t, y_t|y_{1:t-1}) = \mathcal{N}([x_t\ y_t]^\intercal; \boldsymbol{\mu}, \boldsymbol{\Sigma})$, where the mean and covariance are

$$\boldsymbol{\mu} = \begin{bmatrix} \mu_X \\ \mu_Y \end{bmatrix} = \begin{bmatrix} \mathbb{E}[X_t|y_{1:t-1}] \\ \mathbb{E}[Y_t|y_{1:t-1}] \end{bmatrix},$$

$$\boldsymbol{\Sigma} = \begin{bmatrix} \Sigma_X & \Sigma_{XY} \\ \Sigma_{YX} & \Sigma_Y \end{bmatrix} = \begin{bmatrix} \text{var}[X_t|y_{1:t-1}] & \text{cov}(X_t, Y_t|y_{1:t-1}) \\ \text{cov}(Y_t, X_t|y_{1:t-1}) & \text{var}[Y_t|y_{1:t-1}] \end{bmatrix}.$$

Each term of the mean vector and covariance matrix is obtained following

$$
\begin{aligned}
\mathbb{E}[X_t|y_{1:t-1}] &\equiv & \mu_{t|t-1} &= & \mu_{t-1|t-1} \\
\text{var}[X_t|y_{1:t-1}] &\equiv & \sigma^2_{t|t-1} &= & \sigma^2_{t-1|t-1} + \sigma^2_W \\
\mathbb{E}[Y_t|y_{1:t-1}] &= & \mu_{t|t-1} & & \\
\text{var}[Y_t|y_{1:t-1}] &= & \underbrace{\sigma^2_{t-1|t-1} + \sigma^2_W}_{\sigma^2_{t|t-1}} + \sigma^2_V & & \\
\text{cov}(X_t, Y_t|y_{1:t-1}) &= & \overbrace{\sigma^2_{t-1|t-1} + \sigma^2_W}. & &
\end{aligned}
$$

The posterior knowledge at $t$ is given by

$$f(x_t|y_{1:t}) = \frac{f(x_t, y_t|y_{1:t-1})}{f(y_t|y_{1:t-1})} = \mathcal{N}(x_t; \mu_{t|t}, \sigma^2_{t|t}),$$

where posterior mean and variance are

$$
\begin{aligned}
\mu_{t|t} &= \mu_X + \Sigma_{XY}\Sigma_Y^{-1}(y_t - \mu_Y) \\
\sigma^2_{t|t} &= \Sigma_X - \Sigma_{XY}\Sigma_Y^{-1}\Sigma_{XY}^\intercal.
\end{aligned}
$$

The *marginal likelihood* $f(y_t|y_{1:t-1})$ describing the prior probability density of observing $y_t$ given all observations up to time $t-1$ is given by

$$f(y_t|y_{1:t-1}) = \mathcal{N}(y_t; \underbrace{\mu_{t|t-1}}_{\mathbb{E}[Y_t|y_{1:t-1}]}, \underbrace{\sigma^2_{t-1|t-1} + \sigma^2_W + \sigma^2_V}_{\text{var}[Y_t|y_{1:t-1}]}).$$

*Example: Temperature time series*   We now look at an illustrative example where we apply the prediction and update steps recursively to estimate the temperature $x_t$ over a period of three hours using imprecise observations made every hour for $t \in \{1, 2, 3\}$. Our prior knowledge is described by $\mathbb{E}[X_0] = 10°C$ and $\sigma[X_0] = 7°C$. The transition and observation models are

$$\underbrace{y_t = x_t + v_t,}_{\text{observation model}} \ v_t : V \sim \mathcal{N}(0, \overbrace{\underbrace{3^2}_{\sigma^2_V}}^{\text{observation error}}),$$

$$\underbrace{x_t = x_{t-1} + w_t,}_{\text{transition model}} \ w_t : W \sim \mathcal{N}(0, \overbrace{\underbrace{0.5^2}_{\sigma^2_W}}^{\text{model error}}).$$

**Note:** $\text{cov}(X_t, Y_t|y_{1:t-1})$
Following the covariance properties,
$$
\begin{aligned}
\text{cov}(X_1 + X_2, X_3 + X_4) &= \text{cov}(X_1, X_3) \\
&\cdots + \text{cov}(X_1, X_4) \\
&\cdots + \text{cov}(X_2, X_3) \\
&\cdots + \text{cov}(X_2, X_4)
\end{aligned}
$$
therefore,
$$
\begin{aligned}
\text{cov}(X_2, Y_2|y_1) &= \text{cov}(X_{1|1} + W, X_{1|1} + W + V) \\
&= \text{cov}(X_{1|1}, X_{1|1}) \\
&\cdots + \text{cov}(X_{1|1}, W)^{\,0, X_{1|1} \perp\!\!\!\perp W} \\
&\cdots + \text{cov}(X_{1|1}, V)^{\,0, X_{1|1} \perp\!\!\!\perp V} \\
&\cdots + \text{cov}(W, X_{1|1})^{\,0, W \perp\!\!\!\perp X_{1|1}} \\
&\cdots + \text{cov}(W, W) \\
&\cdots + \text{cov}(W, V)^{\,0, W \perp\!\!\!\perp V} \\
&= \text{cov}(X_{1|1}, X_{1|1}) + \text{cov}(W, W) \\
&= \sigma^2_{1|1} + \sigma^2_W.
\end{aligned}
$$

**Note:** $f(y_t|y_{1:t-1}) \equiv f(y_t|y_{1:t-1}; \boldsymbol{\theta})$, that is, we can represent either implicitly or explicitly the dependency on the model parameters $\boldsymbol{\theta}$ that enter in the definition of the transition and observation equations. See §12.1.4 for further details.

We refer to $f(y_t|y_{1:t-1})$ as the *marginal likelihood* because the hidden state $x_t$ was marginalized. In the case of the multivariate Normal, the solution to $\int f(y_t, x_t|y_{1:t-1})dx_t$ is analytically tractable; see §4.1.3.

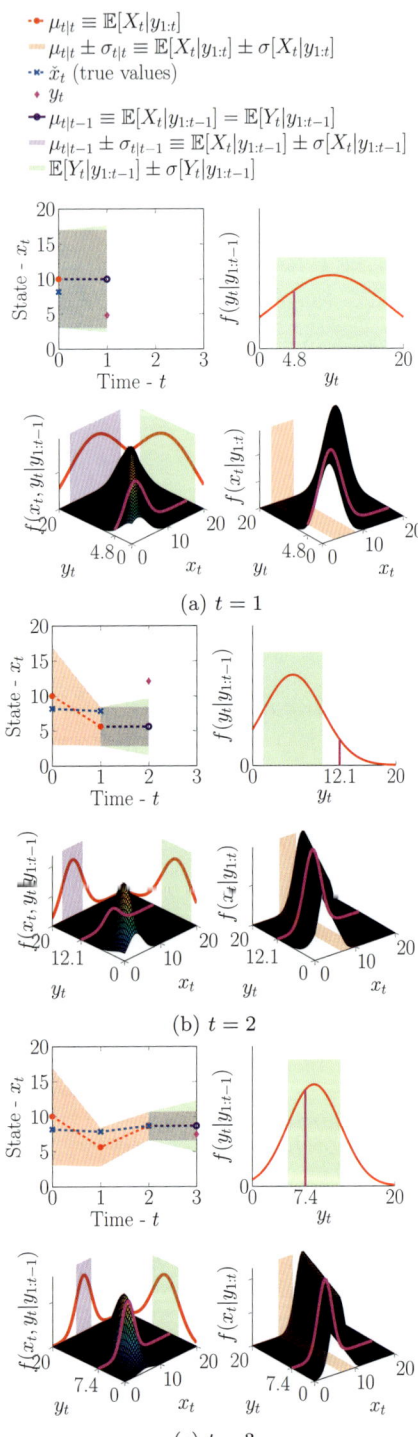

- • -- $\mu_{t|t} \equiv \mathbb{E}[X_t|y_{1:t}]$
- ▬ $\mu_{t|t} \pm \sigma_{t|t} \equiv \mathbb{E}[X_t|y_{1:t}] \pm \sigma[X_t|y_{1:t}]$
- ·✱· $\check{x}_t$ (true values)
- ◆ $y_t$
- •—• $\mu_{t|t-1} \equiv \mathbb{E}[X_t|y_{1:t-1}] = \mathbb{E}[Y_t|y_{1:t-1}]$
- ▬ $\mu_{t|t-1} \pm \sigma_{t|t-1} \equiv \mathbb{E}[X_t|y_{1:t-1}] \pm \sigma[X_t|y_{1:t-1}]$
- ▬ $\mathbb{E}[Y_t|y_{1:t-1}] \pm \sigma[Y_t|y_{1:t-1}]$

(a) $t = 1$

(b) $t = 2$

(c) $t = 3$

Figure 12.5: Step-by-step probabilistic hidden-state estimation. Note that the pink curve overlaid on the surfaces describes either the joint or the conditional PDF evaluated for the observed value $y_t$.

Performing the prediction step from $t = 0 \rightarrow t = 1$ leads to

$$
\begin{aligned}
\mathbb{E}[X_1] &= \mathbb{E}[X_0] &= 10 \\
\mathbb{E}[Y_1] &= \mathbb{E}[X_1] &= 10 \\
\mathrm{var}[X_1] &= \mathrm{var}[X_0] + \sigma_W^2 &= 49.25 \\
\mathrm{var}[Y_1] &= \mathrm{var}[X_1] + \sigma_V^2 &= 58.25 \\
\mathrm{cov}(X_1, Y_1) &= \mathrm{var}[X_1] &= 49.25 \\
\rho_{XY} &= \frac{\mathrm{cov}(X_1,Y_1)}{\sigma[X_1]\sigma[Y_1]} &= 0.92.
\end{aligned}
$$

At the time $t = 1$, the observed temperature is $y_1 = 4.8$, leading to a likelihood $f(y_t|y_{1:t-1}) = 0.04$. Performing the update step leads to

$$
\begin{aligned}
\mathbb{E}[X_{1|1}] &= \mathbb{E}[X_1] + \frac{\mathrm{cov}(X_1,Y_1)(y_1-\mathbb{E}[Y_1])}{\mathrm{var}[Y_1]} &= 5.6 \\
\mathrm{var}[X_{1|1}] &= \mathrm{var}[X_1] - \frac{\mathrm{cov}(X_1,Y_1)^2}{\mathrm{var}[Y_1]} &= 7.6.
\end{aligned}
$$

Following the Gaussian conditional example presented in §4.1.4, figure 12.5a presents graphically each of the quantities associated with the prior knowledge at $t = 1$ obtained by propagating the knowledge of $X_0$ in the transition and observation models, the likelihood of $y_t$, and the posterior knowledge. By repeating the same prediction and update steps for $t = 2$, where $y_2 = 12.1$, we obtain

**Prediction step**

$$
\begin{aligned}
\mathbb{E}[X_{2|1}] &= \mathbb{E}[X_{1|1}] &= 5.6 \\
\mathbb{E}[Y_{2|1}] &= \mathbb{E}[X_{2|1}] &= 5.6 \\
\mathrm{var}[X_{2|1}] &= \mathrm{var}[X_{1|1}] + \sigma_W^2 &= 7.9 \\
\mathrm{var}[Y_{2|1}] &= \mathrm{var}[X_{2|1}] + \sigma_V^2 &= 16.9 \\
\mathrm{cov}(X_{2|1}, Y_{2|1}) &= \mathrm{var}[X_{2|1}] &= 7.9 \\
\rho_{XY} &= \frac{\mathrm{cov}(X_{2|1},Y_{2|1})}{\sigma[X_{2|1}]\sigma[Y_{2|1}]} &= 0.68 \\
f(y_2|y_1) &= 0.03
\end{aligned}
$$

**Update step**

$$
\begin{aligned}
\mathbb{E}[X_{2|2}] &= \mathbb{E}[X_{2|1}] + \frac{\mathrm{cov}(X_{2|1},Y_{2|1})(y_2-\mathbb{E}[Y_{2|1}])}{\mathrm{var}[Y_{2|1}]} &= 8.6 \\
\mathrm{var}[X_{2|2}] &= \mathrm{var}[X_{2|1}] - \frac{\mathrm{cov}(X_{2|1},Y_{2|1})^2}{\mathrm{var}[Y_{2|1}]} &= 4.2.
\end{aligned}
$$

These results are illustrated in figure 12.5b. The details of subsequent steps are not presented because they consist in repeating recursively the same prediction and update steps. Figure 12.5c shows how, in only three time steps, we went from a diffuse knowledge describing $X_0$ to a more precise estimate for $X_2$.

### 12.1.2   General Formulation

We now extend the prediction and update steps with a formulation compatible with multiple hidden states. For example, consider the case where want to use a SSM to estimate the hidden states $\mathbf{x}_t = [x_t \; \dot{x}_t]^\intercal$, that is, the temperature $x_t$ and its *rate of change* $\dot{x}_t$. The observation model then becomes

$$y_t = x_t + v_t \equiv \overbrace{\mathbf{C}}^{\mathbf{C}=[1\ 0]} \mathbf{x}_t + v_t,$$

where the *observation matrix* $\mathbf{C} = [1\ 0]$ indicates that an observation $y_t$ is a function of $x_t$ but not of the rate of change $\dot{x}_t$. $v_t$ is a realization of an observation error $V \sim \mathcal{N}(v; 0, \sigma_V^2)$. In the general case, observation errors are described by a multivariate Normal random variable $\mathbf{V} \sim \mathcal{N}(\mathbf{v}; \mathbf{0}, \mathbf{R})$, where $\mathbf{R}$ is the covariance matrix. The transition model for the two hidden states is

$$\left. \begin{array}{rcl} x_t &=& x_{t-1} + \Delta t \cdot \dot{x}_{t-1} + w_t \\ \dot{x}_t &=& \dot{x}_{t-1} + \dot{w}_t \end{array} \right\} \equiv \mathbf{x}_t = \overbrace{\mathbf{A}}^{\mathbf{A}=\left[\begin{smallmatrix}1 & \Delta t \\ 0 & 1\end{smallmatrix}\right]} \mathbf{x}_{t-1} + \overbrace{\mathbf{w}_t}^{\mathbf{w}_t=\left[\begin{smallmatrix}w_t \\ \dot{w}_t\end{smallmatrix}\right]},$$

where the transition matrix $\mathbf{A}$ describes the model kinematics and $\mathbf{w}_t$ are realizations of the model errors $\mathbf{W} \sim \mathcal{N}(\mathbf{w}; \mathbf{0}, \mathbf{Q})$ that affect each of the hidden states.

In the *prediction step*, we compute our prior knowledge at $t$ from our posterior knowledge at $t-1$, using the transition model and the observation model,

$$\begin{array}{rcl} \boldsymbol{\mu}_{t|t-1} &=& \mathbf{A}\boldsymbol{\mu}_{t-1|t-1} \\ \boldsymbol{\Sigma}_{t|t-1} &=& \mathbf{A}\boldsymbol{\Sigma}_{t-1|t-1}\mathbf{A}^\intercal + \mathbf{Q} \\ \mathbb{E}[\mathbf{Y}_t|\mathbf{y}_{1:t-1}] &=& \mathbf{C}\boldsymbol{\mu}_{t|t-1} \\ \mathrm{cov}(\mathbf{Y}_t|\mathbf{y}_{1:t-1}) &=& \mathbf{C}\boldsymbol{\Sigma}_{t|t-1}\mathbf{C}^\intercal + \mathbf{R} \\ \mathrm{cov}(\mathbf{X}_t, \mathbf{Y}_t|\mathbf{y}_{1:t-1}) &=& \boldsymbol{\Sigma}_{t|t-1}\mathbf{C}^\intercal. \end{array} \tag{12.5}$$

These terms are collectively describing our joint prior knowledge $f(\mathbf{x}_t, \mathbf{y}_t|\mathbf{y}_{1:t-1}) = \mathcal{N}([\mathbf{x}_t \; \mathbf{y}_t]^\intercal; \boldsymbol{\mu}, \boldsymbol{\Sigma})$ with the mean vector and covariance matrix defined as

$$\boldsymbol{\mu} = \left[\begin{array}{c} \boldsymbol{\mu}_\mathbf{X} \\ \boldsymbol{\mu}_\mathbf{Y} \end{array}\right] = \left[\begin{array}{c} \boldsymbol{\mu}_{t|t-1} \\ \mathbb{E}[\mathbf{Y}_t|\mathbf{y}_{1:t-1}] \end{array}\right]$$

$$\boldsymbol{\Sigma} = \left[\begin{array}{cc} \boldsymbol{\Sigma}_\mathbf{X} & \boldsymbol{\Sigma}_\mathbf{XY} \\ \boldsymbol{\Sigma}_\mathbf{XY}^\intercal & \boldsymbol{\Sigma}_\mathbf{Y} \end{array}\right] = \left[\begin{array}{cc} \boldsymbol{\Sigma}_{t|t-1} & \mathrm{cov}(\mathbf{X}_t, \mathbf{Y}_t|\mathbf{y}_{1:t-1}) \\ \mathrm{cov}(\mathbf{X}_t, \mathbf{Y}_t|\mathbf{y}_{1:t-1})^\intercal & \mathrm{cov}(\mathbf{Y}_t|\mathbf{y}_{1:t-1}) \end{array}\right].$$

In the *update step*, the posterior knowledge is obtained using the conditional multivariate Normal PDF,

$$f(\mathbf{x}_t|\mathbf{y}_{1:t}) = \frac{f(\mathbf{x}_t, \mathbf{y}_t|\mathbf{y}_{1:t-1})}{f(\mathbf{y}_t|\mathbf{y}_{1:t-1})} = \mathcal{N}(\mathbf{x}_t; \boldsymbol{\mu}_{t|t}, \boldsymbol{\Sigma}_{t|t}),$$

which is defined by its posterior mean vector and covariance matrix,

$$\begin{array}{rcl} \boldsymbol{\mu}_{t|t} &=& \boldsymbol{\mu}_\mathbf{X} + \boldsymbol{\Sigma}_\mathbf{XY}\boldsymbol{\Sigma}_\mathbf{Y}^{-1}(\mathbf{y}_t - \boldsymbol{\mu}_\mathbf{Y}) \\ \boldsymbol{\Sigma}_{t|t} &=& \boldsymbol{\Sigma}_\mathbf{X} - \boldsymbol{\Sigma}_\mathbf{XY}\boldsymbol{\Sigma}_\mathbf{Y}^{-1}\boldsymbol{\Sigma}_\mathbf{XY}^\intercal. \end{array} \tag{12.6}$$

**Observation model**
$$\mathbf{y}_t = \mathbf{C}\mathbf{x}_t + \mathbf{v}_t, \quad \underbrace{\mathbf{v} : \mathbf{V} \sim \mathcal{N}(\mathbf{v}; \mathbf{0}, \mathbf{R})}_{\text{Observation errors}}$$

$\mathbf{C}$: Observation matrix
$\mathbf{R}$: Observation errors covariance

**Transition model**
$$\mathbf{x}_t = \mathbf{A}\mathbf{x}_{t-1} + \mathbf{w}_t, \quad \underbrace{\mathbf{w} : \mathbf{W} \sim \mathcal{N}(\mathbf{w}; \mathbf{0}, \mathbf{Q})}_{\text{Transition errors}}$$

$\mathbf{A}$: Transition matrix
$\mathbf{Q}$: Transition errors covariance

The posterior knowledge $f(\mathbf{x}_t|\mathbf{y}_{1:t})$ combines the information coming from the prior knowledge $\{\boldsymbol{\mu}_0, \boldsymbol{\Sigma}_0\}$, the sequence of observation $\mathbf{y}_{1:t}$, and the system dynamics described by the transition model. Equations 12.5 and 12.6, obtained using the theory presented in §3.4.1 and §4.1.3 are referred to as the *Kalman filter*[1] where they are reorganized as

[1] Kalman, R. E. (1960). A new approach to linear filtering and prediction problems. *Transactions of the ASME–Journal of Basic Engineering 82*(Series D), 35–45.

**Prediction step**

$$\boldsymbol{\mu}_{t|t-1} = \mathbf{A}\boldsymbol{\mu}_{t-1|t-1} \qquad \text{Prior mean}$$
$$\boldsymbol{\Sigma}_{t|t-1} = \mathbf{A}\boldsymbol{\Sigma}_{t-1|t-1}\mathbf{A}^\mathsf{T} + \mathbf{Q} \quad \text{Prior covariance}$$

**Update step**

$$f(\mathbf{x}_t|\mathbf{y}_{1:t}) = \mathcal{N}(\mathbf{x}_t; \boldsymbol{\mu}_{t|t}, \boldsymbol{\Sigma}_{t|t})$$
$$\boldsymbol{\mu}_{t|t} = \boldsymbol{\mu}_{t|t-1} + \mathbf{K}_t\mathbf{r}_t \qquad \text{Posterior expected value}$$
$$\boldsymbol{\Sigma}_{t|t} = (\mathbf{I} - \mathbf{K}_t\mathbf{C})\boldsymbol{\Sigma}_{t|t-1} \quad \text{Posterior covariance}$$
$$\mathbf{r}_t = \mathbf{y}_t - \hat{\mathbf{y}}_t \qquad \text{Innovation vector}$$
$$\hat{\mathbf{y}}_t = \mathbf{C}\boldsymbol{\mu}_{t|t-1} \qquad \text{Prediction observations vector}$$
$$\mathbf{K}_t = \boldsymbol{\Sigma}_{t|t-1}\mathbf{C}^\mathsf{T}\mathbf{G}_t^{-1} \qquad \text{Kalman gain matrix}$$
$$\mathbf{G}_t = \mathbf{C}\boldsymbol{\Sigma}_{t|t-1}\mathbf{C}^\mathsf{T} + \mathbf{R} \quad \text{Innovation covariance matrix}$$

In the name *Kalman filter*, the term *filter* refers to the action of removing the observation errors from an observed time series. Filtering is an intrinsically recursive operation that is suited for applications where a flow of data can be continuously processed as it gets collected. Note that in the classic Kalman formulation,[2] there are additional control terms for including the effect of external actions on the system. These terms are omitted here because, contrarily to the field of aerospace, where flaps, rudders, or trust vectors are employed to continuously modify the state (i.e., position, speed, and acceleration) of a vehicle traveling through space, civil systems seldom have the possibility of such direct control actions being applied to them.

[2] Simon, D. (2006). *Optimal state estimation: Kalman, H infinity, and nonlinear approaches*. Wiley.

*Example: Temperature time series*   We want to estimate $\mathbf{x}_t = [x_t \ \dot{x}_t]^\mathsf{T}$, the temperature and its rate of change over a period of 5 hours using imprecise observations made every hour for $t \in \{1, 2, \cdots, 5\}$. The observation and transition models are respectively

$$y_t = \overbrace{\mathbf{C}}^{[1\ 0]} \mathbf{x}_t + \overbrace{v_t}^{V\sim\mathcal{N}(0,\sigma_V^2)}, \qquad \overbrace{\begin{bmatrix} x_t \\ \dot{x}_t \end{bmatrix}}^{\mathbf{x}_t} = \overbrace{\begin{bmatrix} 1 & \Delta t \\ 0 & 1 \end{bmatrix}}^{\mathbf{A}} \mathbf{x}_{t-1} + \overbrace{\begin{bmatrix} w_t \\ \dot{w}_t \end{bmatrix}}^{\mathbf{w}_t \sim \mathcal{N}(\mathbf{0},\mathbf{Q})}.$$

Our prior knowledge for the temperature is $\mathbb{E}[X_0] = 10°C$, $\sigma[X_0] =$

$2°C$, and for the rate of change, $\mathbb{E}[\dot{X}_0] = -1°C$, $\sigma[\dot{X}_0] = 1.5°C$. We assume that our knowledge for $X_0$ and $\dot{X}_0$ is independent $(X_0 \perp\!\!\!\perp \dot{X}_0 \rightarrow \rho_{X_0 \dot{X}_0} = 0)$, so the prior mean vector and covariance matrix are

$$\boldsymbol{\mu}_0 = \begin{bmatrix} 10 \\ 0 \end{bmatrix}, \; \boldsymbol{\Sigma}_0 = \begin{bmatrix} 2^2 & 0 \\ 0 & 1.5^2 \end{bmatrix}.$$

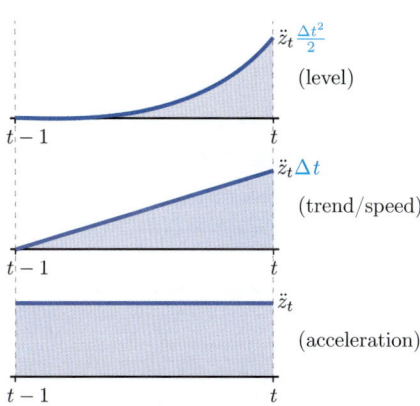

Figure 12.6: The error caused on the level and the trend by a constant acceleration $\ddot{z}_t$ over a time step of length $\Delta t$.

The observation-errors standard deviation is $\sigma_V = 4°C$. Model errors are defined by the covariance matrix $\mathbf{Q}$, which describes the errors on the temperature estimate and its rate of change. We assume that these errors are caused by a constant acceleration that takes place during the interval $\Delta t$, which is not accounted for in the model, that builds on the hypothesis of a constant speed and no acceleration. In order to define the covariance matrix $\mathbf{Q}$, we need to define a new vector $\mathbf{z}_t = [z_t \; \dot{z}_t \; \ddot{z}_t]^\mathsf{T}$ that describes a realization of $\mathbf{Z}_t \sim \mathcal{N}(\mathbf{z}_t; \mathbf{0}, \boldsymbol{\Sigma}_Z)$, that is, the effect of a constant acceleration having a random magnitude such that $\mathbf{Z}_t \perp\!\!\!\perp \mathbf{Z}_{t+i}, \forall i \neq 0$. As illustrated in figure 12.6, having a constant acceleration $\ddot{z}_t$ over a time step of length $\Delta t$ causes a change of $\ddot{z}_t \Delta t$ in the trend (i.e., speed) and a change of $\ddot{z}_t \frac{\Delta t^2}{2}$ in the level. We can express the errors caused by the constant acceleration in $\mathbf{z}_t$ in terms of our transition errors $\mathbf{w}_t$ as

[3] Bar-Shalom, Y., X. Li, and T. Kirubarajan (2001). *Estimation with applications to tracking and navigation: Theory algorithms and software*. John Wiley & Sons.

$$\mathbf{W}_t = \mathbf{A}_z \mathbf{Z}_t \sim \mathcal{N}(\mathbf{w}_t; \mathbf{0}, \underbrace{\mathbf{A}_z \boldsymbol{\Sigma}_z \mathbf{A}_z^\mathsf{T}}_{\mathbf{Q}}),$$

where

$$\mathbf{A}_z = \begin{bmatrix} 0 & 0 & \frac{\Delta t^2}{2} \\ 0 & 0 & \Delta t \end{bmatrix}, \; \boldsymbol{\Sigma}_z = \begin{bmatrix} 0 & 0 & 0 \\ 0 & 0 & 0 \\ 0 & 0 & \sigma_W^2 \end{bmatrix}.$$

Therefore, the definition of the matrix $\mathbf{Q}$ for the transition errors $\mathbf{W}_t$ is

$$\mathbf{Q} = \mathbf{A}_z \boldsymbol{\Sigma}_z \mathbf{A}_z^\mathsf{T} = \sigma_W^2 \begin{bmatrix} \frac{\Delta t^4}{4} & \frac{\Delta t^3}{2} \\ \frac{\Delta t^3}{2} & \Delta t^2 \end{bmatrix}.$$

For further details regarding the derivation of the process noise covariance matrices, the reader should consult the textbook by Bar-Shalom, Li, and Kirubarajan.[3] In this example, the noise parameter is equal to $\sigma_W^2 = 0.1$ and $\Delta t = 1$. Figure 12.7 presents the state estimates for the temperature and its rate of change along with the true values that were employed to generate simulated observations.

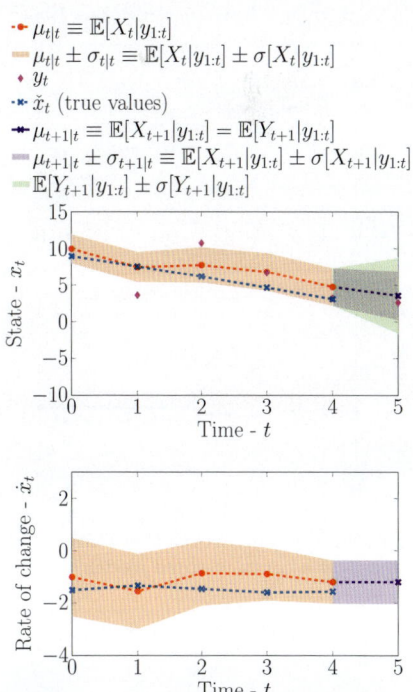

Figure 12.7: Example of hidden-state estimation for a constant-speed model.

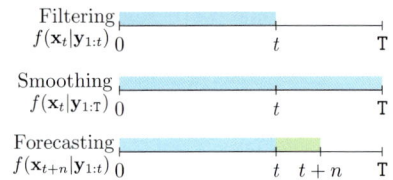

Figure 12.8: Comparison of *filtering*, *smoothing*, and *forecasting*, where the blue region corresponds to the data employed to estimate $x_t$ or forecast $x_{t+n}$.

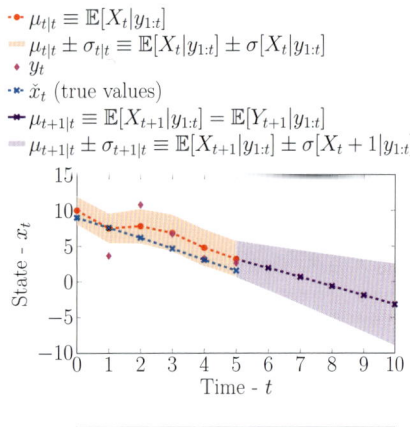

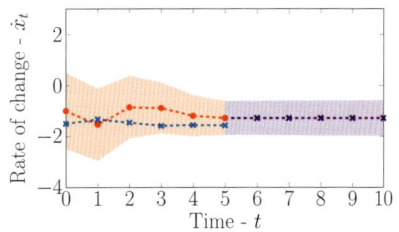

Figure 12.9: Example of hidden-state forecast for a constant-speed model.

### 12.1.3   Forecasting and Smoothing

The *filtering* method presented in the previous section estimates $f(\mathbf{x}_t|\mathbf{y}_{1:t})$, that is, the posterior probability density of hidden states $\mathbf{x}_t$, given all the observations up to time $t$. With this procedure, acquiring new data at a time $t+1$ does not modify our knowledge about $\mathbf{x}_t$. This second task of estimating the posterior probability of hidden states $\mathbf{x}_t$ given all the observations up to time $\mathsf{T} > t$, $f(\mathbf{x}_t|\mathbf{y}_{1:\mathsf{T}})$, is referred to as *smoothing*. A third task consists in *forecasting* the hidden states $\mathbf{x}_{t+n}$, $n$ time steps in the future, given all the observations up to time $t$, that is, $f(\mathbf{x}_{t+n}|\mathbf{y}_{1:t})$. The three tasks of *filtering*, *smoothing*, and *forecasting* are illustrated schematically in figure 12.8.

*Forecasting:* $f(\mathbf{x}_{t+n}|\mathbf{y}_{1:t})$   In the context of SSMs, forecasting is trivial because it only involves recursively propagating the current knowledge $\boldsymbol{\mu}_{t|t}, \boldsymbol{\Sigma}_{t|t}$ at time $t$ through the prediction step. In short, forecasting from a time $t$ to $t+1$ follows

$$f(\mathbf{x}_{t+1}|\mathbf{y}_{1:t}) = \mathcal{N}(\mathbf{x}_{t+1}; \boldsymbol{\mu}_{t+1|t}, \boldsymbol{\Sigma}_{t+1|t})$$

$$\boldsymbol{\mu}_{t+1|t} = \mathbf{A}\boldsymbol{\mu}_{t|t}$$

$$\boldsymbol{\Sigma}_{t+1|t} = \mathbf{A}\boldsymbol{\Sigma}_{t|t}\mathbf{A}^{\mathsf{T}} + \mathbf{Q}.$$

Note that whenever *data is missing* at a given time step, only the prediction step is recursively applied until new data is obtained. Therefore, missing data is treated in the same way as forecasting using the transition model. This is an interesting feature for SSMs, which do not require any special treatment for cases where data is missing.

*Example: Temperature time series (continued)*   We build on the temperature estimation example presented in figure 12.7 to illustrate the concept of forecasting. The goal here is to forecast the temperature for five hours after the last time step for which data is available. The result is presented in figure 12.9. Note how, because no observation is available to update the predictions, the standard deviation of the estimated hidden states keeps increasing because at each time step, the model error covariance matrix $\mathbf{Q}$ is added recursively to the hidden-state covariance matrix.

*Smoothing:* $f(\mathbf{x}_t|\mathbf{y}_{1:\mathsf{T}})$   With smoothing, we perform the reverse process of updating the knowledge at a time $t$, using the knowledge at $t+1$. Smoothing starts from the last time step $\mathsf{T}$ estimated using the filtering procedure described in §12.1.2 and moves recursively backward until the initial time step is reached. The information flow during filtering and smoothing is illustrated in figure 12.10. The

| $t$ | 0 | 1 | $t$ | T − 1 | T |
|---|---|---|---|---|---|

Filter: $\boldsymbol{\mu}_0$ $\boldsymbol{\Sigma}_0$ → $\boldsymbol{\mu}_{1|1}$ $\boldsymbol{\Sigma}_{1|1}$ → $\cdots$ → $\boldsymbol{\mu}_{t|t}$ $\boldsymbol{\Sigma}_{t|t}$ → $\cdots$ → $\boldsymbol{\mu}_{T-1|T-1}$ $\boldsymbol{\Sigma}_{T-1|T-1}$ → $\boldsymbol{\mu}_{T|T}$ $\boldsymbol{\Sigma}_{T|T}$

Smooth: $\boldsymbol{\mu}_{0|T}$ $\boldsymbol{\Sigma}_{0|T}$ ← $\boldsymbol{\mu}_{1|T}$ $\boldsymbol{\Sigma}_{1|T}$ ← $\cdots$ ← $\boldsymbol{\mu}_{t|T}$ $\boldsymbol{\Sigma}_{t|T}$ ← $\cdots$ ← $\boldsymbol{\mu}_{T-1|T}$ $\boldsymbol{\Sigma}_{T-1|T}$

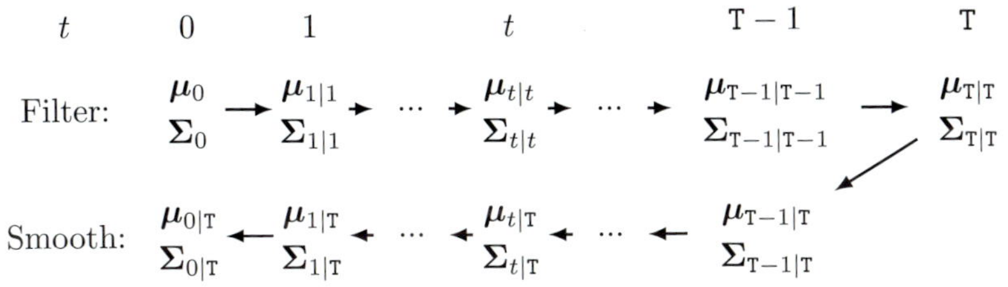

Figure 12.10: Information flow for the filtering and smoothing processes.

posterior knowledge $f(\mathbf{x}_t|\mathbf{y}_{1:T})$ combines the information coming from the prior knowledge $\{\boldsymbol{\mu}_0, \boldsymbol{\Sigma}_0\}$, the entire sequence of observation $\mathbf{y}_{1:T}$, and the transition model. The smoothing procedure can be employed to learn initial hidden states $\{\boldsymbol{\mu}_0, \boldsymbol{\Sigma}_0\} \to \{\boldsymbol{\mu}_{0|T}, \boldsymbol{\Sigma}_{0|T}\}$. Updating initial values using the smoother typically improves the model quality. The most common algorithm for performing smoothing is the *RTS Kalman smoother*,[4] which is defined following

$$
\begin{aligned}
f(\mathbf{x}_t|\mathbf{y}_{1:T}) &= \mathcal{N}(\mathbf{x}_t; \boldsymbol{\mu}_{t|T}, \boldsymbol{\Sigma}_{t|T}) \\
\boldsymbol{\mu}_{t|T} &= \boldsymbol{\mu}_{t|t} + \mathbf{J}_t \left( \boldsymbol{\mu}_{t+1|T} - \boldsymbol{\mu}_{t+1|t} \right) \\
\boldsymbol{\Sigma}_{t|T} &= \boldsymbol{\Sigma}_{t|t} + \mathbf{J}_t \left( \boldsymbol{\Sigma}_{t+1|T} - \boldsymbol{\Sigma}_{t+1|t} \right) \mathbf{J}_t^{\mathsf{T}} \\
\mathbf{J}_t &= \boldsymbol{\Sigma}_{t|t} \mathbf{A}^{\mathsf{T}} \boldsymbol{\Sigma}_{t+1|t}^{-1}.
\end{aligned}
$$

[4] Rauch, H. E., C. T. Striebel, and F. Tung (1965). Maximum likelihood estimates of linear dynamic systems. *AIAA Journal 3*(8), 1445–1450.

As illustrated in figure 12.10, the RTS Kalman smoother is performed iteratively starting from the Kalman filter step $t = T − 1$, where the mean vector $\boldsymbol{\mu}_{t+1|T} = \boldsymbol{\mu}_{T|T}$ and covariance matrix $\boldsymbol{\Sigma}_{t+1|T} = \boldsymbol{\Sigma}_{T|T}$ are already available from the last Kalman filter step.

*Example: Smoothing temperature estimates*   Figure 12.11 presents the RTS smoothing procedure applied to the temperature estimation example initially presented in figure 12.5. These results show how, between the filtering and smoothing procedures, the mean at $t = 2$ went from 8.6 to 8.4, and the variance from 4.2 to 1.8. The uncertainty is reduced with smoothing because the estimate $\mathrm{var}[X_{2|3}]$ is based on one more data point than $\mathrm{var}[X_{2|2}]$. Figure 12.11 presents the entire smoothed estimate, which almost coincides with the true values and has a narrower uncertainty confidence region than the filtering estimates.

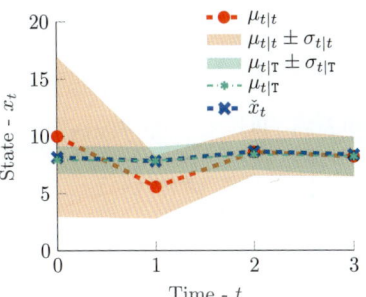

Figure 12.11: Comparison of filtering, smoothing, and the true states.

### 12.1.4  Parameter Estimation

The definition of matrices $\mathbf{A}$, $\mathbf{C}$, $\mathbf{Q}$, and $\mathbf{R}$ involved in the observation and transition models requires estimating several parameters $\boldsymbol{\theta}$ using observations $\mathcal{D} = \{y_{1:\mathsf{T}}\}$. As for other methods presented in previous chapters, the posterior for parameters is

$$f(\boldsymbol{\theta}|\mathcal{D}) = \frac{f(\mathcal{D}|\boldsymbol{\theta})f(\boldsymbol{\theta})}{f(\mathcal{D})}.$$

In this Bayesian formulation for estimating the posterior PDF for parameters, the key component is the marginal likelihood, $f(\mathcal{D}|\boldsymbol{\theta}) \equiv f(y_{1:\mathsf{T}}|\boldsymbol{\theta})$. With the hypothesis that observations are conditionally independent given the states $\mathbf{x}_t$, the joint marginal likelihood (i.e., the joint prior probability density of observations given a set of parameters) is

$$f(\mathbf{y}_{1:\mathsf{T}}|\boldsymbol{\theta}) = \prod_{t=1}^{\mathsf{T}} f(\mathbf{y}_t|\mathbf{y}_{1:t-1},\boldsymbol{\theta}),$$

where the marginal prior probability of each observation is obtained using the filtering procedure,

$$f(\mathbf{y}_t|\mathbf{y}_{1:t-1},\boldsymbol{\theta}) = \mathcal{N}(\mathbf{y}_t; \mathbf{C}\boldsymbol{\mu}_{t|t-1}, \mathbf{C}\boldsymbol{\Sigma}_{t|t-1}\mathbf{C}^{\mathsf{T}} + \mathbf{R}).$$

The posterior $f(\boldsymbol{\theta}|\mathcal{D})$ can be estimated using the sampling techniques[5] presented in chapter 7. However, because evaluating $f(\mathbf{y}_{1:\mathsf{T}}|\boldsymbol{\theta})$ can be computationally demanding, most practical applications found in the literature only employ the maximum likelihood estimate (MLE) rather than estimating the full posterior with a Bayesian approach. For an MLE, the optimal set of parameters is then defined as

$$\boldsymbol{\theta}^* = \arg\max_{\boldsymbol{\theta}} \overbrace{\ln f(\mathbf{y}_{1:\mathsf{T}}|\boldsymbol{\theta})}^{\text{log-likelihood}},$$

where because often $f(\mathbf{y}_t|\mathbf{y}_{1:t-1},\boldsymbol{\theta}) \ll 1$, the product of the marginals $\prod_{t=1}^{\mathsf{T}} f(\mathbf{y}_t|\mathbf{y}_{1:t-1},\boldsymbol{\theta}) \approx 0$ is sensitive to zero underflow issues. When the contrary happens, so that $f(\mathbf{y}_t|\mathbf{y}_{1:t-1},\boldsymbol{\theta}) \gg 1$, the problem of numerical overflow arises. In order to mitigate these issues, the log-likelihood is employed instead of the likelihood,

$$\ln f(\mathbf{y}_{1:\mathsf{T}}|\boldsymbol{\theta}) = \sum_{t=1}^{\mathsf{T}} \ln f(\mathbf{y}_t|\mathbf{y}_{1:t-1},\boldsymbol{\theta}).$$

As illustrated in figure 12.12, the optimal set of parameters $\boldsymbol{\theta}^*$ corresponds to the location where the derivative of the log-likelihood equals zero. The values $\boldsymbol{\theta}^*$ can be found using a gradient

We refer to $f(\mathbf{y}_{1:\mathsf{T}}|\boldsymbol{\theta})$ as the joint *marginal likelihood* because the hidden variables $\mathbf{x}_{1:\mathsf{T}}$ are marginalized. In the remainder of this chapter, for the sake of brevity, we use the term *likelihood* instead of *marginal likelihood*.

[5] Nguyen, L. H., I. Gaudot, and J.-A Goulet (2019). Uncertainty quantification for model parameters and hidden state variables in Bayesian dynamic linear models. *Structural Control and Health Monitoring* 26(3), e2309.

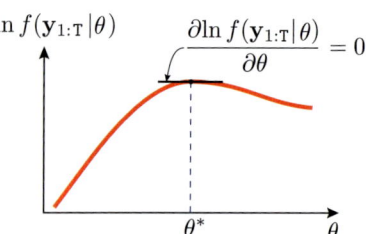

Figure 12.12: The MLE optimal parameters correspond to the location where the derivative of the log-likelihood equals zero.

ascent method such as the Newton-Raphson algorithm presented in §5.2. Note that because Newton-Raphson is a convex optimization method, and because SSMs typically involve a *non-convex* (see chapter 5) and *non-identifiable* (see §6.3.4) likelihood function, the optimization is particularly sensitive to initial values $\boldsymbol{\theta}_0$. It is thus essential to initialize parameters using engineering knowledge and to try several initial starting locations.

Note that a closed-form *expectation maximization* method exists for estimating the parameters of SSMs.[6] However, even if it is mathematically elegant, in the context of civil engineering applications such as those presented in the following sections, it is still sensitive to the selection of the initial values and is thus easily trapped in local maxima.

### 12.1.5  Limitations and Practical Considerations

There are some limitations to the Kalman filter formulation. First, it is sensitive to numerical errors when the covariance is rapidly reduced in the measurement step, for example, when accurate measurements are used or after a period with missing data, or when there are orders of magnitude separating the eigenvalues of the covariance matrix. Two alternative formulations that are numerically more robust are the UD filter and square-root filter. The reader interested in these methods should consult specialized textbooks such as the one by Simon[7] or Gibbs.[8]

A second key limiting aspect is that the formulation of the Kalman filter and smoother are only applicable for linear observation and transition models. When problems at hand require nonlinear models, there are two main types of alternatives: First, there are the *unscented Kalman filter*[9] and *extended Kalman filter*,[10] which allow for an approximation of the Kalman algorithm while still describing the hidden states using multivariate Normal random variables. The second option is to employ sampling methods such as the *particle filter*,[11] where the hidden-state variables are described by particles (i.e., samples) that are propagated through the transition model. The weight of each particle is updated using the likelihood computed from the observation model. Particle filtering is suited to estimate PDFs that are not well represented by the multivariate Normal. The reader interested in these nonlinear-compatible formulations can consult Murphy's[12] textbook for further details.

[6] Ghahramani, Z. and G. E. Hinton (1996). Parameter estimation for linear dynamical systems. Technical Report CRG-TR-96-2, University of Toronto, Dept. of Computer Science.

[7] Simon, D. (2006). *Optimal state estimation: Kalman, H infinity, and nonlinear approaches.* Wiley.

[8] Gibbs, B. P. (2011). *Advanced Kalman filtering, least-squares and modeling: A practical handbook.* Wiley.

[9] Wan, E. A. and R. van der Merwe (2000). The unscented Kalman filter for nonlinear estimation. In *Adaptive Systems for Signal Processing, Communications, and Control Symposium,* 153–158. IEEE.

[10] Ljung, L. (1979). Asymptotic behavior of the extended Kalman filter as a parameter estimator for linear systems. *IEEE Transactions on Automatic Control 24*(1), 36–50.

[11] Del Moral, P. (1997). Non-linear filtering: interacting particle resolution. *Comptes Rendus de l'Académie des Sciences—Series I—Mathematics 325*(6), 653–658.

[12] Murphy, K. P. (2012). *Machine learning: A probabilistic perspective.* MIT Press.

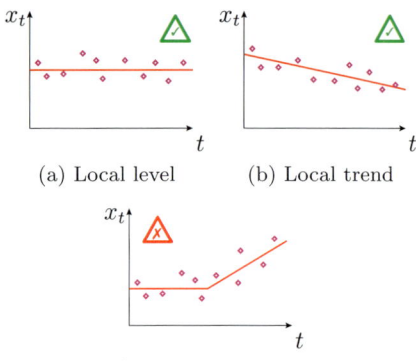

(a) Local level          (b) Local trend

(c) Regime switching

Figure 12.13: Comparison between single- and multiple-regime models.

[13] Blom, H. A. P. and Y. Bar-Shalom (1988). The interacting multiple model algorithm for systems with Markovian switching coefficients. *IEEE Transactions on Automatic Control 33*(8), 780–783.
[14] Murphy, K. P. (1998). Switching Kalman filters. Citeseer; and Kim, C.-J. and C. R. Nelson (1999). *State-space models with regime switching: Classical and Gibbs-sampling approaches with applications.* MIT Press.

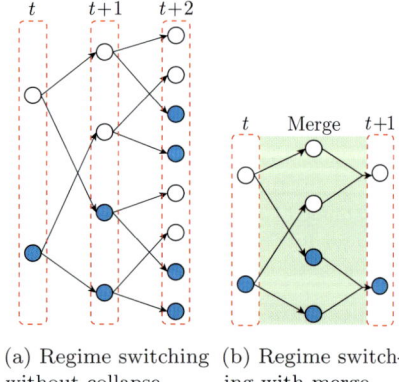

(a) Regime switching without collapse    (b) Regime switching with merge (SKF)

Figure 12.14: Comparison of regime switching with and without merging the state at each time step. (Adapted from Nguyen and Goulet (2018).)

## 12.2   State-Space Models with Regime Switching

The state-space model theory we have covered so far is applicable for a set of system responses following a same regime. Figures 12.13a–b are both examples of behaviors where SSMs are applicable. Figure 12.13c presents an example where a standard SSM is not applicable because there is a switch from a zero-speed regime to a constant-speed regime.

Estimating the probability of hidden states along with the probability of each regime involves computational challenges. For example, if we consider a case with $\mathtt{S} = 2$ possible regimes, we then have 2 possible transitions along with 2 models to describe each regime. We start from a time step $t$, where the system can be in regime 1 with probability $\pi_1^1$ and in regime 2 with probability $\pi_1^2 = 1 - \pi_1^1$, as illustrated in figure 12.14a. There are 2 regimes × 2 possible transition paths, leading to 4 possible combinations to describe the joint probability at times $t$ and $t + 1$. Going from $t + 1$ to $t + 2$, each of the 4 paths at $t + 1$ branches into 2 new ones at $t + 2$, leading to a total of now 8 paths. We see through this example that the number of paths to consider increases exponentially ($\mathtt{S}^{\mathtt{T}}$) with the number of time steps $\mathtt{T}$. Even for short time series, this approach becomes computationally prohibitive. In this section, we present the *generalized pseudo Bayesian algorithm of order two*,[13] which is behind the *switching Kalman filter*.[14] The key aspect of this method is to collapse the possible transition paths at each time step, as illustrated in figure 12.14b.

### 12.2.1   Switching Kalman Filter

We first explore the switching Kalman filter (SKF) while considering only $\mathtt{S} = 2$ possible regimes. The goal is to estimate at each time step $t$ the probability of each regime, along with the estimates for the hidden states corresponding to each of them. The notation that will be employed in the SKF is

$$
\begin{aligned}
\boldsymbol{\mu}_{t-1|t-1}^i &= \mathbb{E}[\mathbf{X}_{t-1}|\mathbf{y}_{1:t-1}, s_{t-1} = i] \\
\boldsymbol{\Sigma}_{t-1|t-1}^i &= \mathrm{cov}[\mathbf{X}_{t-1}, \mathbf{X}_{t-1}|\mathbf{y}_{1:t-1}, s_{t-1} = i] \\
\pi_{t-1|t-1}^i &= \mathrm{Pr}(S_{t-1} = i|\mathbf{y}_{1:t-1}),
\end{aligned}
$$

where the superscript on the mean vector and covariance matrix refers to the regime number, which is described by the variable $s \in \{1, 2, \cdots, \mathtt{S}\}$. $\pi_{t|t}^i$ is the shorthand notation for the probability of the regime $i$ at time $t$, given all the available observations up to time $t$, $\mathrm{Pr}(S_t = i|\mathbf{y}_{1:t})$.

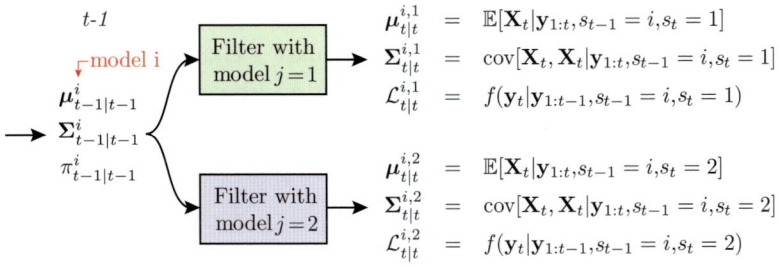

$$\mu_{t|t}^{i,1} = \mathbb{E}[\mathbf{X}_t | \mathbf{y}_{1:t}, s_{t-1} = i, s_t = 1]$$
$$\Sigma_{t|t}^{i,1} = \text{cov}[\mathbf{X}_t, \mathbf{X}_t | \mathbf{y}_{1:t}, s_{t-1} = i, s_t = 1]$$
$$\mathcal{L}_{t|t}^{i,1} = f(\mathbf{y}_t | \mathbf{y}_{1:t-1}, s_{t-1} = i, s_t = 1)$$

$$\mu_{t|t}^{i,2} = \mathbb{E}[\mathbf{X}_t | \mathbf{y}_{1:t}, s_{t-1} = i, s_t = 2]$$
$$\Sigma_{t|t}^{i,2} = \text{cov}[\mathbf{X}_t, \mathbf{X}_t | \mathbf{y}_{1:t}, s_{t-1} = i, s_t = 2]$$
$$\mathcal{L}_{t|t}^{i,2} = f(\mathbf{y}_t | \mathbf{y}_{1:t-1}, s_{t-1} = i, s_t = 2)$$

Figure 12.15: Switching Kalman filter: How initial states transition through two distinct models.

*Filtering step*   As presented in figure 12.15, starting from regime 1, at $t - 1$, there are two possible paths, each of them corresponding to one of the regime switches. The two possible paths $j \in \{1, 2\}$ are explored by estimating the posterior state estimates $\{\mu_{t|t}^{1,j}, \Sigma_{t|t}^{1,j}\}$, along with their likelihood $\mathcal{L}_{t|t}^{1,j}$, using the filtering procedure presented in §12.1.2,

$$(\mu_{t|t}^{i,j}, \Sigma_{t|t}^{i,j}, \mathcal{L}_{t|t}^{i,j}) = \text{Filter}(\mu_{t-1|t-1}^{i}, \Sigma_{t-1|t-1}^{i}, \mathbf{y}_t; \mathbf{A}_j, \mathbf{C}_j, \mathbf{Q}_j, \mathbf{R}_j). \tag{12.7}$$

In equation 12.7, the initial states are obtained from the regime $i$, and the matrices $\{\mathbf{A}_j, \mathbf{C}_j, \mathbf{Q}_j, \mathbf{R}_j\}$ are those from the model $j$.

Once the procedure presented in figure 12.15 is repeated for each model, the next step consists in merging the state estimates that landed in a same regime. Figure 12.16 presents an example where the state estimates from the model starting from regime 1 at $t - 1$ and transiting through regime 1 are merged with the state estimates from the model starting from regime 2 at $t - 1$ and transiting through regime 1. This merge step is performed by considering that the final state estimates $\{\mu_{t|t}^{i}, \Sigma_{t|t}^{i}\}$, along with its probability $\pi_{t|t}^{i}$, describe a multivariate Normal, itself obtained by mixing the state estimates originating from all paths that transited from $t - 1 \to t$ through model $j$.

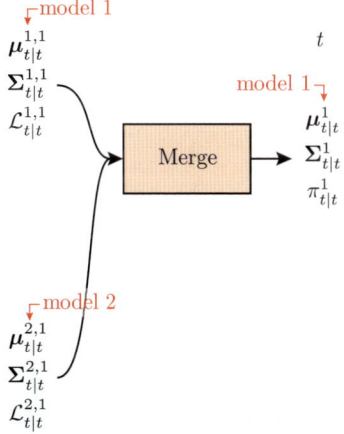

Figure 12.16: The switching Kalman filter *merge step.*

*Gaussian mixture*   We now take a quick detour from the switching Kalman filter in order to review the mathematical details related to *Gaussian mixtures* (see §10.1.1) that are behind the merge step. For example, we have two states $X_1$ and $X_2$ described by Gaussian random variables so that

$$X_1 \sim \mathcal{N}(x_1; \mu_1, \sigma_1^2)$$
$$X_2 \sim \mathcal{N}(x_2; \mu_2, \sigma_2^2).$$

The outcome of the random process, where $\pi_1$ and $\pi_2 = 1 - \pi_1$ correspond to the probability of each state, is described by a Gaussian mixture,

$$X_{12} \sim f(x_{12}) = \pi_1 \mathcal{N}(x_{12}; \mu_1, \sigma_1^2) + \pi_2 \mathcal{N}(x_{12}; \mu_2, \sigma_2^2).$$

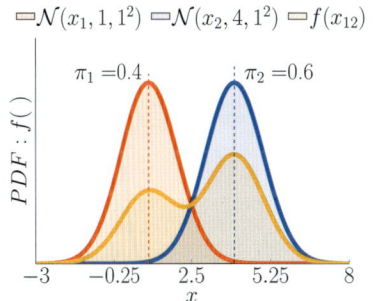

Figure 12.17: Example of a Gaussian mixture PDF.

[15] Runnalls, A. R. (2007). Kullback-Leibler approach to Gaussian mixture reduction. *IEEE Transactions on Aerospace and Electronic Systems 43*(3), 989–999.

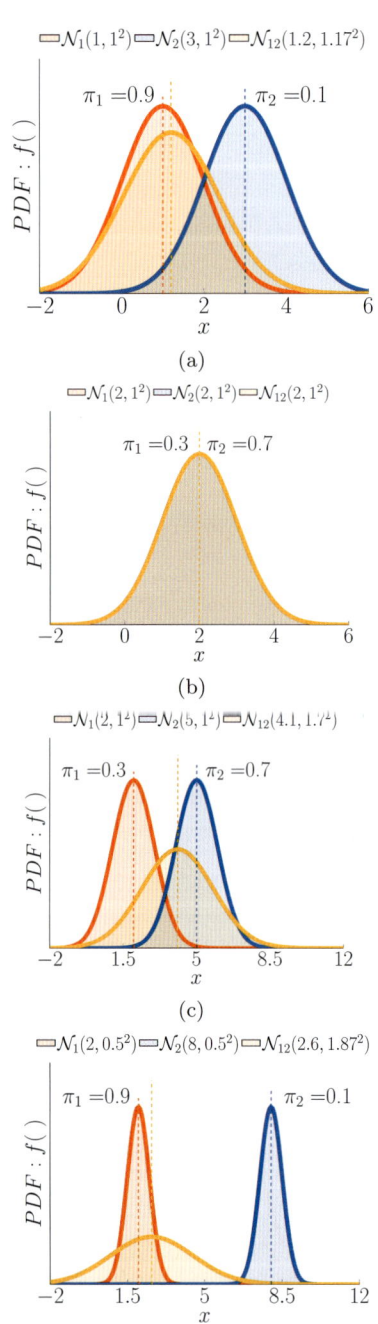

Figure 12.18: Examples of Gaussian mixtures.

Figure 12.17 presents an example of two Gaussian PDFs combining in a Gaussian mixture. We can approximate a Gaussian mixture by a single Gaussian[15] $f(x_{12}) \approx \mathcal{N}(x_{12}; \mu_{12}, \sigma_{12}^2)$ having a mean and variance given by

$$
\begin{aligned}
\mu_{12} &= \pi_1 \mu_1 + \pi_2 \mu_2 \\
\sigma_{12}^2 &= \pi_1 \sigma_1^2 + \pi_2 \sigma_2^2 + \pi_1 \pi_2 (\mu_1 - \mu_2)^2.
\end{aligned}
$$

Figure 12.18 presents four examples of Gaussian mixtures approximated by a single Gaussian PDF. Notice how in (b), when the PDFs of $X_1$ and $X_2$ have the same parameters, the result coincides with the two distributions. Also, by comparing figures (c) and (d), we see that the greater the distance $|\mu_1 - \mu_2|$ is, the greater $\sigma_{12}$ is.

*Merge step*   Going back to the switching Kalman filter, the merge step is based on the approximation of a Gaussian mixture by a single Gaussian PDF. The joint posterior probability of the regimes $S_{t-1} = i$ and $S_t = j$ is proportional to the product of the likelihood $\mathcal{L}_{t|t}^{i,j}$, the prior probability $\pi_{t-1|t-1}^i$, and the probability $Z^{i,j}$ of transiting from $s_{t-1} = i$ to $s_t = j$,

$$
\begin{aligned}
\mathbf{M}_{t-1,t|t}^{i,j} &= \Pr(S_{t-1} = i, S_t = j | \mathbf{y}_{1:t}) \\
&= \frac{\mathcal{L}_{t|t}^{i,j} \cdot z^{i,j} \cdot \pi_{t-1|t-1}^i}{\sum_i \sum_j \mathcal{L}_{t|t}^{i,j} \cdot Z^{i,j} \cdot \pi_{t-1|t-1}^i}.
\end{aligned}
\tag{12.8}
$$

The posterior probability of the regime $j$ at time $t$ is obtained by marginalizing the joint posterior in equation 12.8. For a regime $j$, this operation corresponds to summing the probability of all paths transiting through model $j$, irrespective of their origin,

$$
\pi_{t|t}^j = \sum_{i=1}^{s} \mathbf{M}_{t-1,t|t}^{i,j}.
$$

The posterior states for each regime $j$, $\boldsymbol{\mu}_{t|t}^j, \boldsymbol{\Sigma}_{t|t}^j$, are estimated using

$$
\begin{aligned}
\mathbf{W}_{t-1|t}^{i,j} &= \Pr(S_{t-1} = i | s_t = j, \mathbf{y}_{1:t}) = \frac{\mathbf{M}_{t-1,t|t}^{i,j}}{\pi_{t|t}^j} \\
\boldsymbol{\mu}_{t|t}^j &= \sum_{i=1}^{s} \boldsymbol{\mu}_{t|t}^{i,j} \mathbf{W}_{t-1|t}^{i,j} \\
\boldsymbol{\mu} &= \boldsymbol{\mu}_{t|t}^{i,j} - \boldsymbol{\mu}_{t|t}^j \\
\boldsymbol{\Sigma}_{t|t}^j &= \sum_{i=1}^{s} \left[ \mathbf{W}_{t-1|t}^{i,j} (\boldsymbol{\Sigma}_{t|t}^{i,j} + \boldsymbol{\mu}\boldsymbol{\mu}^\intercal) \right].
\end{aligned}
$$

The general overview of the switching Kalman filter is presented in figure 12.19. Note that in addition to the parameters defining each filtering model, the SKF requires defining the prior probability of each regime $\pi_0^j$ along with the $\mathtt{S} \times \mathtt{S}$ transition matrix $\mathbf{Z}$ defining the probability of transiting from a state at time $t - 1$ to another at time $t + 1$.

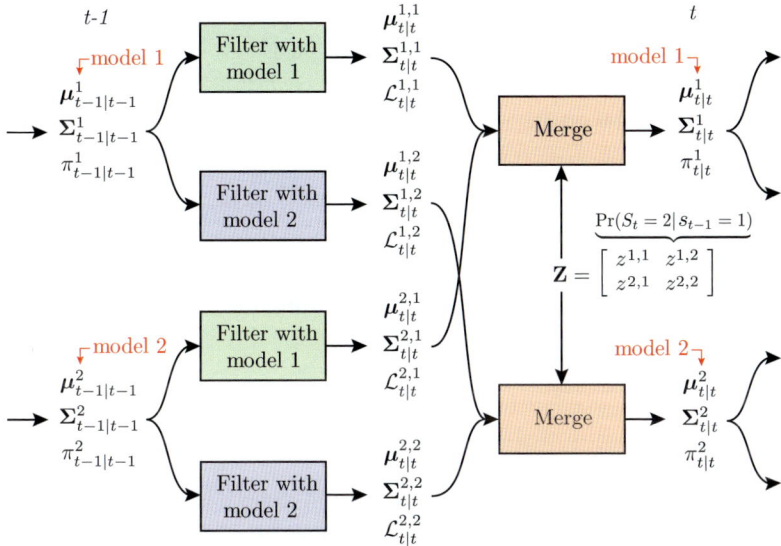

Figure 12.19: General overview of the switching Kalman filter procedure.

### 12.2.2  Example: Temperature Data with Regime Switch

This example explores an application of the switching Kalman filter to temperature data subject to a regime switch. Figure 12.20 presents the data set where there is a switch between a zero-speed regime and a constant-speed regime. Note that with the SKF, the filtering model associated with each regime must have the same hidden states. In this example, the two states $\mathbf{x}_t = [x_t \ \dot{x}_t]^\mathsf{T}$ are the temperature and its rate of change, with initial states for both models initialized as

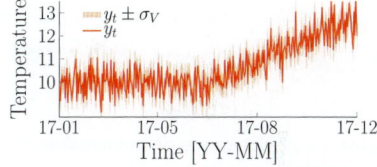

Figure 12.20: Example of data set where there is a switch between a zero-speed regime and a constant-speed regime.

$$\boldsymbol{\mu}_0 = \begin{bmatrix} 10 \\ 0 \end{bmatrix}, \ \boldsymbol{\Sigma}_0 = \begin{bmatrix} 0.01^2 & 0 \\ 0 & 0.001^2 \end{bmatrix}, \ \boldsymbol{\pi}_0 = \begin{bmatrix} 0.95 \\ 0.05 \end{bmatrix}.$$

Note that the initial states $\{\boldsymbol{\mu}_0, \boldsymbol{\Sigma}_0, \boldsymbol{\pi}_0\}$ can be estimated using the smoothing algorithm. The first regime is described by a local level where the rate of change is forced to be equal to 0 in the transition matrix $[\mathbf{A}]_{:,2} = 0$,

$$\overbrace{\begin{bmatrix} x_t \\ \dot{x}_t \end{bmatrix}}^{} \begin{bmatrix} 1 & 0 \\ 0 & 0 \end{bmatrix} \quad \overbrace{\begin{bmatrix} w_t \\ \dot{w}_t \end{bmatrix}}^{\mathbf{W} \sim \mathcal{N}(\mathbf{0}, \mathbf{Q}_{11/12})}$$

$$\underbrace{\mathbf{x}_t}_{} = \underbrace{\mathbf{A}}_{} \ \mathbf{x}_{t-1} + \underbrace{\mathbf{w}_t}_{} \quad \text{(Regime 1)}.$$

The second regime is described by a local trend model where

$$\underbrace{\begin{bmatrix} x_t \\ \dot{x}_t \end{bmatrix}}_{\mathbf{x}_t} = \underbrace{\begin{bmatrix} 1 & \Delta t \\ 0 & 1 \end{bmatrix}}_{\mathbf{A}} \mathbf{x}_{t-1} + \overbrace{\underbrace{\begin{bmatrix} w_t \\ \dot{w}_t \end{bmatrix}}_{}}^{\mathbf{W}\sim\mathcal{N}(\mathbf{0},\mathbf{Q}_{22/21})} \quad \text{(Regime 2)}.$$

Describing the four possible paths, that is, $\{1 \rightarrow 1, \ 1 \rightarrow 2, \ 2 \rightarrow 1, \ 2 \rightarrow 2\}$, requires the definition of four covariance matrices for the transition-model errors,

$$\mathbf{Q}_{11} = \mathbf{Q}_{21} = \mathbf{Q}_{22} = \mathbf{0}, \ \mathbf{Q}_{12} = \begin{bmatrix} 0 & 0 \\ 0 & \sigma_{12}^2 \end{bmatrix}.$$

In $[\mathbf{Q}_{12}]_{2,2}$, the MLE optimal parameter estimated using empirical data is $\sigma_{12} = 2.3 \times 10^{-4}$. There is a single observation model for both regimes defined following

$$y_t = \overbrace{\mathbf{C}}^{[1,0]} \mathbf{x}_t + \overbrace{v_t}^{V\sim\mathcal{N}(0,0.5^2)}.$$

The MLE for terms on the main diagonal are $z_{ii} = 0.997$ so that the complete transition matrix is defined as

$$\mathbf{Z} = \begin{bmatrix} 0.997 & 0.003 \\ 0.003 & 0.997 \end{bmatrix}.$$

Figure 12.21 presents the outcome from the switching Kalman filter: (a) presents the probability of the local trend regime, which marks the transition between the zero-speed and constant-speed regimes; (b) presents the level describing the filtered system responses where the observation errors were removed; and (c) presents the rate of change of the system responses, that is, the trend.

## 12.3 Linear Model Structures

This section presents how to build linear models from generic components. This approach is referred to as *Bayesian dynamic linear models* by Goulet and Nguyen,[16] and more commonly as *Bayesian forecasting* or *dynamic linear models* in the literature[17]. Note that the term *linear* does not mean that these models can only be employed to describe linear relationships with respect to time; it refers to the linearity of the observation and transition equations.

Five main components are reviewed here: local level, local trend, local acceleration, periodic, and autoregressive. As illustrated in figure 12.22, these components can be seen as the building

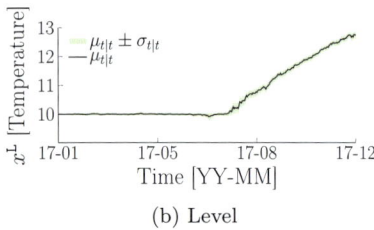

(a) Probability of the *local trend* regime

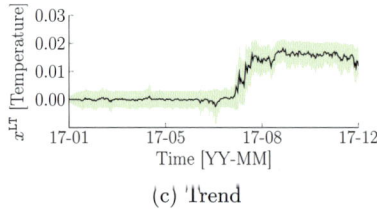

(b) Level

(c) Trend

Figure 12.21: Example of hidden-states estimation using the switching Kalman filter.

[16] Nguyen, L. H. and J.-A. Goulet (2018). Anomaly detection with the switching Kalman filter for structural health monitoring. *Structural Control and Health Monitoring* 25(4), e2136; and Goulet, J.-A. (2017). Bayesian dynamic linear models for structural health monitoring. *Structural Control and Health Monitoring* 24(12), 1545–2263.

[17] Harrison, P. J. and C. F. Stevens (1976). Bayesian forecasting. *Journal of the Royal Statistical Society: Series B (Methodological)* 38(3), 205–228; West, M. and J. Harrison (1999). *Bayesian forecasting and dynamic models*. Springer; and West, M. (2013). Bayesian dynamic modelling. In *Bayesian theory and applications*, 145–166. Oxford University Press.

Figure 12.22: Generic components that can be seen as the building blocks, which, once assembled, allow modeling complex empirical responses. From left to right: local level, local trend, local acceleration, periodic, and autoregressive.

blocks that, once assembled, allow modeling complex time series. Each component has its own specific mathematical formulation and is intended to model a specific behavior. We will first see the mathematical formulation associated with each component; then we present how to assemble components to form $\{\mathbf{A}, \mathbf{C}, \mathbf{Q}, \mathbf{R}\}$ matrices.

### 12.3.1   Generic Components

*Local level*   A local level describes a quantity that is *locally constant* over a time step. It is described by a single hidden variable $x^{\text{LL}}$. Because of its locally constant nature, its transition model predicts that the hidden state at $t$ is equal to the one at $t-1$, so the transition coefficient is 1. The standard deviation describing the prediction error for the locally constant model is $\sigma_W^{\text{LL}}$. Local levels are part of the hidden variables that contribute to observations, so the coefficient in the observation matrix equals 1. The hidden-state variables as well at the formulation for the transition and observation model matrices are

$$\mathbf{x}^{\text{LL}} = x^{\text{LL}}, \mathbf{A}^{\text{LL}} = 1,\ \mathbf{C}^{\text{LL}} = 1,\ \mathbf{Q}^{\text{LL}} = (\sigma^{\text{LL}})^2.$$

Local level ($x^{\text{LL}}$)

Figure 12.23 presents realizations of a stochastic process defined by the transition model of a local level. We can see that the process variability increases as the standard deviation $\sigma^{\text{LL}}$ increases.

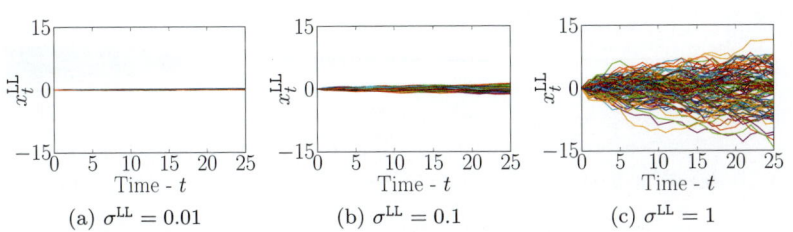

(a) $\sigma^{\text{LL}} = 0.01$    (b) $\sigma^{\text{LL}} = 0.1$    (c) $\sigma^{\text{LL}} = 1$

Figure 12.23: Examples of local level components. The initial state $x_0^{\text{LL}} = 0$ is the same for all realizations.

*Local trend*   A local trend describes a quantity that has a *locally constant* rate of change over a time step. It acts jointly with a hidden variable describing the level. It is thus defined by two hidden variables $[x^{\text{L}}\ x^{\text{LT}}]^{\intercal}$. Because of its locally constant nature, its transition model predicts that the hidden state at $t$ is equal to the current one, $x_t^{\text{LT}} = x_{t-1}^{\text{LT}}$, so the transition coefficient is 1. The

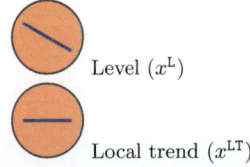

Level ($x^{\text{L}}$)

Local trend ($x^{\text{LT}}$)

hidden state $x_t^{\text{L}} = x_{t-1}^{\text{L}} + x_{t-1}^{\text{LT}}\Delta t$ is described by the current level value plus the change brought by the local trend, multiplied by the time step length $\Delta t$. As we have seen in §12.1.2, the definition of the process-noise covariance matrix $\mathbf{Q}^{\text{LT}}$ requires considering the uncertainty that is caused by a constant acceleration with a random magnitude that can occur over the span of a time step. Analytically, it corresponds to

$$\mathbf{Q}^{\text{LT}} = (\sigma^{\text{LT}})^2 \begin{bmatrix} \frac{\Delta t^4}{4} & \frac{\Delta t^3}{2} \\ \frac{\Delta t^3}{2} & \Delta t^2 \end{bmatrix},$$

where $\sigma^{\text{LT}}$ is the process noise parameter. In most cases involving a local trend, only the hidden variable representing the level contributes to observations, so the coefficient in the observation matrix for $x^{\text{LT}}$ equals 0. The hidden-state estimation as well as the formulation for the transition and observation model matrices are

$$\mathbf{x}^{\text{LT}} = \begin{bmatrix} x^{\text{L}} \\ x^{\text{LT}} \end{bmatrix}, \ \mathbf{A}^{\text{LT}} = \begin{bmatrix} 1 & \Delta t \\ 0 & 1 \end{bmatrix}, \ \mathbf{C}^{\text{LT}} = \begin{bmatrix} 1 \\ 0 \end{bmatrix}^{\text{T}}.$$

Figure 12.24 presents realizations of a stochastic process defined by the transition model of a local trend. We can see that the hidden-state variable $x^{\text{LT}}$ is locally constant and that $x^{\text{L}}$ has a locally constant speed. A local trend is employed to model a system response that changes over time with a locally constant speed.

Figure 12.24; Examples of local trend components. The initial state $[x_0^{\text{L}} \ x_0^{\text{LT}}] = [0 \ -1]^{\text{T}}$ is the same for all realizations.

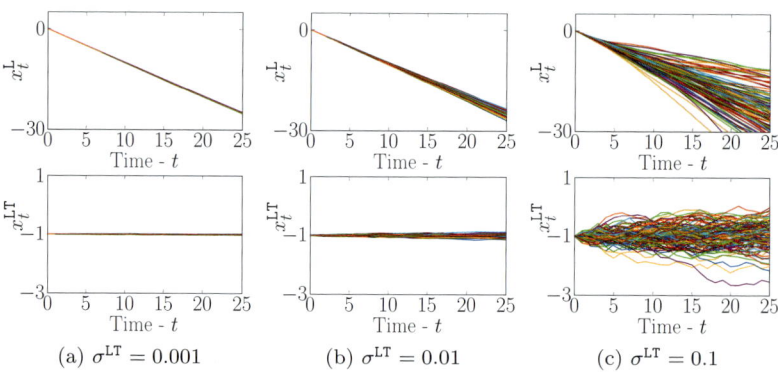

(a) $\sigma^{\text{LT}} = 0.001$   (b) $\sigma^{\text{LT}} = 0.01$   (c) $\sigma^{\text{LT}} = 0.1$

*Local acceleration*   A local acceleration describes a quantity that has a *locally constant* acceleration over a time step. It acts jointly with a level hidden variable and a trend hidden variable. It is thus defined by three hidden-state variables $[x^{\text{L}} \ x^{\text{T}} \ x^{\text{LA}}]^{\text{T}}$. Because of its locally constant nature, its transition model predicts that the hidden state at $t$ is equal to the current one, $x_t^{\text{LA}} = x_{t-1}^{\text{LA}}$, so the transition coefficient is 1. The hidden state $x_t^{\text{T}} = x_{t-1}^{\text{T}} + x_{t-1}^{\text{LA}}\Delta t$ is

Level ($x^{\text{L}}$)

Trend ($x^{\text{T}}$)

Local acceleration ($x^{\text{LA}}$)

described by the current value plus the change brought by the local acceleration multiplied by the time step length $\Delta t$. The hidden state $x_t^{\mathrm{L}} = x_{t-1}^{\mathrm{L}} + x_{t-1}^{\mathrm{T}}\Delta t + x_{t-1}^{\mathrm{LA}}\frac{\Delta t^2}{2}$. The hidden-state variables as well as the formulation for the transition and observation model matrices are

$$
\mathbf{x}^{\mathrm{LA}} = \begin{bmatrix} x^{\mathrm{L}} \\ x^{\mathrm{T}} \\ x^{\mathrm{LA}} \end{bmatrix}, \quad \mathbf{A}^{\mathrm{LA}} = \begin{bmatrix} 1 & \Delta t & \frac{\Delta t^2}{2} \\ 0 & 1 & \Delta t \\ 0 & 0 & 1 \end{bmatrix}, \quad \mathbf{C}^{\mathrm{LA}} = \begin{bmatrix} 1 \\ 0 \\ 0 \end{bmatrix}^{\mathsf{T}},
$$

$$
\mathbf{Q}^{\mathrm{LA}} = (\sigma^{\mathrm{LA}})^2 \begin{bmatrix} \frac{\Delta t^4}{4} & \frac{\Delta t^3}{2} & \frac{\Delta t^2}{2} \\ \frac{\Delta t^3}{2} & \Delta t^2 & \Delta t \\ \frac{\Delta t^2}{2} & \Delta t & 1 \end{bmatrix}.
$$

Figure 12.25 presents realizations of a stochastic process defined by the transition model of a local acceleration. We can see that the $x^{\mathrm{LA}}$ hidden variable is locally constant, $x^{\mathrm{T}}$ has a locally constant speed, and $x^{\mathrm{L}}$ has a locally constant acceleration. A local acceleration is employed to model a system response that changes over time with a locally constant acceleration.

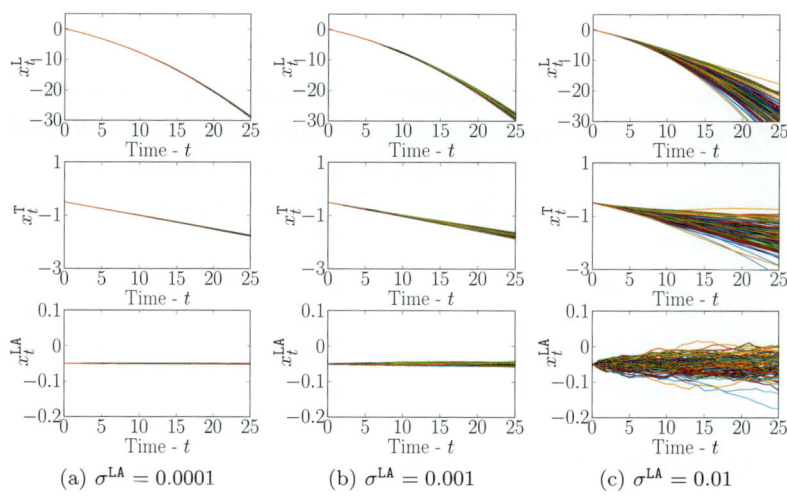

(a) $\sigma^{\mathrm{LA}} = 0.0001$    (b) $\sigma^{\mathrm{LA}} = 0.001$    (c) $\sigma^{\mathrm{LA}} = 0.01$

Figure 12.25: Examples of local acceleration components. The initial state $[x_0^{\mathrm{L}} \; x_0^{\mathrm{T}} \; x_0^{\mathrm{LA}}] = [0 \; -0.5 \; -0.5]^{\mathsf{T}}$ is the same for all realizations.

*Periodic*  Periodic components are expressed in their *Fourier form* where they can be described by two hidden variables $\mathbf{x}^{\mathrm{S}} = [x^{\mathrm{S1}} \; x^{\mathrm{S2}}]^{\mathsf{T}}$. Here, only the first hidden variable contributes to observations, so the coefficients in the observation matrix are 1 for $x^{\mathrm{S1}}$ and 0 for $x^{\mathrm{S2}}$. The hidden-state variables as well as the formulation for the transition and observation model matrices are

Periodic ($x^{\mathrm{S1}}, x^{\mathrm{S2}}$)

$$
\mathbf{x}^{\mathrm{S}} = \begin{bmatrix} x^{\mathrm{S1}} \\ x^{\mathrm{S2}} \end{bmatrix}, \quad \mathbf{A}^{\mathrm{S}} = \begin{bmatrix} \cos\omega & \sin\omega \\ -\sin\omega & \cos\omega \end{bmatrix}, \quad \mathbf{C}^{\mathrm{S}} = \begin{bmatrix} 1 \\ 0 \end{bmatrix}^{\mathsf{T}},
$$

$$\mathbf{Q}^{\mathrm{S}} = (\sigma^{\mathrm{S}})^2 \begin{bmatrix} 1 & 0 \\ 0 & 1 \end{bmatrix},$$

where the frequency $\omega = \frac{2\pi \cdot \Delta t}{p}$ is defined as a function of the period $p$ and the time step length. Figure 12.26 presents realizations of the hidden variables describing the periodic process. A periodic component is employed to model sine-like periodic phenomena with known period $p$. For example, in the context of civil engineering, it can be employed to model the effect of daily and seasonal temperature changes.

Figure 12.26: Examples of periodic components. The initial state $[x_0^{\mathrm{S1}}\ x_0^{\mathrm{S2}}] = [0\ 1]^{\mathsf{T}}$ and the period $p = 10$ are the same for all realizations.

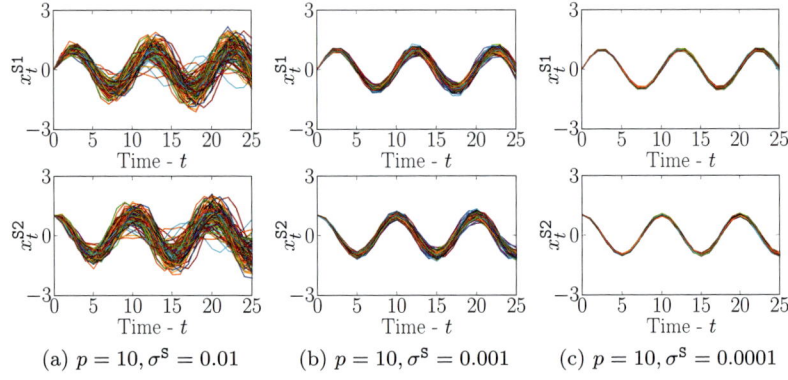

(a) $p = 10, \sigma^{\mathrm{S}} = 0.01$    (b) $p = 10, \sigma^{\mathrm{S}} = 0.001$    (c) $p = 10, \sigma^{\mathrm{S}} = 0.0001$

*Autoregressive*   The autoregressive (AR) component is described by a single hidden variable $x^{\mathrm{AR}}$ that presents a dependence on itself over consecutive time steps. Here, we only present the autoregressive process of order 1, where the current state only depends on the previous one,

$$x_t^{\mathrm{AR}} = \phi^{\mathrm{AR}} x_{t-1}^{\mathrm{AR}} + w^{\mathrm{AR}}, \quad w^{\mathrm{AR}} : W^{\mathrm{AR}} \sim \mathcal{N}(w^{\mathrm{AR}}; 0, (\sigma^{\mathrm{AR}})^2).$$

$\phi^{\mathrm{AR}}$ is the autoregression coefficient defining the transition matrix. The autoregressive hidden-state variable contributes to the observation, so the coefficient in the observation matrix equals 1. The state variable as well as the formulation for the transition and observation model matrices are

$$\mathbf{x}^{\mathrm{AR}} = x^{\mathrm{AR}}, \ \mathbf{A}^{\mathrm{AR}} = \phi^{\mathrm{AR}}, \ \mathbf{C}^{\mathrm{AR}} = 1, \ \mathbf{Q}^{\mathrm{AR}} = (\sigma^{\mathrm{AR}})^2.$$

Figure 12.27 presents examples of realizations of an autoregressive process for different parameters. Like for other components, the variability increases with the process-error standard deviation $\sigma^{\mathrm{AR}}$. A key property of the autoregressive process is that for $0 < \phi^{\mathrm{AR}} < 1$, it leads to a stationary process for which the standard deviation is

$$(\sigma^{\mathrm{AR},0})^2 = \frac{(\sigma^{\mathrm{AR}})^2}{1 - (\phi^{\mathrm{AR}})^2}.$$

Autoregressive ($x^{\mathrm{AR}}$)

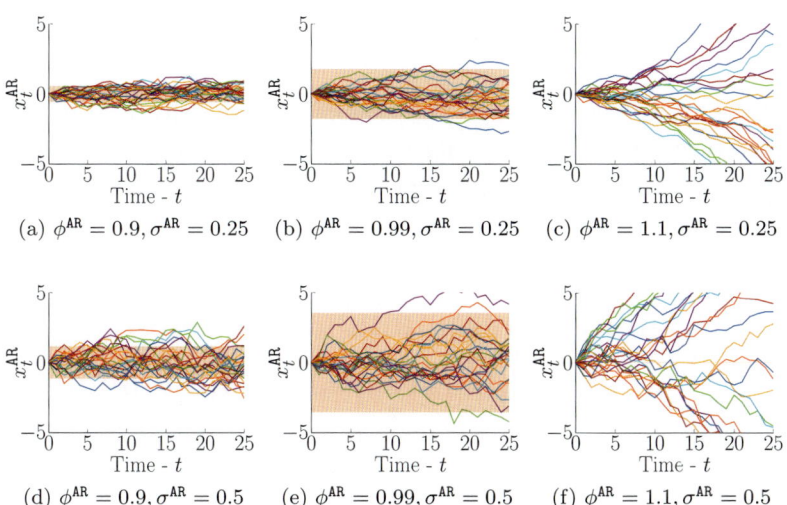

Figure 12.27: Examples of autoregressive components. The shaded region corresponds to the $\pm\sigma^{\mathrm{AR},0}$ confidence region for the process's stationary parameters. Note that cases (c) and (f) are nonstationary because $\phi^{\mathrm{AR}} > 1$. The initial state $x_0^{\mathrm{AR}} = 0$ is the same for all realizations.

(a) $\phi^{\mathrm{AR}} = 0.9, \sigma^{\mathrm{AR}} = 0.25$    (b) $\phi^{\mathrm{AR}} = 0.99, \sigma^{\mathrm{AR}} = 0.25$    (c) $\phi^{\mathrm{AR}} = 1.1, \sigma^{\mathrm{AR}} = 0.25$

(d) $\phi^{\mathrm{AR}} = 0.9, \sigma^{\mathrm{AR}} = 0.5$    (e) $\phi^{\mathrm{AR}} = 0.99, \sigma^{\mathrm{AR}} = 0.5$    (f) $\phi^{\mathrm{AR}} = 1.1, \sigma^{\mathrm{AR}} = 0.5$

The confidence region corresponding to $\pm\sigma^{\mathrm{AR},0}$ is represented by the shaded area. Notice that for $\phi^{\mathrm{AR}} \geq 1$ the process is nonstationary, which explains why examples (c) and (f) do not display the $\pm\sigma^{\mathrm{AR},0}$ region. The special case where $\phi^{\mathrm{AR}} = 1$ corresponds to a *random walk*, which is equivalent to the local level component because $x_t = x_{t-1} + w_t$.

The autoregressive (AR) component plays a key role in the model. It is there to describe the model errors that have inherent dependencies over time steps; the definition of a transition model involves some approximation errors, and because the transition model is employed to make predictions over successive time steps, the prediction errors are correlated across time. Moreover, note that this correlation tends to one as the time-step length tends to zero. If we do not employ an AR process to describe these time-dependent model errors, our ability to correctly estimate other hidden-state variables can be diminished.

### 12.3.2   Component Assembly

A time series of observations $y_{1:\mathtt{T}}$ is modeled by selecting the relevant components and then assembling them to form the global matrices $\{\mathbf{A}, \mathbf{C}, \mathbf{Q}, \mathbf{R}\}$. The global hidden-state variable vector $\mathbf{x}$ is obtained by concatenating in a column vector the state variables defining each selected component. The observation matrix $\mathbf{C}$ is obtained by concatenating in a row vector the specific $\mathbf{C}$ vectors defining each selected component. The transition matrix $\mathbf{A}$ and the process noise covariance $\mathbf{Q}$ are obtained by forming *block diagonal* matrices ($\mathrm{blkdiag}(\cdot, \cdot)$). $\mathbf{A}$ and $\mathbf{Q}$ each employs the specific terms defining the selected components. For a single time series, the observation error covariance matrix $\mathbf{R}$ contains only one term, $\sigma_V^2$.

*Example: Climate change*    This example studies daily average temperature data recorded at Montreal's YUL airport for the period 1953–2013, as presented in figure 12.28. The goal is to quantify the current yearly rate of change in the average temperature. Because

Figure 12.28: Montreal YUL airport daily average temperature observations for the period 1953–2013.

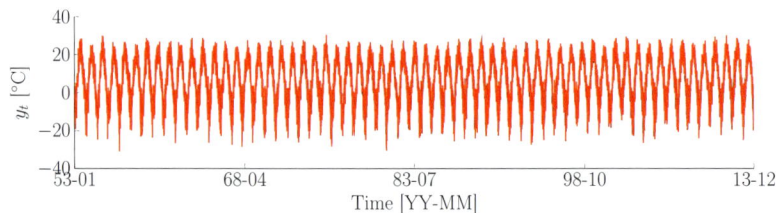

data is recorded daily, the time-step length is $\Delta t = 1\,\text{day}$. Figure 12.28 shows a yearly seasonal pattern with frequency $\omega = \frac{2\pi}{365.2422}$. In addition to the periodic component, this model requires a local trend to quantify the rate of change and an autoregressive one to capture the time-correlated model prediction errors. For these components, the hidden-state variables are

$$\mathbf{x}_t = \begin{bmatrix} x^{\text{L}} & x^{\text{LT}} & x^{\text{S1}} & x^{\text{S2}} & x^{\text{AR}} \end{bmatrix}^{\mathsf{T}}.$$

**Note:** The length of a solar year is 365.2422 days.

The specific formulation for each component is described below:
Local trend component - LT

$$\mathbf{A}^{\text{LT}} = \begin{bmatrix} 1 & 1 \\ 0 & 1 \end{bmatrix}$$

$$\mathbf{Q}^{\text{LT}} = (\sigma^{\text{LT}})^2 \cdot \begin{bmatrix} 1/4 & 1/2 \\ 1/2 & 1 \end{bmatrix}$$

$$\mathbf{C}^{\text{LT}} = \begin{bmatrix} 1 & 0 \end{bmatrix}$$

Periodic component - S

$$\mathbf{A}^{\text{S}} = \begin{bmatrix} \cos\omega & \sin\omega \\ -\sin\omega & \cos\omega \end{bmatrix}, \quad \omega = \frac{2\pi}{365.2422}$$

$$\mathbf{Q}^{\text{S}} = \mathbf{0}\ (2 \times 2\ \text{matrix})$$

$$\mathbf{C}^{\text{S}} = \begin{bmatrix} 1 & 0 \end{bmatrix}$$

Autoregressive component - AR

$$\mathbf{A}^{\text{AR}} = [\phi^{\text{AR}}]$$

$$\mathbf{Q}^{\text{AR}} = [(\sigma^{\text{AR}})^2]$$

$$\mathbf{C}^{\text{AR}} = [1]$$

The global matrices describing the model are obtained by assembling the components so that

$$\begin{aligned}
\mathbf{A} &= \text{blkdiag}(\mathbf{A}^{\text{LT}}, \mathbf{A}^{\text{S}}, \mathbf{A}^{\text{AR}}) \\
\mathbf{Q} &= \text{blkdiag}(\mathbf{Q}^{\text{LT}}, \mathbf{Q}^{\text{S}}, \mathbf{Q}^{\text{AR}}) \\
\mathbf{C} &= [\mathbf{C}^{\text{LT}} \ \mathbf{C}^{\text{S}} \ \mathbf{C}^{\text{AR}}] \\
\mathbf{R} &= \sigma_V^2.
\end{aligned}$$

The prior knowledge defining the initial states is defined as

$$\begin{aligned}
\mathbb{E}[X_0] &= [5\ 0\ 0\ 0\ 0]^{\mathsf{T}} \\
\text{cov}[X_0] &= \text{diag}([100\ 10^{-6}\ 100\ 100\ 100]).
\end{aligned}$$

The unknown parameters and their MLE optimal values learned from data using the Newton-Raphson algorithm are

$$\boldsymbol{\theta}^* = [\phi^{\text{AR}}, \sigma^{\text{LT}}, \sigma^{\text{AR}}, \sigma_V]^{\mathsf{T}} = [0.69\ 2.2 \times 10^{-6}\ 3.4\ 2.1 \times 10^{-4}]^{\mathsf{T}}.$$

Figure 12.29 presents the smoothed estimates for hidden-state variables $\mathbf{x}_t$. Notice in figure 12.29a how the long-term average temperature drops from the 1950s to the 1970s and then increases steadily. Figure 12.29b presents the rate of change in °C/day. When reported on an annual basis, the annual rate of change in 2013 is estimated to be $\mu_{\text{T}|\text{T}} = 0.06\,°\text{C/year}$ with a standard deviation $\sigma_{\text{T}|\text{T}} = 0.05\,°\text{C/year}$.

### 12.3.3   Modeling Dependencies Between Observations

State-space models can be employed to model the dependencies between several observations. Take, for example, the seasonal variations in the air temperature affecting the displacement of a structure. If we observe both quantities, that is, the temperature and the displacement, we can perform their joint estimation.

In the general case where there are D observed time series, we have to select relevant components for each time series and then assemble them together. The global state vector $\mathbf{x}_t$ for all time series and all components is obtained by concatenating in a column vector the state vectors $\mathbf{x}_t^j$, $\forall j \in \{1, 2, \cdots, \text{D}\}$ from each time series. The global observation ($\mathbf{C}$), transition ($\mathbf{A}$), process noise covariance ($\mathbf{Q}$), and observation covariance ($\mathbf{R}$) matrices are all built in the same way by concatenating block diagonal matrices from each time series,

$$\begin{aligned}
\mathbf{C} &= \text{blkdiag}(\mathbf{C}^1, \mathbf{C}^2, \cdots, \mathbf{C}^{\text{D}}) \\
\mathbf{A} &= \text{blkdiag}(\mathbf{A}^1, \mathbf{A}^2, \cdots, \mathbf{A}^{\text{D}}) \\
\mathbf{Q} &= \text{blkdiag}(\mathbf{Q}^1, \mathbf{Q}^2, \cdots, \mathbf{Q}^{\text{D}}) \\
\mathbf{R} &= \text{blkdiag}(\mathbf{R}^1, \mathbf{R}^2, \cdots, \mathbf{R}^{\text{D}}).
\end{aligned}$$

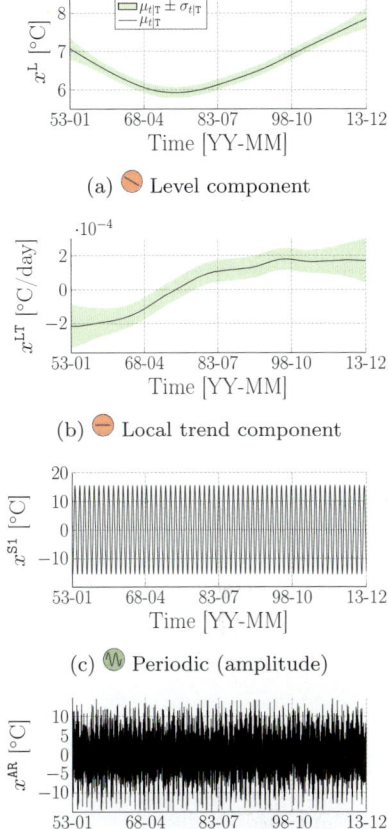

(a) 🌑 Level component

(b) ⊖ Local trend component

(c) 🌘 Periodic (amplitude)

(d) 🌔 Autoregressive component

Figure 12.29: Hidden-state estimates for the temperature time series.

The global matrix $\mathbf{C}$ requires special attention because it contains the information regarding the time-series dependencies. In the $\mathbf{C}$ matrix, each row corresponds to a time series. The component-wise representation of the observation matrix is

$$\mathbf{C} = \begin{bmatrix} \mathbf{C}^1 & \mathbf{C}_{1,2}^c & \cdots & \mathbf{C}_{1,j}^c & \cdots & \mathbf{C}_{1,\mathsf{D}}^c \\ \mathbf{C}_{2,1}^c & \mathbf{C}^2 & \cdots & \mathbf{C}_{2,j}^c & \cdots & \mathbf{C}_{2,\mathsf{D}}^c \\ \vdots & \vdots & \vdots & \vdots & \ddots & \vdots \\ \mathbf{C}_{i,1}^c & \mathbf{C}_{i,2}^c & \cdots & \mathbf{C}_{i,j}^c & \cdots & \mathbf{C}_{i,\mathsf{D}}^c \\ \vdots & \vdots & \vdots & \vdots & \ddots & \vdots \\ \mathbf{C}_{\mathsf{D},1}^c & \mathbf{C}_{\mathsf{D},2}^c & \cdots & \mathbf{C}_{\mathsf{D},j}^c & \cdots & \mathbf{C}^{\mathsf{D}} \end{bmatrix},$$

where the off-diagonal terms $\mathbf{C}_{i,j}^c$ are row vectors with the same number of elements as the matrix $\mathbf{C}^j$. When there is no dependence between observations, the off-diagonal matrices $\mathbf{C}_{i,j}^c$ contain only zeros. When dependence exists between observations, it can be encoded in a dependence matrix such as

$$\mathbf{D} = \begin{bmatrix} 1 & d_{1,2} & \cdots & d_{1,j} & \cdots & d_{1,\mathsf{D}} \\ d_{2,1} & 1 & \cdots & d_{2,j} & \cdots & d_{2,\mathsf{D}} \\ \vdots & \vdots & \vdots & \vdots & \ddots & \vdots \\ d_{i,1} & d_{i,2} & \cdots & 1 & \cdots & d_{i,\mathsf{D}} \\ \vdots & \vdots & \vdots & \vdots & \ddots & \vdots \\ d_{\mathsf{D},1} & d_{\mathsf{D},2} & \cdots & d_{\mathsf{D},j} & \cdots & 1 \end{bmatrix},$$

where $d_{i,j} \in \{0,1\}$ is employed to indicate whether (1) dependence does exist or (0) does not. When $d_{i,j} = 0$, $\mathbf{C}_{i,j}^c = [\mathbf{0}]$; when dependence exists and $d_{i,j} = 1$, the off-diagonal matrix $\mathbf{C}_{i,j}^c = \left[ \phi_1^{i|j} \ \phi_2^{i|j} \ \cdots \ \phi_{k_j}^{i|j} \right] \in \mathbb{R}^{k_j}$ contains $k_j$ regression coefficients $\phi_k^{i|j}$ describing the linear dependence between the $k^{\text{th}}$ component from the $j^{\text{th}}$ time series and the $i^{\text{th}}$ time series. All regression coefficients $\phi^{i|j}$ are treated as unknown parameters to be learned from data.

*Example: Bridge displacement versus temperature*   We consider the joint estimation of the hidden-state variables for two time series: the displacement and temperature measured on a bridge from 2013 until 2015. Both time series are presented in figure 12.30, where we can see a negative correlation between the observed displacement and temperature seasonal cycles.

The *temperature* is modeled using a local level, two periodic components with respective periods of 365.2422 and 1 day, and an autoregressive component,

$$\mathbf{x}_t^{\mathsf{T}} = [\underbrace{x_t^{\mathsf{T,LL}}}\ \underbrace{x_t^{\mathsf{T1,S1}}\ x_t^{\mathsf{T1,S2}}}_{p=365.24}\ \underbrace{x_t^{\mathsf{T2,S1}}\ x_t^{\mathsf{T2,S2}}}_{p=1}\ \underbrace{x_t^{\mathsf{T,AR}}}]^{\mathsf{T}}.$$

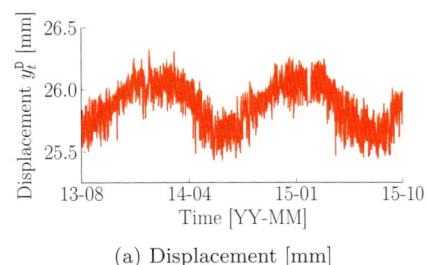

(a) Displacement [mm]

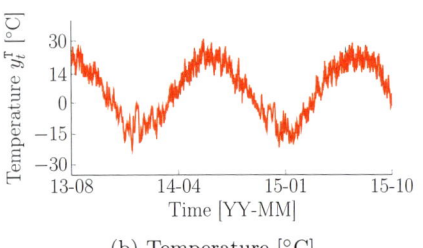

(b) Temperature [°C]

Figure 12.30: Observed time series for the displacement and temperature of a bridge.

All matrices defining the transition and the observation models are obtained by assembling the specific mathematical formulation associated with each component,

$$\mathbf{A}^{\mathrm{T}} = \mathrm{block\,diag}\left(1, \begin{bmatrix} \cos\omega^{\mathrm{T1}} & \sin\omega^{\mathrm{T1}} \\ -\sin\omega^{\mathrm{T1}} & \cos\omega^{\mathrm{T1}} \end{bmatrix}, \begin{bmatrix} \cos\omega^{\mathrm{T2}} & \sin\omega^{\mathrm{T2}} \\ -\sin\omega^{\mathrm{T2}} & \cos\omega^{\mathrm{T2}} \end{bmatrix}, \phi^{\mathrm{T,AR}}\right)$$

$$\mathbf{Q}^{\mathrm{T}} = \mathrm{block\,diag}\left(0,0,0,0,0,(\sigma^{\mathrm{T,AR}})^2\right)$$

$$\mathbf{C}^{\mathrm{T}} = [1\ 1\ 0\ 1\ 0\ 1]$$

$$\mathbf{R}^{\mathrm{T}} = (\sigma^{\mathrm{T}}_V)^2,$$

where $\omega^{\mathrm{T1}} = \frac{2\pi}{365.2422}$ and $\omega^{\mathrm{T2}} = \frac{2\pi}{1}$.

The *displacement* is modeled using a local level and an autoregressive component,

$$\mathbf{x}^{\mathrm{D}}_t = [x^{\mathrm{D,LL}}_t\ \underbrace{x^{\mathrm{D,AR}}_t}]^{\mathsf{T}}.$$

Matrices defining the transition and the observation models are again obtained by assembling the specific mathematical formulation associated with each component,

$$\mathbf{A}^{\mathrm{D}} = \mathrm{block\,diag}\left(1, \phi^{\mathrm{D,AR}}\right)$$

$$\mathbf{Q}^{\mathrm{D}} = \mathrm{block\,diag}\left(0, (\sigma^{\mathrm{D,AR}})^2\right)$$

$$\mathbf{C}^{\mathrm{D}} = [1\ 1]$$

$$\mathbf{R}^{\mathrm{D}} = (\sigma^{\mathrm{D}}_V)^2.$$

The global model matrices for the joint data set $\mathbf{y}_{1:\mathrm{T}} = \begin{bmatrix} \mathbf{y}^{\mathrm{D}}_{1:\mathrm{T}} \\ \mathbf{y}^{\mathrm{T}}_{1:\mathrm{T}} \end{bmatrix}$ are assembled following

$$\mathbf{x} = \begin{bmatrix} \mathbf{x}^{\mathrm{D}} \\ \mathbf{x}^{\mathrm{T}} \end{bmatrix}$$

$$\mathbf{A} = \mathrm{block\,diag}\left(\mathbf{A}^{\mathrm{D}}, \mathbf{A}^{\mathrm{T}}\right)$$

$$\mathbf{C} = \mathrm{block\,diag}\left(\mathbf{C}^{\mathrm{D}}, \mathbf{C}^{\mathrm{T}}\right)$$

$$\mathbf{R} = \mathrm{block\,diag}\left(\mathbf{R}^{\mathrm{D}}, \mathbf{R}^{\mathrm{T}}\right)$$

$$\mathbf{Q} = \mathrm{block\,diag}\left(\mathbf{Q}^{\mathrm{D}}, \mathbf{Q}^{\mathrm{T}}\right).$$

The dependence matrix is defined as $\mathbf{D} = \begin{bmatrix} 1 & 1 \\ 0 & 1 \end{bmatrix}$, which indicates that the displacement observations depend on the hidden states defined for the temperature. The assembled global observation matrix is thus

$$\mathbf{C} = \begin{bmatrix} \mathbf{C}^{\mathrm{D}} & \mathbf{C}^{c}_{\mathrm{D|T}} \\ \mathbf{0} & \mathbf{C}^{\mathrm{T}} \end{bmatrix} = \begin{bmatrix} 1 & 1 & 0 & \phi^{\mathrm{D|T_1}} & 0 & \phi^{\mathrm{D|T_2}} & 0 & \phi^{\mathrm{D|T_2}} \\ 0 & 0 & 1 & 1 & 0 & 1 & 0 & 1 \end{bmatrix},$$

where the first row corresponds to the displacement and the second to the temperature. Note that there are two regression coefficients describing the dependence between the displacement observations

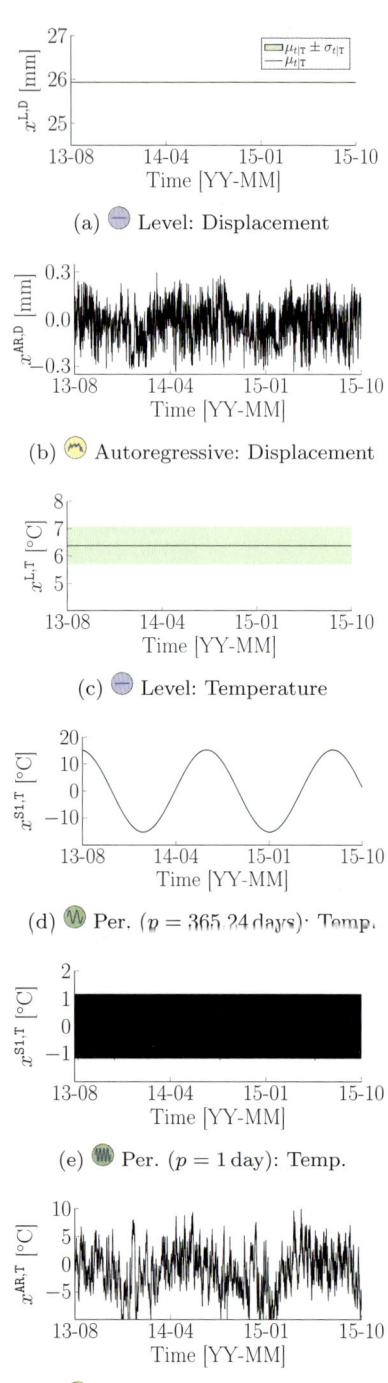

(a) ⬤ Level: Displacement

(b) ⬤ Autoregressive: Displacement

(c) ⬤ Level: Temperature

(d) ⬤ Per. ($p = 365.24$ days): Temp.

(e) ⬤ Per. ($p = 1$ day): Temp.

(f) ⬤ Autoregressive: Temperature

Figure 12.31: Hidden-state estimates for the displacements and temperatures.

and the temperature hidden-state variables: $\phi^{D|T_1}$, the long-term seasonal effects described by the first periodic component, and $\phi^{D|T_2}$, the short-term ones described by the second periodic component and the autoregressive term. MLE optimal model parameter values estimated from data are

$$\boldsymbol{\theta} = [\quad \phi^{D|T_1} \quad \phi^{D|T_2} \quad \phi^{D,AR} \quad \phi^{T,AR} \quad \sigma^{D,AR} \quad \sigma^{T,AR} \quad \sigma_V^D \quad \sigma_V^T]^\mathsf{T}$$
$$= [-0.017 \quad 0.01 \quad 0.982 \quad 0.997 \quad 0.024 \quad 0.28 \quad 10^{-5} \quad 10^{-4}]^\mathsf{T}.$$

The resulting smoothed hidden-state estimations are presented in figure 12.31. The periodic behavior displayed in figure 12.30a is not present in any of the displacement components presented in figures 12.31a–b. This is because the periodic behavior displayed in the displacement is modeled through the dependence over the temperature's hidden states.

## 12.4   Anomaly Detection

It is possible to detect anomalies in time series using the regime-switching method presented in §12.2. The idea is to create two models representing, respectively, a normal and an abnormal regime. The key strength of this approach is to consider not only the likelihood of each model at each time step but also the information about the prior probability of each regime as well as the probability of switching from one regime to another. In the context of anomaly detection, it allows reducing the number of false alarms while increasing the detection capacity. In a general context, the number of regimes is not limited, so several models can be tested against each other.

*Example: Anomaly detection*   This example illustrates the concept of anomaly detection for observations made on a dam displacement. Figure 12.32 presents observations showing a downward trend as well as a yearly periodic pattern. Because no temperature observations are available for a joint estimation, this case is thus treated as a single time series. The components employed are a local acceleration, two periodic components with respective periods $p = 365.2422$ and $p = 162.1211$ days, and one autoregressive component. In this case, we want to employ two different models, where one is constrained to have a constant speed and the other a constant acceleration. The vector of hidden-state variables for both regimes is

$$\mathbf{x}_t = [\quad x_t^L \quad x_t^T \quad x_t^{LA} \quad \underbrace{x_t^{T1,S1} \quad x_t^{T1,S2}}_{p=365.24} \quad \underbrace{x_t^{T2,S1} \quad x_t^{T2,S2}}_{p=162.12} \quad x_t^{AR}]^\mathsf{T}.$$

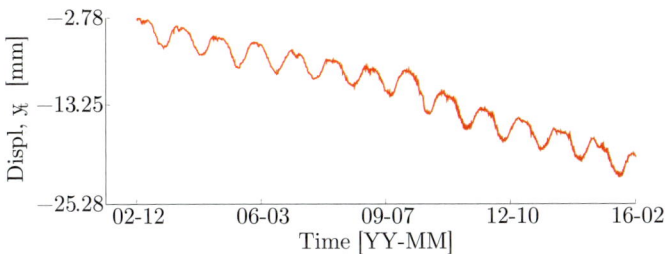

Figure 12.32: Example of a dam displacement time series that contains an anomaly that requires using the switching Kalman filter.

Global model matrices $\mathbf{A}^{(j)}, \mathbf{C}^{(j)}, \mathbf{Q}^{(j)}, \mathbf{R}^{(j)}$ for regimes $j \in \{1, 2\}$ are assembled following the procedure described in §12.3.2. The only things differentiating the two models are the transition matrices and the process noise covariance for the local acceleration component,

$$\mathbf{A}^{(1)} = \begin{bmatrix} 1 & \Delta t & 0 \\ 0 & 1 & 0 \\ 0 & 0 & 0 \end{bmatrix}, \quad \mathbf{A}^{(2)} = \begin{bmatrix} 1 & \Delta t & \frac{\Delta t^2}{2} \\ 0 & 1 & \Delta t \\ 0 & 0 & 1 \end{bmatrix}.$$

The model associated with the first regime has the particularity that the terms in the transition matrix that corresponds to the acceleration are set equal to zero. The covariance matrices for the transition errors from regimes $1 \rightarrow 2$ and $2 \rightarrow 1$ are respectively

$$\mathbf{Q}_{12} = \begin{bmatrix} (\sigma^{\mathrm{LA}})^2 \cdot \frac{\Delta t^2}{20} & 0 & 0 \\ 0 & (\sigma^{\mathrm{LTT}})^2 \cdot \frac{\Delta t^3}{3} & 0 \\ 0 & 0 & (\sigma^{\mathrm{LA}})^2 \cdot \Delta t \end{bmatrix}$$

$$\mathbf{Q}_{21} = \begin{bmatrix} (\sigma^{\mathrm{LT}})^2 \cdot \frac{\Delta t^3}{3} & 0 & 0 \\ 0 & (\sigma^{\mathrm{LTT}})^2 \cdot \Delta t & 0 \\ 0 & 0 & 0 \end{bmatrix}.$$

The transition matrix is parameterized by $z^{11}$ and $z^{22}$ as defined in

$$\mathbf{Z} = \begin{bmatrix} z^{11} & 1 - z^{11} \\ 1 - z^{22} & z^{22} \end{bmatrix}.$$

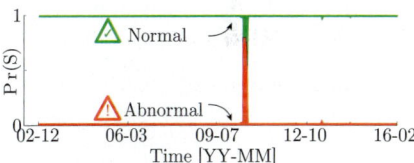

Figure 12.33: Probability of regimes 1 and 2 obtained using the SKF algorithm.

Figure 12.33 presents the probability of each regime as a function of time; and figure 12.34 presents the hidden-state estimates for the level, trend, and acceleration. This example displays a switch between the normal and the abnormal regimes during a short time span. These results illustrate how anomalies in time series can be detected using regime switching. They also illustrate how the baseline response of a system can be isolated from the raw observations contaminated by observation errors and external effects caused by environmental factors. The model allows estimating that before the regime switch, the rate of change was -2 mm/year, and it increased in magnitude to -2.9mm/year after the switch.

Figure 12.34: Expected values $\mu_{t|t}$ and uncertainty bound $\mu_{t|t} \pm \sigma_{t|t}$ for hidden states resulting from a combination of regimes 1 and 2 using the SKF algorithm.

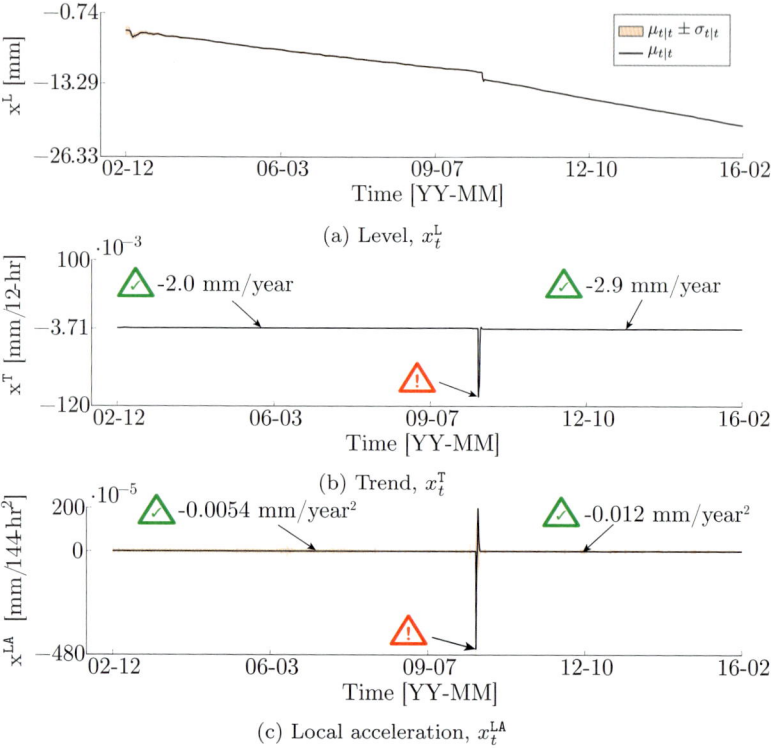

(a) Level, $x_t^{\mathrm{L}}$

(b) Trend, $x_t^{\mathrm{T}}$

(c) Local acceleration, $x_t^{\mathrm{LA}}$

[18] Nguyen, L. H. and J.-A. Goulet (2018). Anomaly detection with the switching Kalman filter for structural health monitoring. *Structural Control and Health Monitoring* 25(4), e2136.

[19] Gaudot, I., L. H. Nguyen, S. Khazaeli, and J.-A. Goulet (2019, May). OpenBDLM, an open-source software for structural health monitoring using bayesian dynamic linear models. In *13th Proceedings from the 13th International Conference on Applications of Statistics and Probability in Civil Engineering (ICASP).*

All the details related to this example can be found in the study performed by Nguyen and Goulet.[18]

The package OpenBDLM[19] allows using the state-space model's theory and includes the formulation for the various components defined for Bayesian dynamic linear models.

## Exercises

P12.1   What is the difference between filtering, forecasting, and smoothing?

P12.2   Why is the Kalman filter not applicable for nonlinear transition or observation models? What are the alternatives?

P12.3   From prior knowledge $\{\mu_0 = 5, \sigma_0^2 = 2\}$ and the transition and observation models below, compute $\{\mu_1, \sigma_1^2\}$ as well as $\{\mathbb{E}[Y_1], \mathrm{var}[Y_1]\}$.

$$
\begin{aligned}
x_t &= x_{t-1} + w_t, & w_t : W \sim \mathcal{N}(w; 0, 0.5^2) \\
y_t &= x_t + v_t, & v_t : V \sim \mathcal{N}(v; 0, 0.25^2)
\end{aligned}
$$

P12.4   How do we handle missing data with the Kalman filter?

P12.5   What is the role of the Markov hypothesis in the Kalman filter?

P12.6   Explain what circumstances require employing a Kalman filter with regime switching?

P12.7   Implement the Kalman filter to reproduce the example in figure 12.9 using the data $\mathbf{y}_{1:5} = [3.6\ 10.8\ 6.7\ 3.2\ 2.6]^\mathsf{T}$.

P12.8   Building on the Kalman filter you implemented in P12.7, implement the Kalman smoother to reproduce the results presented in figure 12.28 using the data set found in the file YUL_1953_2013.csv for Montreal's airport.

P12.9   Repeat the analysis from P12.8 for the data from Toronto's airport (YZZ_1938_2013.csv), located 550 km to the southwest of Montreal, and compare both results.

# 13
# Model Calibration

In civil engineering, *model calibration* is employed for the task of estimating the parameters of *hard-coded physics-based models* using empirical observations $\mathcal{D} = \{(x_i, y_i), \forall i \in \{1 : \mathtt{D}\}\} = \{\mathcal{D}_x, \mathcal{D}_y\}$. The term *hard-coded physics-based* refers to mathematical models made of constitutive laws and rules that are themselves based on physics or domain-specific knowledge. In contrast, models presented in chapters 8, 9, and 12 are referred to as *empirical models* because they are built on empirical data with little or no hard-coded rules associated with the underlying physics behind the phenomena studied. The hard-coded model of a system is described by a function $g(\boldsymbol{\theta}, \mathbf{x})$, which depends on the covariates $\mathbf{x} = [x_1 \ x_2 \ \cdots \ x_{\mathtt{X}}]^{\mathsf{T}} \in \mathbb{R}^{\mathtt{X}}$ as well as its model parameters $\boldsymbol{\theta} = [\theta_1 \ \theta_2 \ \cdots \ \theta_{\mathtt{P}}]^{\mathsf{T}}$. The *observed system responses* $\mathcal{D}_y = \{y_1, y_2, \cdots, y_{\mathtt{D}}\}$ depend on covariates $\mathcal{D}_x = \{\mathbf{x}_1, \mathbf{x}_2, \cdots, \mathbf{x}_{\mathtt{D}}\}$. Figure 13.1 presents an example of a model where the parameters $\boldsymbol{\theta}$ describe the boundary conditions and the initial tension in cables. For this example, the covariates $\mathbf{x}$ and $\mathbf{x}^*$ describe the locations where predictions are made, as well as the type of predictions, that is, displacement, rotation, strain, and so on. The vector $\mathbf{x}$ describes covariates for observed locations, and $\mathbf{x}^*$ those for the unobserved locations.

Model calibration is not in itself a subfield of machine learning. Nevertheless, in this chapter, we explore how the machine learning methods presented in previous chapters can be employed to address the challenges associated with the probabilistic inference of model parameters for hard-coded models. This application is classified under the umbrella of *unsupervised learning* because, as we will see, the main task consists in inferring hidden-state variables and parameters that are themselves not observed.

*Problem setups*    There are several typical setups where we want to employ empirical observations in conjunction with hard-coded

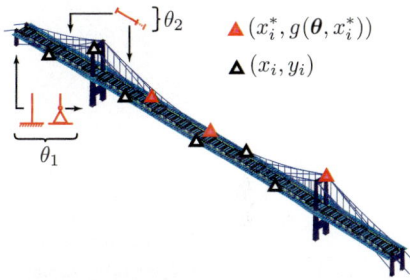

Figure 13.1: Example of hard-coded finite-element model for which we want to infer the parameter values $\boldsymbol{\theta} = [\theta_1 \ \theta_2]^{\mathsf{T}}$ from sets of observations $(x_i, y_i)$.

**Nomenclature**

*Observed system responses*
$\mathcal{D}_y = \{y_1, y_2, \cdots, y_{\mathtt{D}}\}$
*Covariates*
$\mathcal{D}_x = \{\mathbf{x}_1, \mathbf{x}_2, \cdots, \mathbf{x}_{\mathtt{D}}\}$
*Parameter values*
$\boldsymbol{\theta} = [\theta_1 \ \theta_2 \ \cdots \ \theta_{\mathtt{P}}]^{\mathsf{T}}$
*Model predictions*
$\mathbf{g}(\boldsymbol{\theta}, \mathbf{x}) = [g(\boldsymbol{\theta}, \mathbf{x}_1) \ \cdots \ g(\boldsymbol{\theta}, \mathbf{x}_{\mathtt{D}})]^{\mathsf{T}}$

models. For the example presented in figure 13.1, if we have a prior knowledge for the joint distribution of model parameters $f(\boldsymbol{\theta})$, we can propagate this prior knowledge through the model in order to quantify the uncertainty associated with predictions. This uncertainty propagation can be, for instance, performed using either Monte Carlo sampling (see §6.5) or first-order linearization (see §3.4.2). When the uncertainty in the prior knowledge $f(\boldsymbol{\theta})$ is weakly informative, it may lead to a large variability in model predictions. In that situation, it becomes interesting to employ empirical observations to reduce the uncertainty related to the prior knowledge of model parameters. The key with model calibration is that model parameters are typically not directly observable. For instance, in figure 13.1 it is often not possible to directly measure the boundary condition properties or to measure the cable internal tension. These properties have to be inferred from the observed structural responses $\mathcal{D}_y$. Another key aspect is that we typically build physics-based models $\mathbf{g}(\boldsymbol{\theta}, \mathbf{x})$ because they can predict quantities that cannot be observed. For example, we may want to learn about model parameters $\boldsymbol{\theta}$ using observations of the static response of a structure defined by the covariates $\mathbf{x}$ and then employ the model to predict the unobserved responses $\mathbf{g}(\boldsymbol{\theta}, \mathbf{x}^*)$, defined by other covariates $\mathbf{x}^*$. The vector $\mathbf{x}^*$ may describe unobserved locations or quantities such as the dynamic behavior instead of the static one employed for parameter estimation. One last possible problem setup is associated with the selection of model classes for describing a phenomenon.

There are three main challenges associated with model calibration: *observation errors*, *prediction errors*, and *model complexity*. The first challenge arises because the observations available $\mathcal{D}_y$ are in most cases contaminated by observation errors. The second is due to the discrepancy between the prediction of hard-coded physics-based models $\mathbf{g}(\boldsymbol{\theta}, \mathbf{x})$ and the real system responses; that is, even when parameter values are known, the model remains an approximation of the reality. The third challenge is related to the difficulties associated with choosing a model structure having the right complexity for the task at hand. In §13.1, we explore the impact of these challenges on the least-squares model calibration approach, which is still extensively used in practice. Then, the subsequent sections explore how to address some of these challenges using a hierarchical Bayesian approach combining concepts presented in chapters 6 and 8.

## 13.1   Least-Squares Model Calibration

This section presents the pitfalls and limitations of the common least-squares deterministic model calibration. With *least-squares* model calibration, the goal is to find the model parameters $\boldsymbol{\theta}$ that minimize the sum of the square of the differences between predicted and measured values. The difference between predicted and observed values at one observed location $\mathbf{x}_i$ is defined as the residual

$$r(\boldsymbol{\theta}, (\mathbf{x}_i, y_i)) = y_i - g(\boldsymbol{\theta}, \mathbf{x}_i).$$

For the entire data set containing D data points, the least-squares loss function is defined as

$$
\begin{aligned}
J(\boldsymbol{\theta}, \mathcal{D}) &= \sum_{i=1}^{D} r(\boldsymbol{\theta}, (\mathbf{x}_i, y_i))^2 \\
&= \|\mathbf{r}(\boldsymbol{\theta}, \mathcal{D})\|^2.
\end{aligned}
$$

Because of the square exponent in the loss function, its value is positive $J(\boldsymbol{\theta}, \mathcal{D}) \geq 0$. The task of identifying the optimal parameters $\boldsymbol{\theta}^*$ is formulated as a minimization problem where the goal is to minimize the sum of the square of the residual at each observed location,

$$
\begin{aligned}
\boldsymbol{\theta}^* &= \underset{\boldsymbol{\theta}}{\arg\min}\ J(\boldsymbol{\theta}, \mathcal{D}) \\
&\equiv \underbrace{\underset{\boldsymbol{\theta}}{\arg\max}\ -J(\boldsymbol{\theta}, \mathcal{D})}_{\text{Optimization problem}}.
\end{aligned}
$$

Gradient-based optimization methods such as those presented in chapter 5 can be employed to identify $\boldsymbol{\theta}^*$.

### 13.1.1   Illustrative Examples

This section presents three examples based on simulated data using the *reference beam model* $\check{g}(\check{E}, x)$ presented in figure 13.2 along with the values for its parameters. Observations consist in displacement measurements made for positions $x_1 = L/2$ and $x_2 = L$. The first example corresponds to the idealized context where there is no observation or prediction errors because the model employed to infer parameter values is the same as the reference model employed to generate synthetic data. The second example employs a simplified model to infer parameters. Finally, the third example employs an overcomplex model to infer parameters. Note that the last two examples include observation errors. For all three cases, there is a single unknown parameter: the elastic modulus $E$, which characterizes the flexural stiffness of the beam.

What we will see in §13.1 is an example of what *not to do* when calibrating a hard-coded physics-based model as well as why one should avoid it.

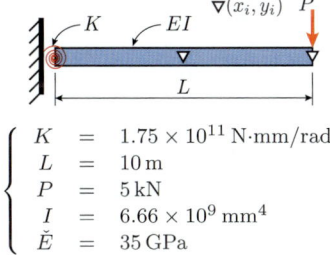

$$
\left\{
\begin{aligned}
K &= 1.75 \times 10^{11}\ \text{N·mm/rad} \\
L &= 10\ \text{m} \\
P &= 5\ \text{kN} \\
I &= 6.66 \times 10^{9}\ \text{mm}^4 \\
\check{E} &= 35\ \text{GPa}
\end{aligned}
\right.
$$

Figure 13.2: The reference model $\check{g}(\check{E}, x)$ that is employed to generate simulated observations.

**Example #1**
**Reference model**
$\breve{g}(\breve{E},x)$

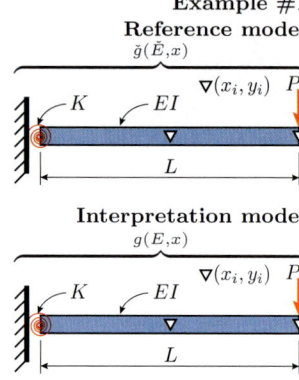

**Interpretation model**
$g(E,x)$

*Example #1: Exact model without observation errors*   The first example illustrates an idealized context where there are no observation or prediction errors because the model employed to infer parameter values $g(E,x)$ is the same as the one employed to generate synthetic data $\breve{g}(\breve{E},x)$. Figure 13.3 compares the predicted and true deflection curves along with observed values. Here, because of the absence of model or measurement errors, the optimal parameter also coincides with its true value $\theta^* = E^* = \breve{E}$, and the predicted deflection obtained using the optimal parameter value $\theta^*$ coincides with the true deflection. Note that the loss function evaluated for $\theta^*$ equals zero. Even if least-squares model calibration worked perfectly in this example, this good performance cannot be generalized to real case studies because in this oversimplified example, there are no model or measurement errors.

Figure 13.3: Least-squares model calibration where the interpretation model is identical to the reference model, $\breve{g}(\breve{E},x) = g(\breve{E},x)$, and where there are no measurement errors.

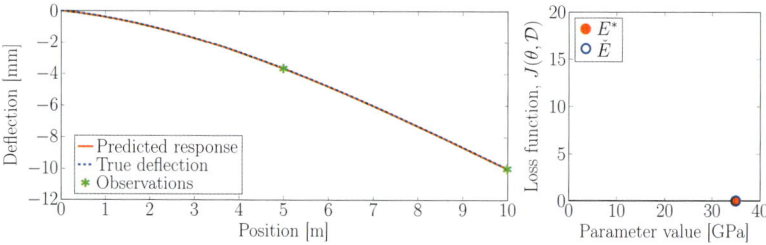

**Example #2**
**Reference model**
$\breve{g}(\breve{E},x)$

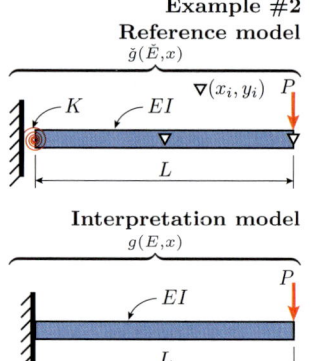

**Interpretation model**
$g(E,x)$

*Example #2: Simplified model with observation errors*   The second example presents a more realistic case, where the model $g(E,x)$ employed to infer parameters contains a simplification in comparison with the real system described by the reference model $\breve{g}(\breve{E},x)$. In this case, the simplification consists in omitting the rotational spring that is responsible for the nonzero initial rotation at the support. In addition, observations are affected by observation errors $v_i : V \sim \mathcal{N}(v;0,1)\,\text{mm}$, where the observation model is defined as

$$y_i = \breve{g}(\breve{E},x_i) + v_i,\ \ V_i \perp\!\!\!\perp V_j, \forall i \neq j.$$

Figure 13.4 compares the predicted and true deflection curves along with observed values. We see that because of the combination of model simplification and observation errors, we obtain a higher loss for the correct parameter value $\breve{E} = 35\,\text{GPa}$ than for the optimal value $E^* = 22\,\text{GPa}$. As a result, the model predictions are also biased for any prediction location $x$. This second example illustrates how deterministic least-squares model calibration may identify wrong parameter values as well as lead to inaccurate predictions for measured and unmeasured locations.

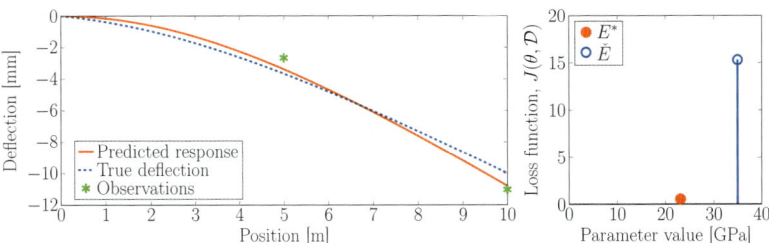

Figure 13.4: Least-squares model calibration where the interpretation model is simplified in comparison with the reference model, $\breve{g}(\breve{E}, x) \neq g(\breve{E}, x)$, and where there are measurement errors.

*Example #3: Overcomplex model with observation errors*    The third example illustrates the pitfalls of performing model calibration using a simplified yet overcomplex interpretation model, that is, a model that contains simplifications yet also contains too many parameters for the problem at hand. Here, overcomplexity is caused by the use of two parameters for describing the elastic modulus, $E_1$ and $E_2$. Like in the second example, observations are affected by observation errors $v_i : V \sim \mathcal{N}(v; 0, 1)$ mm.

Figure 13.5 compares the predicted and true deflection curves along with observed values. Note how the over-parameterization allowed to wrongly match both observations. Parameter values identified are biased in comparison with the true value, and the predictions made by the model are also biased. Because in practice the true deflection and parameter values remain unknown, the biggest danger is *overfitting* where we could believe that the model is capable of providing accurate predictions because the calibration wrongly appears to be perfect. In the context of least-squares model calibration, this issue can be mitigated using cross-validation (see §8.1.2), where only subsets of observations are employed to train the model, and then the remaining data is employed to test the model's predictive capacity. This procedure would reveal that the interpretation model is poorly defined. In a more general way, we can also use Bayesian model selection (see §6.8) for quantifying the probability of each model class.

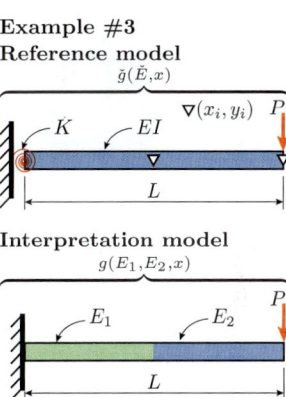

**Example #3**
**Reference model**
$\breve{g}(\breve{E}, x)$

**Interpretation model**
$g(E_1, E_2, x)$

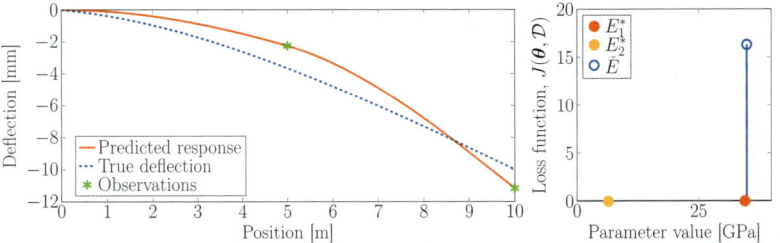

Figure 13.5: Least-squares model calibration where the interpretation model is overcomplex in comparison with the reference model, $\breve{g}(\breve{E}, x) \neq g(\breve{E}_1, \breve{E}_2, x)$, and where there are measurement errors.

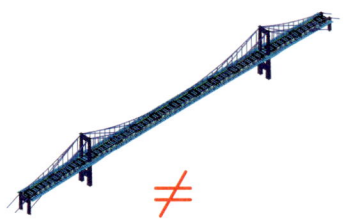

Figure 13.6: In complex engineering contexts, no matter how good a hard-coded physics-based model is, it never perfectly represents the reality. (Photo: Tamar Bridge, adapted from Peter Edwards)

### 13.1.2   Limitations of Deterministic Model Calibration

Deterministic model calibration using methods such as least-squares residual minimization performs well when model and measurement errors are negligible and when data sets are large; the issue is that in practical engineering applications, the errors associated with measurements and especially hard-coded models are not negligible, and data sets are typically small. As illustrated in figure 13.6, for engineering systems, no matter how much effort is devoted to building a hard-coded physics-based model, it will always remain a simplified version of the reality. Therefore, model simplifications introduce prediction errors that need to be considered explicitly as hidden variables to be inferred in addition to model parameters $\boldsymbol{\theta}$.

### 13.2   Hierarchical Bayesian Estimation

This section presents a Bayesian *hierarchical* method for jointly estimating model parameters as well as prediction errors. The justification for this approach is that no matter how good a hard-coded model is, it remains an approximation of the reality. Therefore, in addition to the model parameters, we need to infer prediction errors. There is typically an issue of non-identifiability (see §6.3.4) between model parameters and prediction errors. As we will see, the problem is made identifiable using prior knowledge. In addition, note that because of observation errors, observations $y_i$ are not exact representations of the reality either, that is,

$$\underbrace{\mathrm{model}(\boldsymbol{\theta})}_{g(\boldsymbol{\theta})} \quad \neq \quad \mathrm{reality}$$

$$\mathrm{model}(\boldsymbol{\theta}) + \underbrace{\mathrm{error}_w}_{w} \quad = \quad \mathrm{reality} \qquad \neq \quad \mathrm{observation}$$

$$\underbrace{\mathrm{model}(\boldsymbol{\theta})}_{g(\boldsymbol{\theta})} + \underbrace{\mathrm{error}_w}_{w} + \underbrace{\mathrm{error}_v}_{v} \quad = \quad \mathrm{reality} + \underbrace{\mathrm{error}_v}_{v} \quad = \quad \underbrace{\mathrm{observation}}_{y}.$$

In the previous equation, observations $\mathbf{y} = [y_1 \ y_2 \ \cdots \ y_D]^\mathsf{T}$ are modeled by the sum of *model predictions* $\mathbf{g}(\boldsymbol{\theta}, \mathbf{x}) \in \mathbb{R}^D$, which are a function of covariates $\mathbf{x} = [\mathbf{x}_1 \ \mathbf{x}_2 \ \cdots \ \mathbf{x}_D]^\mathsf{T} \in \mathbb{R}^{X \times D}$ and *model parameters* $\boldsymbol{\theta} = [\theta_1 \ \theta_2 \ \cdots \ \theta_P]^\mathsf{T} \in \mathbb{R}^P$; *prediction errors* $\mathbf{w}(\mathbf{x}) = [w(\mathbf{x}_1) \ w(\mathbf{x}_2) \ \cdots \ w(\mathbf{x}_D)]^\mathsf{T} \in \mathbb{R}^D$, which are also a function of covariates $\mathbf{x}$; and *measurement errors* $\mathbf{v} = [v_1 \ v_2 \ \cdots \ v_D]^\mathsf{T} \in \mathbb{R}^D$.

### 13.2.1   Joint Posterior Formulation

The model predictions $\mathbf{g}(\boldsymbol{\theta}, \mathbf{x})$ depend on the known covariates $\mathbf{x}$ describing the prediction locations and unknown model parameters $\boldsymbol{\theta}$ describing physical properties of the system. The observed structural responses are described by

$$\overset{\text{Prediction errors}}{\mathbf{y} = \mathbf{g}(\boldsymbol{\theta}, \mathbf{x}) + \mathbf{w}(\mathbf{x}) + \mathbf{v},} \tag{13.1}$$

where the annotations indicate:
- Observations
- Deterministic predictions from a hard-coded model
- Model parameters
- Covariates
- Measurement errors
- Prediction errors

where measurement errors are assumed to be independent of the covariates $\mathbf{x}$. All the terms in equation 13.1 are considered as deterministic quantities because, for a given set of observations $\mathcal{D} = \{\mathcal{D}_x, \mathcal{D}_y\} = \{(\mathbf{x}_i, y_i), \forall i \in \{1 : \mathtt{D}\}\}$, quantities $\mathbf{x}, \mathbf{y}, \boldsymbol{\theta}$, $\mathbf{w}(\mathbf{x})$, and $\mathbf{v}$ are not varying, yet values for $\boldsymbol{\theta}, \mathbf{w}(\mathbf{x})$, and $\mathbf{v}$ remain unknown.

The prior knowledge for model parameters is described by the random variable $\boldsymbol{\Theta} \sim f(\boldsymbol{\theta}; \mathbf{z}_p)$, where $\mathbf{z}_p$ are the *hyperparameters*, that is, parameters of the prior probability density function (PDF). When hyperparameters $\mathbf{z}_p$ are unknown, the random variable $\mathbf{Z}_p$ is described by a *hyper-prior* $f(\mathbf{z}_p)$ and a *hyper-posterior* $f(\mathbf{z}_p|\mathcal{D})$, that is, the prior and posterior PDFs of hyperparameters. Unlike the parameter values $\boldsymbol{\theta}$, for which true values may exist, hyperparameters $\mathbf{z}_p$ are parameters of our prior knowledge, for which, by nature, no true value exists. When hyperparameters $\mathbf{z}_p$ are assumed to be fixed constants that are not learned from data, they can be implicitly included in the prior PDF formulation $f(\boldsymbol{\theta}; \mathbf{z}_p) = f(\boldsymbol{\theta})$.

The concept of *prior, posterior, hyper-prior*, and *hyper-posterior* not only applies to unknown parameters $\boldsymbol{\theta}$ but also to prediction errors $\mathbf{w}(\mathbf{x})$ and measurement errors $\mathbf{v}$. Note that, unlike for hyperparameters $\mathbf{z}_p$ and $\mathbf{z}_w$, a true value can exist for $\mathbf{z}_v$. When it exists, the true value for $\mathbf{z}_v$ can typically be inferred from the statistical precision of the measuring instruments employed to obtain $\mathbf{y}$. In this section, in order to simplify the notation and to allow focusing the attention on model parameters $\boldsymbol{\theta}$ and prediction errors $\mathbf{w}(\mathbf{x})$, we assume that $\mathbf{z}_p$ and $\mathbf{z}_v$ are known constants.

The formulation for the joint posterior PDF for unknown model parameters $\boldsymbol{\theta}$, prediction errors $\mathbf{w}(\mathbf{x})$, and prediction errors prior PDF hyperparameters $\mathbf{z}_w$ follows:

$$\underset{\text{posterior}}{\underbrace{f(\boldsymbol{\theta}, \mathbf{w}(\mathcal{D}_x), \mathbf{z}_w|\mathcal{D})}} = \frac{\overset{\text{likelihood}}{\overbrace{f(\mathcal{D}_y|\boldsymbol{\theta}, \mathbf{w}(\mathcal{D}_x), \mathcal{D}_x)}} \cdot \overset{\text{prior}}{\overbrace{f(\boldsymbol{\theta}) \cdot f(\mathbf{w}(\mathcal{D}_x)|\mathbf{z}_w)}} \cdot \overset{\text{hyper-prior}}{\overbrace{f(\mathbf{z}_w)}}}{f(\mathcal{D}_y)}.$$

with annotations:
- Model parameters
- Prediction errors at location $\mathcal{D}_x$
- Hyperparameters for the prediction errors prior PDF
- Data

$$\tag{13.2}$$

*Likelihood*    The formulation for the joint posterior in equation 13.2 allows the explicit quantification of the dependence between model parameters and prediction errors at observed locations. In the case where several sets of model parameters and prediction errors $\{\boldsymbol{\theta}, \mathbf{w}(\mathcal{D}_x)\}$ can equally explain the observations, then the problem is non-identifiable (see §6.3.4) and several sets of values will end up having an equal posterior probability, given that the prior probability of each set is also equal. Here, the likelihood of data $\mathcal{D}_y$ is formulated considering that $\mathbf{g}(\boldsymbol{\theta}, \mathcal{D}_x)$ and $\mathbf{w}(\mathcal{D}_x)$ are unknown constants, so the only aleatory quantities are the observation errors $\mathbf{V} \sim \mathcal{N}(\mathbf{v}; \mathbf{0}, \boldsymbol{\Sigma}_v)$. The formulation of the likelihood is then

$$
\begin{aligned}
f(\mathcal{D}_y | \boldsymbol{\theta}, \mathbf{w}(\mathcal{D}_x), \mathcal{D}_x) &= f\big(\overbrace{(\mathbf{g}(\boldsymbol{\theta}, \mathcal{D}_x) + \mathbf{w}(\mathcal{D}_x)}^{\text{constants}} + \overbrace{\mathbf{V}}^{\mathcal{N}(\mathbf{v};\mathbf{0},\boldsymbol{\Sigma}_v)}) = \mathcal{D}_y\big) \\
&= \mathcal{N}(\mathbf{y} = \mathcal{D}_y; \overbrace{\mathbf{g}(\boldsymbol{\theta}, \mathcal{D}_x) + \mathbf{w}(\mathcal{D}_x)}^{\boldsymbol{\mu}_y}, \overbrace{\boldsymbol{\Sigma}_v}^{\boldsymbol{\Sigma}_y}) \\
&= \frac{\exp\left(-\frac{1}{2}\big(\mathcal{D}_y - (\mathbf{g}(\boldsymbol{\theta}, \mathcal{D}_x) + \mathbf{w}(\mathcal{D}_x))\big)^{\mathsf{T}} \boldsymbol{\Sigma}_v^{-1} \big(\mathcal{D}_y - (\mathbf{g}(\boldsymbol{\theta}, \mathcal{D}_x) + \mathbf{w}(\mathcal{D}_x))\big)\right)}{(2\pi)^{\mathsf{D}/2}\sqrt{\det \boldsymbol{\Sigma}_v}} \\
&\propto \exp\left(-\frac{1}{2}\big(\mathcal{D}_y - (\mathbf{g}(\boldsymbol{\theta}, \mathcal{D}_x) + \mathbf{w}(\mathcal{D}_x))\big)^{\mathsf{T}} \boldsymbol{\Sigma}_v^{-1} \big(\mathcal{D}_y - (\mathbf{g}(\boldsymbol{\theta}, \mathcal{D}_x) + \mathbf{w}(\mathcal{D}_x))\big)\right) \\
&= \exp\left(-\frac{1}{2}\sum_{i=1}^{\mathsf{D}} \frac{(y_i - (g(\boldsymbol{\theta}, \mathbf{x}_i) + w(\mathbf{x}_i)))^2}{\sigma_v^2}\right), \text{ for } V_i \perp\!\!\!\perp V_j, \forall i \neq j.
\end{aligned}
$$

*Prior*    It is necessary to define a mathematical formulation for the prior PDF of prediction errors, $\mathbf{W} \sim f(\mathbf{w}|\mathbf{z}_w)$. A convenient choice consists in describing the prior PDF for prediction errors using a Gaussian process analogous to the one described in §8.2. With the assumption that our model is unbiased, that is, the prior expected value for prediction errors is $\mathbf{0}$, this Gaussian process is expressed as

$$
\begin{aligned}
\mathbf{W} &\sim \mathcal{N}(\mathbf{w}; \mathbf{0}, \boldsymbol{\Sigma}_w), \text{ where} \\
[\boldsymbol{\Sigma}_w]_{ij} &= \sigma_w(\mathbf{x}_i) \cdot \sigma_w(\mathbf{x}_j) \cdot \rho(\mathbf{x}_i, \mathbf{x}_j, \boldsymbol{\ell}), \text{ and} \\
\sigma_w(\mathbf{x}) &= \mathrm{fct}(\mathbf{x}, \mathbf{z}_s), \\
\rho(\mathbf{x}_i, \mathbf{x}_j, \boldsymbol{\ell}) &= \exp\left(-\tfrac{1}{2}(\mathbf{x}_i - \mathbf{x}_j)^{\mathsf{T}} \boldsymbol{\Sigma}_{\boldsymbol{\ell}}^{-1}(\mathbf{x}_i - \mathbf{x}_j)\right), \\
\mathbf{z}_w &= [\mathbf{z}_s\ \boldsymbol{\ell}]^{\mathsf{T}}.
\end{aligned}
$$

The prediction error standard deviation $\sigma_w(\mathbf{x})$ is typically covariate dependent, so that it needs to be represented by a problem-specific function $\mathrm{fct}(\mathbf{x}, \mathbf{z}_s)$, where $\mathbf{z}_s$ are its parameters. The correlation structure can be represented by a square-exponential covariance function, where $\boldsymbol{\Sigma}_{\boldsymbol{\ell}}$ is a diagonal matrix, parameterized by the length-scale parameters $\boldsymbol{\ell} = [\ell_1\ \ell_2\ \cdots\ \ell_{\mathsf{X}}]^{\mathsf{T}}$. $\ell_k$ is the hyperparameter describing how the correlation decreases as the distance

$(x_{k,i} - x_{k,j})^2$ increases. Note that this choice for the correlation structure is not exclusive and others can be employed as described in §8.2. The hyperparameters regrouped in $\mathbf{z}_w$ need to be learned from data.

*Example #4: Joint posterior estimation*   We are revisiting example #2, where the model employed to infer parameters $g(E, x)$ contains simplifications in comparison with the reference model $\check{g}(\check{E}, x)$ describing the real system studied. In this case, the simplification consists in omitting the rotational spring allowing for a nonzero initial rotation at the support. In addition, observations are affected by independents and identically distributed observation errors $v_i : V \sim \mathcal{N}(v; 0, 1)\,\mathrm{mm}$.

Figure 13.7a presents the true deflection obtained from the reference model as well as from the interpretation model using the true parameter value $\check{E} = 35\,\mathrm{GPa}$. We can see that for any location other than $x = 0$, there is a systematic prediction error. The prior knowledge for these theoretically unknown prediction errors is modeled using a zero-mean Gaussian process,

$$f(\mathbf{w}) = \mathcal{N}(\mathbf{w}; \mathbf{0}, \boldsymbol{\Sigma}_w),$$

where each term of the covariance matrix is defined as

$$[\boldsymbol{\Sigma}_w]_{ij} = \overbrace{(a \cdot x_i)}^{\sigma_i} \cdot \overbrace{(a \cdot x_j)}^{\sigma_j} \cdot \overbrace{\exp\left(-\frac{1}{2}\frac{(x_i - x_j)^2}{\ell^2}\right)}^{\rho_{ij}}. \qquad (13.3)$$

The prior knowledge for prediction errors is parameterized by $\mathbf{z}_w = [a\ \ell]^{\mathsf{T}}$. Figure 13.7b presents the marginal $\pm\sigma_w$ confidence interval around the predicted deflection. The prior knowledge for each parameter and hyperparameter is elastic modulus $\mathcal{N}(E; 20, 20^2)\,\mathrm{GPa}$, model error prior scale factor $\mathcal{N}(a; 10^{-4}, (5 \times 10^{-4})^2)$, and length-scale $\mathcal{N}(\ell; 5, 50^2)\,\mathrm{m}$. Note that all of those priors are assumed to be independent from each other, so the joint prior can be described by the product of the marginals. The analytic formulation of the posterior PDF for the unknown model parameter $\theta = E$, the prediction errors $w(5), w(10)$, and the hyperparameters $a, \ell$ is known up to a constant, so that

$$
\begin{aligned}
f(E, w(5), w(10), a, \ell | \mathcal{D}_y) \quad \propto \quad & \mathcal{N}(y_1; g(E, 5) + w(5), 1^2) & \cdots \\
& \cdot\mathcal{N}(y_2; g(E, 10) + w(10), 1^2) & \cdots \\
& \cdot\mathcal{N}([w(5)\ w(10)]^{\mathsf{T}}; \mathbf{0}, \boldsymbol{\Sigma}_w) & \cdots \\
& \cdot\mathcal{N}(E; 20, 20^2) & \cdots \\
& \cdot\mathcal{N}(a; 10^{-4}, (5 \times 10^{-4})^2) & \cdots \\
& \cdot\mathcal{N}(\ell; 5, 50^2).
\end{aligned}
$$

**Example #4**
**Reference model**
$$\check{g}(\check{E}, x)$$

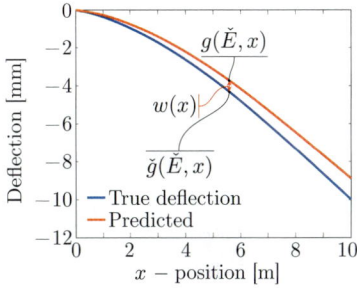

**Interpretation model**
$$g(E, x)$$

$$
\begin{aligned}
\mathcal{D}_x &= \{5, 10\}\,m \\
\mathcal{D}_y &= \{-4.44, -11.28\}\,mm
\end{aligned}
$$

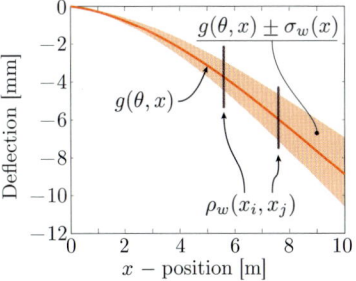

(a) Comparison of the real and predicted deflection obtained using the true parameter value $\check{E}$.

(b) The prior knowledge for prediction errors is described using a Gaussian process.

Figure 13.7: Graphical representation of the prior knowledge for prediction errors arising from the use of a simplified interpretation model.

Using the Metropolis sampling method presented in §7.1, three parallel chains, containing a total of $S = 10^5$ joint samples, are taken from the unnormalized posterior $f(E, w(5), w(10), a, \ell | \mathcal{D}_y)$, where each joint sample is described by $\{E, w(5), w(10), a, \ell\}_s$. The first half of each chain are discarded as burn-in samples; the Metropolis acceptance rate is approximately 0.25, and the estimated potential scale reduction (EPSR, i.e., chain stationarity metric; see §7.3.3) is below 1.1 for all parameters and hyperparameters. The Markov chain Monte Carlo (MCMC) posterior samples are presented in figure 13.8, where, when it exists, the true value is represented by the symbol $*$. Note the almost perfect correlation in the posterior between $w_1(5)$ and $w_2(10)$. It happens because, as illustrated in figure 13.7, whenever the model either under- or overestimates the displacement, it does for both locations. Notice that there is also a clear dependence between prediction errors $w_i$ and the elastic modulus $E$, where small (large) values for $E$ are compensated by large positive (negative) prediction errors. These are examples of non-identifiability regularized by prior knowledge, as described in §6.3.4.

Figure 13.8: Bivariate scatter plots and histograms describing the marginal posterior PDFs. The symbol $*$ indicates the true values.

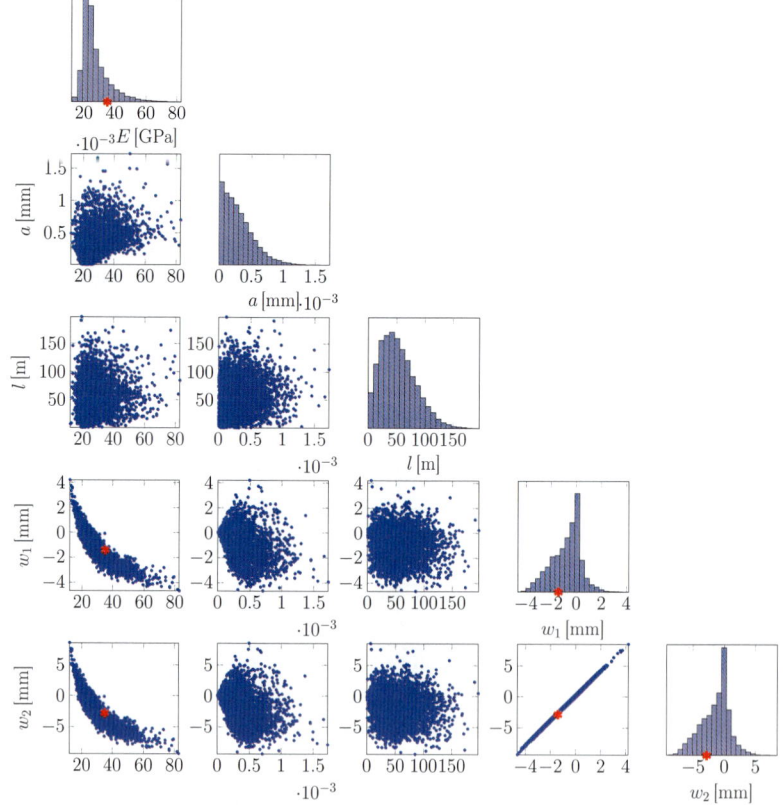

### 13.2.2 Predicting at Unobserved Locations

In practical applications, we are typically interested in predicting the structure's responses $u$ at unobserved locations $\mathbf{x}_*$. Analogous to equation 13.1, the predicted values at unobserved locations are defined as

$$u = g(E, \mathbf{x}_*) + \mathbf{w}(\mathbf{x}_*).$$

When predicting at unobserved locations, we want to consider the knowledge obtained at observed locations, as described by the posterior samples displayed in figure 13.8. When using MCMC samples (see §7.5), we can approximate the posterior predictive expected values and covariances by following

$$
\begin{aligned}
\mathbb{E}[\mathbf{U}|\mathcal{D}] &= \int \mathbf{u} \cdot f(\mathbf{u}|\mathcal{D})d\mathbf{u} \\
&\approx \frac{1}{\mathsf{S}}\sum_{s=1}^{\mathsf{S}} \mathbf{u}_s \qquad \text{(Posterior predictive mean)} \\
\mathrm{cov}(\mathbf{U}|\mathcal{D}) &= \mathbb{E}[(\mathbf{U} - \mathbb{E}[\mathbf{U}|\mathcal{D}])(\mathbf{U} - \mathbb{E}[\mathbf{U}|\mathcal{D}])^{\mathsf{T}}] \\
&\approx \frac{1}{\mathsf{S}-1}\sum_{s=1}^{\mathsf{S}}(\mathbf{u}_s - \mathbb{E}[\mathbf{U}|\mathcal{D}])(\mathbf{u}_s - \mathbb{E}[\mathbf{U}|\mathcal{D}])^{\mathsf{T}}. \quad \text{(Posterior predictive covariance)}
\end{aligned}
$$

In order to compute $\mathbb{E}[\mathbf{U}|\mathcal{D}]$ and $\mathrm{cov}(\mathbf{U}|\mathcal{D})$, we have to generate realizations of $\mathbf{u}_s$ from $f(\mathbf{u}|\mathcal{D})$ using the MCMC samples $\{E, w(5), w(10), a, \ell\}_s$ obtained from the unnormalized posterior $f(E, w(5), w(10), a, \ell|\mathcal{D}_y)$. Through that process, we must first generate realizations for the covariance matrices $\{\boldsymbol{\Sigma}_*^s, \boldsymbol{\Sigma}_{\mathbf{w}*}^s, \boldsymbol{\Sigma}_{\mathbf{w}}^s\}$ describing the model errors prior PDF. These realizations are then used to update the posterior mean vector and covariance matrix describing the model errors at prediction locations, so that

$$
\begin{Bmatrix} w_s(5) \\ w_s(10) \\ a_s \\ \ell_s \end{Bmatrix} \rightarrow \begin{Bmatrix} \boldsymbol{\Sigma}_*^s \\ \boldsymbol{\Sigma}_{\mathbf{w}*}^s \\ \boldsymbol{\Sigma}_{\mathbf{w}}^s \end{Bmatrix} \rightarrow \begin{cases} \boldsymbol{\mu}_{*|\mathbf{w}}^s = \boldsymbol{\Sigma}_{\mathbf{w}*}^{s\mathsf{T}}(\boldsymbol{\Sigma}_{\mathbf{w}}^s)^{-1}[w_s(5)\ w_s(10)]^{\mathsf{T}} \\ \boldsymbol{\Sigma}_{*|\mathbf{w}}^s = \boldsymbol{\Sigma}_*^s - \boldsymbol{\Sigma}_{\mathbf{w}*}^{s\mathsf{T}}(\boldsymbol{\Sigma}_{\mathbf{w}}^s)^{-1}\boldsymbol{\Sigma}_{\mathbf{w}*}^s. \end{cases}
$$

Realizations of the model errors at prediction locations $\mathbf{w}_{s*}$ can be sampled from the multivariate Gaussian PDF defined by

$$\mathbf{w}_{s*} : \mathbf{W}_* \sim \mathcal{N}(\mathbf{w}_*; \boldsymbol{\mu}_{*|\mathbf{w}}^s, \boldsymbol{\Sigma}_{*|\mathbf{w}}^s).$$

Finally, we can obtain a realization of the structure's behavior at predicted locations by summing the realization of model errors $\mathbf{w}_{s*}$ with the model predictions evaluated for the model parameter MCMC sample $E_s$,

$$\mathbf{u}_s = g(E_s, \mathbf{x}_*) + \mathbf{w}_{s*}.$$

**Note:** From equation 13.3,

$$[\boldsymbol{\Sigma}_*^s]_{ij} = \overset{\sigma_i}{\overbrace{(a_s \cdot x_i^*)}} \cdot \overset{\sigma_j}{\overbrace{(a_s \cdot x_j^*)}} \cdot \overset{\rho_{ij}}{\overbrace{\exp\left(-\frac{1}{2}\frac{(x_i^* - x_j^*)^2}{\ell_s^2}\right)}}$$

$$[\boldsymbol{\Sigma}_{\mathbf{w}*}^s]_{ij} = (a_s \cdot x_i) \cdot (a_s \cdot x_j^*) \cdot \exp\left(-\frac{1}{2}\frac{(x_i - x_j^*)^2}{\ell_s^2}\right)$$

$$[\boldsymbol{\Sigma}_{\mathbf{w}}^s]_{ij} = (a_s \cdot x_i) \cdot (a_s \cdot x_j) \cdot \exp\left(-\frac{1}{2}\frac{(x_i - x_j)^2}{\ell_s^2}\right)$$

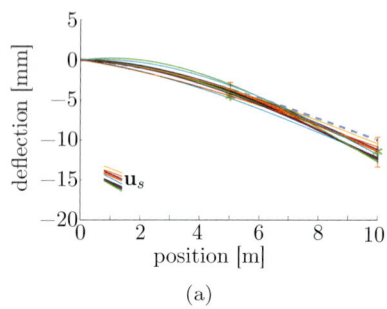

(a)

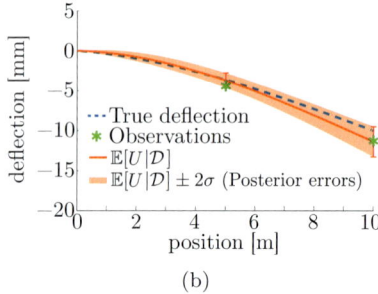

(b)

Figure 13.9: Comparison of the posterior predictive model predictions with the displacement of the reference model.

[1] Kennedy, M. C. and A. O'Hagan (2001). Bayesian calibration of computer models. *Journal of the Royal Statistical Society: Series B (Statistical Methodology) 63*(3), 425–464.

[2] Brynjarsdóttir, J. and A. O'Hagan (2014). Learning about physical parameters: The importance of model discrepancy. *Inverse Problems 30*(11), 114007.

Figure 13.9a presents a comparison of the posterior predictive model prediction samples $\mathbf{u}_s$ with the deflection of the reference model. Figure 13.9b presents the same comparison, this time using the posterior predictive mean vector and confidence interval. Notice how the confidence interval and the smoothness of the realizations match the true deflection profile.

The scatter plots in figure 13.8 show that significant uncertainty remains regarding the parameter $E$ and prediction errors $w(5), w(10)$. However, because there is a strong dependence in the joint posterior PDF describing these variables, the posterior uncertainty for the deflection remains limited. It happens because abnormally small (large) values for the elastic modulus $E_s$ result in large (small) negative displacements, which are compensated for by large positive (negative) prediction errors $w_s(5), w_s(10)$. This example illustrates why, when calibrating model parameters, it is also necessary to infer model prediction errors because parameters and prediction errors are typically dependent on each other.

The most difficult part of such a hierarchical Bayesian estimation procedure is to define an appropriate function for describing the prior knowledge of model prediction errors. This aspect is what is limiting its practical application to complex case studies where the prediction errors depend not only on the prediction locations but also on the prediction types, for example, rotation, strain, acceleration, and so on. When several formulations for the prior knowledge of prediction errors are available, the probability of each of them can be estimated using Bayesian model selection (see §6.8) or cross-validation (see §8.1.2). Further details about hierarchical Bayesian estimation applied to model calibration can be found in Kennedy and O'Hagan;[1] Brynjarsdóttir and O'Hagan.[2]

*Exercises*

P13.1  Explain the issues with the least-square calibration of hard-coded physics-based models.

P13.2  Explain why there are dependencies between unknown model parameters and model prediction errors.

P13.3  Explain the differences between a prior and a hyper-prior.

P13.4  Reproduce the results presented in figure 13.4 using least-square model calibration. Use the data set $\mathcal{D}_x = \{-3.3, 6.6, 10\}$ m, $\mathcal{D}_y = \{-2.8, -6.9, -12.6\}$ mm, along with the beam displacement governed by the equation

$$u(x) = \frac{Px^2}{6EI} \cdot (3L - x) + \frac{PLx}{K}.$$

P13.5  Repeat the problem in P13.4, this time using the hierarchical Bayesian approach presented in §13.2.

# Part V

# Reinforcement Learning

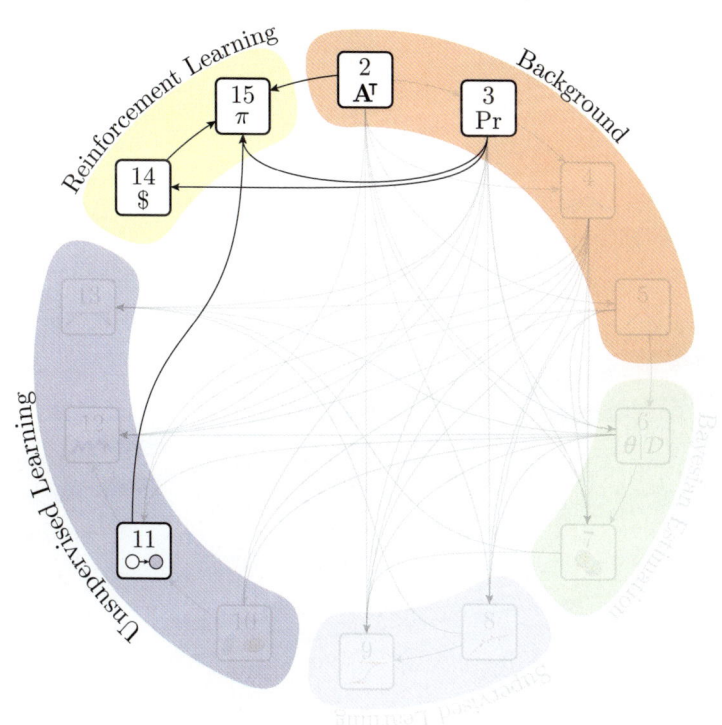

# 14
# *Decisions in Uncertain Contexts*

This chapter explores how to make rational decisions regarding actions to be taken in a context where the state variables on which a decision is based are themselves uncertain. Basic contextual notions will be presented through an illustrative example before introducing the formal mathematical notation associated with *rational decisions* and *value of information*.

## 14.1  *Introductory Example*

In order to introduce the notions associated with *utility theory*, we revisit the soil contamination example presented in §8.2.5, where we have a cubic meter of potentially contaminated soil. The two possible states $x \in \{⚠, \triangle\}$ are either *contaminated* or *not contaminated*, and the two possible actions $a \in \{🗒, ♻\}$ are either to *do nothing* or to *send the soil to a recycling plant* where it is going to be decontaminated. A soil sample is defined as contaminated if the pollutant concentration $c$ exceeds a threshold value $c > \phi_{\mathrm{adm.}}$. The issue is that for most m$^3$ in an industrial site, there are no observations made in order to verify whether or not it is contaminated. Here, we rely on a regression model built from a set of discrete observations to predict for any $x, y, z$ coordinates what our knowledge of the contaminant concentration is as described by its expected value and variance. In this example, our current knowledge indicates that, for a specific m$^3$,

$$\begin{aligned} \Pr(C \leq \phi_{\mathrm{adm.}}) &\equiv \Pr(X = \triangle) &= 0.9 \\ \Pr(C > \phi_{\mathrm{adm.}}) &\equiv \Pr(X = ⚠) &= 0.1. \end{aligned}$$

The optimal decision regarding the action to choose depends on the value incurred by taking an action given the state $x$. In this example we have two states and two actions, so there are four possibilities to be defined, as illustrated in figure 14.1. The incurred

(a) Given an $m^3$ of possibly contaminated soil.

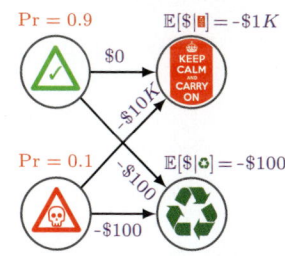

(b) State probabilities, values, and expected values.

Figure 14.1: Decision context for a soil contamination example.

**Actions**
Do nothing 🔋
Recycle ♻

**States**
Not contaminated △
Contaminated ⚠

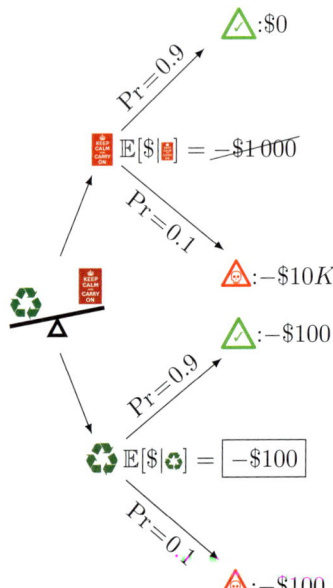

Figure 14.2: Decision tree for the soil contamination example.

value (\$) for each pair of states $x$ and actions taken $a$ are

$$\$(a,x) = \$ \begin{bmatrix} \text{🔋,△} & \text{🔋,⚠} \\ \text{♻,△} & \text{♻,⚠} \end{bmatrix} \equiv \begin{bmatrix} \$0 & -\$10K \\ -\$100 & -\$100 \end{bmatrix},$$

where doing nothing when the soil is not contaminated incurs no cost, decontaminating incurs a cost of \$100/m³ whether or not the soil is contaminated, and omitting the decontamination when it is actually contaminated (i.e., $c > \phi_{\text{adm.}}$) incurs a cost of \$10 K in future legal fees and compensatory damages.

Figure 14.2 depicts a *decision tree* illustrating the actions along with the probability of each outcome. The optimal action to be taken must be selected based on *the expected* value conditional on each action,

$$\begin{aligned} \mathbb{E}[\$|\text{🔋}] &= \$(\text{🔋,△}) \times \Pr(X = \text{△}) + \$(\text{🔋,⚠}) \times \Pr(X = \text{⚠}) \\ &= (\$0 \times 0.9) + (-\$10K \times 0.1) \\ &= -\$1\,000 \\ \mathbb{E}[\$|\text{♻}] &= \$(\text{♻,△}) \times \Pr(X = \text{△}) + \$(\text{♻,⚠}) \times \Pr(X = \text{⚠}) \\ &= (-\$100 \times 0.9) + (-\$100 \times 0.1) \\ &= \boxed{-\$100}. \end{aligned}$$

Because $\mathbb{E}[\$|\text{♻}] > \mathbb{E}[\$|\text{🔋}]$, the optimal action is thus to send the soil to a recycling plant where it will be decontaminated. As you can expect, changing the probability of each state or changing the incurred relative value (\$) for a pair of states $x$ and actions taken $a$ will influence the optimal action to be taken. The next section presents the *utility theory*, which formalizes the method employed in this introductory example.

## 14.2   Utility Theory

*Utility theory* defines how rational decisions are made. This section presents the nomenclature associated with utility theory, it formalizes what a rational decision is, and it presents its fundamental axioms.

### 14.2.1   Nomenclature

A decision consists in the task of choosing an action $\mathsf{a}_i$ from a set of possible actions $\mathcal{A} = \{\mathsf{a}_1, \mathsf{a}_2, \cdots, \mathsf{a}_A\}$. This decision is based on our knowledge of one or several state variables $x$, which can either be discrete or continuous. Our knowledge of a state variable is either described by its probability density $X \sim f(x)$ or mass function $X \sim p(x) = \Pr(X = x)$. The *utility* $\mathbb{U}(a, x)$ quantifies the relative

| | |
|---|---|
| $\mathcal{A} = \{\mathsf{a}_1, \cdots, \mathsf{a}_A\}$ | A set of possible actions |
| $x \in \mathcal{X} \subseteq \mathbb{Z}$ or $\subseteq \mathbb{R}$ | An outcome from a set of possible states |
| $\Pr(X = x) \equiv \Pr(x)$ | Probability of a state $x$ |
| $\mathbb{U}(a, x)$ | Utility given a state $x$ and an action $a$ |

preferences we have for the joint result of taking an action $a$ while being in a state $x$.

*Soil contamination example*   For the example presented in the introduction of this chapter, the actions and states can be formulated as binary discrete variables; $a \in \{\text{■}, \text{♲}\} \equiv \{0, 1\}$ and $x \in \{\triangle, \blacktriangle\} \equiv \{0, 1\}$. The probability of each state is $\Pr(X = x) = \{0.9, 0.1\}$, and the utility of each pair of actions, and states, is

$$\mathbb{U}(a, x) = \mathbb{U} \begin{bmatrix} \text{■},\triangle & \text{■},\blacktriangle \\ \text{♲},\triangle & \text{♲},\blacktriangle \end{bmatrix} \equiv \begin{bmatrix} \$0 & -\$10K \\ -\$100 & -\$100 \end{bmatrix}.$$

### 14.2.2   Rational Decisions

In the context of the utility theory, a *rational decision* is defined as a decision that maximizes the *expected utility*. In an uncertain context, the perceived benefit of an action $\mathsf{a}_i$ is measured by the expected utility, $\overline{\mathbb{U}}(a) \equiv \mathbb{E}[\mathbb{U}(a, X)]$. When $X$ is a discrete random variable so that $x \in \mathcal{X} = \{1, 2, \cdots, \mathsf{X}\}$, the expected utility is

$$\mathbb{E}[\mathbb{U}(a, X)] = \sum_{x \in \mathcal{X}} \mathbb{U}(a, x) \cdot p(x).$$

Instead, when $X$ is a continuous random variable, the expected utility is

$$\mathbb{E}[\mathbb{U}(a, X)] = \int \mathbb{U}(a, x) \cdot f(x) dx.$$

The *optimal* action $a^*$ is the one that *maximizes the expected utility*,

$$\begin{aligned} a^* &= \underset{a \in \mathcal{A}}{\arg \max} \; \mathbb{E}[\mathbb{U}(a, X)], \\ \overline{\mathbb{U}}(a^*) \equiv \mathbb{E}[\mathbb{U}(a^*, X)] &= \underset{a \in \mathcal{A}}{\max} \; \mathbb{E}[\mathbb{U}(a, X)]. \end{aligned}$$

### 14.2.3   Axioms of Utility Theory

The axioms of utility theory are based on the concepts of *lotteries*. A lottery $L$ is defined by a set of possible outcomes $x \in \mathcal{X} = \{1, 2, \cdots, \mathsf{X}\}$, each having its own probability of occurrence $p_X(x) = \Pr(X = x)$, so that

$$L = [\{p_X(1), x{=}1\}; \{p_X(2), x{=}2\}; \cdots ; \{p_X(\mathsf{X}), x{=}\mathsf{X}\}]. \qquad (14.1)$$

A decision is a choice made between several lotteries. For the soil contamination example, there are two lotteries, each one corresponding to a possible action. If we choose to send the soil to a recycling facility, we are certain of the outcome, that is, the soil is not contaminated after its treatment. If we choose to do nothing, there are two possible outcomes; either no significant contaminant was present with a probability of 0.9, or the soil was wrongly left without treatment with a probability of 0.1. These two lotteries can be summarized as

$$L_{\text{♻}} = [\{1.0, (\text{♻}, \triangle)\}; \{0.0, (\text{♻}, \triangle)\}]$$
$$L_{\text{▮}} = [\{0.9, (\text{▮}, \triangle)\}; \{0.1, (\text{▮}, \triangle)\}].$$

The nomenclature employed for ordering preferences is $L_i \succ L_j$ if we prefer $L_i$ over $L_j$, $L_i \sim L_j$ if we are indifferent about $L_i$ and $L_j$, and $L_i \succeq L_j$ if either we prefer $L_i$ over $L_j$ or are indifferent. The axioms of utility theory[1] define a rational behavior. Here, we review these axioms following the nomenclature employed by Russell and Norvig:[2]

[1] Von Neumann, J. and O. Morgenstern (1947). *The theory of games and economic behavior.* Princeton University Press.
[2] Russell, S. and P. Norvig (2009). *Artificial Intelligence: A Modern Approach* (3rd ed.). Prentice-Hall.

| | |
|---|---|
| $L_i \succ L_j$ | $L_i$ is preferred over $L_j$ |
| $L_i \sim L_j$ | $L_i$ and $L_j$ are indifferent |
| $L_i \succeq L_j$ | $L_i$ is preferred over $L_j$ or are indifferent |

*Orderability (Completeness)* — A decision maker has well-defined preferences for lotteries so that one of $(L_i \succ L_j)$, $(L_j \succ L_i)$, $(L_i \sim L_j)$ holds.

*Transitivity* — Preferences over lotteries are *transitive*, so that if $(L_i \succ L_j)$ and $(L_j \succ L_k)$, then $(L_i \succ L_k)$.

*Continuity* — If the preferences for lotteries are ordered following $(L_i \succ L_j \succ L_k)$, then a probability $p$ exists so that $[\{p, L_i\}; \{1 - p, L_k\}] \sim L_j$.

*Substitutability (Independence)* — If two lotteries are equivalent $(L_i \sim L_j)$, then the lottery $L_i$ or $L_j$ can be substituted by another equivalent lottery $L_k$ following $[\{p, L_i\}; \{1 - p, L_k\}] \sim [\{p, L_j\}; \{1 - p, L_k\}]$.

*Monotonicity* — If the lottery $L_i$ is preferred over $L_j$, $L_i \succ L_j$, then for a probability $p$ greater than $q$, $[\{p, L_i\}; \{1 - p, L_j\}] \succ [\{q, L_i\}; \{1 - q, L_j\}]$.

*Decomposability* — The decomposability property assumes that there is no reward associated with the decision-making process itself, that is, there is no fun in gambling.

If the preferences over lotteries can be defined following all these axioms, then a rational decision maker should choose the lottery associated with the action that maximizes the expected utility.

## 14.3   Utility Functions

The axioms of utility theory are used to define a utility function so that if a lottery $L_i$ is preferred over $L_j$, then the lottery $L_i$ must have a greater utility than $L_j$. If we are indifferent about two lotteries, their utilities must be equal. These properties are

summarized as

$$\begin{aligned} \mathbb{U}(L_i) &> \mathbb{U}(L_j) &\Leftrightarrow& \quad L_i &\succ& \quad L_j \\ \mathbb{U}(L_i) &= \mathbb{U}(L_j) &\Leftrightarrow& \quad L_i &\sim& \quad L_j. \end{aligned}$$

The *expected utility* of a lottery (see equation 14.1) is defined as the sum of the utility of each possible outcome in the lottery multiplied by its probability,

$$\mathbb{E}[\mathbb{U}(L)] = \sum_{x \in \mathcal{X}} p(x) \cdot \mathbb{U}(x).$$

Keep in mind that a utility function defines the relative preferences between outcomes and not the absolute one. It means that we can transform a utility function through an affine function,

$$\mathbb{U}^{\text{tr}}(x) = w\mathbb{U}(x) + b, \quad w > 0,$$

without changing the preferences of a decision maker. In the case where we multiply a utility function by a negative constant $w < 0$, this is no longer true; instead of maximizing the expected utility, the problem would then consist in *minimizing the expected loss*,

$$a^* = \arg\min_a \mathbb{E}[\mathbb{L}(a, X)].$$

*Nonlinear utility functions*    We now look at the example of a literal lottery as the term is commonly employed outside the field of utility theory. This lottery is defined like so: you receive \$200 if a coin toss lands on heads, and you pay \$100 if a coin toss lands on tails. This situation is formalized by the two lotteries respectively corresponding to taking the bet (⬤) or passing (🔴),

$$\begin{aligned} L_{\text{⬤}} &= [\{\tfrac{1}{2}, + \text{💵💵}\}; \{\tfrac{1}{2}, -\text{💵}\}] \\ L_{\text{🔴}} &= [\{1, \$0\}], \end{aligned}$$

where you are only certain of the outcome if you choose to pass. The question is, *Do you want to take the bet?* In practice, when people are asked, most would not accept the bet, which contradicts the utility theory principle requiring that one chooses the lottery with the highest expected utility. For this problem,

$$\begin{aligned} \mathbb{E}[\$(L_{\text{⬤}})] &= \tfrac{1}{2} \times +\$200 + \tfrac{1}{2} \times -\$100 = +\$50 \\ \mathbb{E}[\$(L_{\text{🔴}})] &= \$0, \end{aligned}$$

which indicates that a rational decision maker should take the bet. Does this mean that the utility theory does not work or that people are acting irrationally? The answer is no; the reason for this behavior is that utility functions for quantities such as monetary value are typically nonlinear.

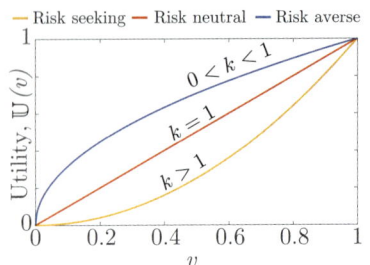

Figure 14.3: Comparison between risk
-*seeking*, -*neutral*, and -*averse* behaviors for
utility functions.

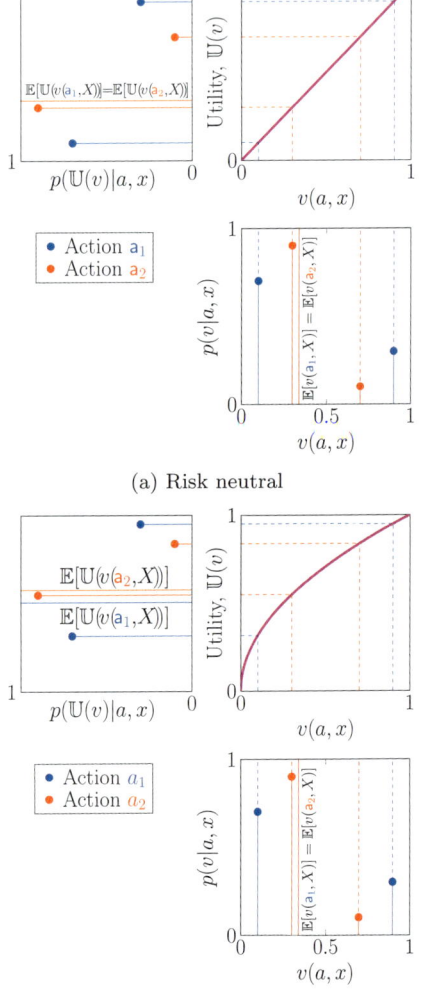

(a) Risk neutral

(b) Risk averse

Figure 14.4: Discrete case: Comparison of
the effect of a risk-neutral and a risk-averse
behavior on the expected utility.

*Risk averse versus risk seeking*   Instead of directly defining the
utility for being in a state $x$ while having taken the action $a$, we
can define it for continuous quantities such as a monetary value.
We define a *value* $v(a, x)$ associated with the outcome of being in
a state $x$ while having taken the action $a$, and $\mathbb{U}(v(a, x)) \equiv \mathbb{U}(v)$ is
the utility for the value associated with $x$ and $a$.

Figure 14.3 presents examples of utility functions expressing different risk behaviors. When a utility function is linear with respect
to a value $v$, we say that it represents a *risk-neutral* behavior, that
is, doubling $v$ also doubles $\mathbb{U}(v)$. In common cases, individuals are
not displaying risk-neutral behavior because, for example, gaining
or losing \$1 will impact behavior differently depending on whether
a person has \$1 or \$1,000,000. A *risk-averse* behavior is characterized by a utility function having a negative second derivative
so that the change in utility for a change of value $v$ decreases as
$v$ increases. The consequence is that given the choice between a
certain event of receiving \$100 and a lottery for which the expected
gain is $\mathbb{E}[L] = \$100$, a risk-averse decision maker would prefer the
certain event. The opposite is a *risk-seeking* behavior, where there
is a small change in the utility function for small values and large
changes in the utility function for large values. When facing the
same previous choice, a risk-seeking decision maker would prefer the
lottery over the certain event.

Let us consider a generic utility function for $v \in (0, 1)$ so that,

$$\mathbb{U}(v) = v^k, \quad \text{where} \begin{cases} 0 < k < 1 & \text{Risk averse} \\ k = 1 & \text{Neutral} \\ k > 1 & \text{Risk seeking.} \end{cases}$$

Figure 14.4 compares the effect of a risk-neutral and a risk-seeking
behavior on the conditional expected utilities $\mathbb{E}[\mathbb{U}(v(\mathsf{a}_i, X))]$. In
this example, there is a binary random variable $X$ describing
the possible state $x \in \{1, 2\}$, where the probability of each outcome depends on an action $\mathsf{a}_i$, and where $v(\mathsf{a}_i, x)$ is the value
of being in a state $x$ while the action $\mathsf{a}_i$ was taken. This illustrative example is designed so that the expected value for both
actions are equal, $\mathbb{E}[v(\mathsf{a}_1, X)] = \mathbb{E}[v(\mathsf{a}_2, X)]$, but not their variance, $\text{var}[v(\mathsf{a}_1, X)] > \text{var}[v(\mathsf{a}_2, X)]$. With a risk-neutral behavior in (a), the expected utilities remain equal for both actions,
$\mathbb{E}[\mathbb{U}(v(\mathsf{a}_1, X))] = \mathbb{E}[\mathbb{U}(v(\mathsf{a}_2, X))]$. For the risk-averse behavior displayed in (b), the expected utility is higher for action 2 than for
action 1, $\mathbb{E}[\mathbb{U}(v(\mathsf{a}_2, X))] > \mathbb{E}[\mathbb{U}(v(\mathsf{a}_1, X))]$, because the variability in
value is greater for action 1 than for action 2.

Figure 14.5 presents the same example, this time for a continuous random variable. The transformation of the probability density

function (PDF) from the value space to the utility space is done following the change of variable rule presented in §3.4. Again, for a risk-neutral behavior (a), the expected utility is equal, and for a risk-averse behavior (b), the expected utility is higher for action 2 than for action 1, $\mathbb{E}[\mathbb{U}(v(\mathsf{a}_2, X))] > \mathbb{E}[\mathbb{U}(v(\mathsf{a}_1, X))]$, because the variability in the value is greater for action 1 than for action 2.

People and organizations typically display a risk-averse behavior, which must be considered if we want to properly model optimal decisions. In such a case, an optimal decision $\mathsf{a}_i$ is the one that maximizes $\mathbb{E}[\mathbb{U}(v(\mathsf{a}_i, X))]$. Although such a decision is optimal in the sense of utility theory, it is not the one with the highest expected monetary value. Only a neutral attitude toward risks maximizes the expected value.

For the bet example presented at the beginning of the section, we discussed that the action chosen is typically to not take the bet, even if taking it would result in the highest expected gain. Part of it has to do with the repeatability of the bet; it is a certainty that the gain per play will tend to the expected value as the number of times played increases. Nevertheless, $100 might appear as an important sum to gamble because if a person plays and loses, he or she might not afford to play again to make up for the loss. In the civil engineering context, risk aversion can be seen through a similar scenario, where if someone loses a large amount of money as the consequence of a decision that was rightfully maximizing the expected utility, that person might not keep his or her job long enough to compensate for the current loss with subsequent profitable decisions.

Risk-averse behavior is the reason of being for insurance companies, who have a neutral attitude toward the risks they are providing insurance for. What they do is cover the risks associated with events that would incur small monetary losses in comparison with the size of the company. Insurers seek to diversify the events covered in order to maximize the independence of the probability of each event. The objective is to avoid being exposed to payments so large that it would jeopardize the solvency of the company. Individuals who buy the insurance are risk averse, so they are ready to pay a premium over the expected value in order not to be put in a risk-neutral position. In other words, they accept paying an amount higher than the expected costs to an insurer, who itself has to pay the expected cost because its exposure to losses is spread over thousands if not millions of customers.

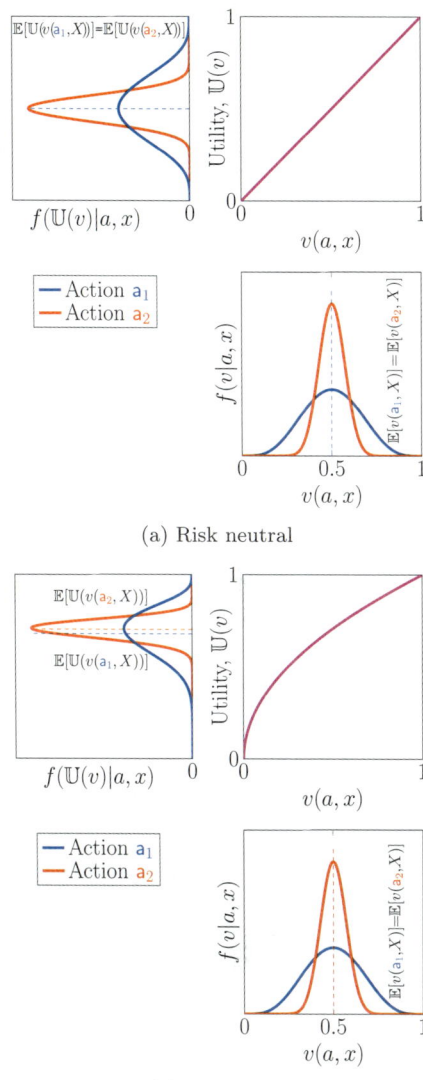

(a) Risk neutral

(b) Risk averse

Figure 14.5: Continuous case: Comparison of the effect of a linear (risk-neutral) and a nonlinear (risk-averse) utility function on the expected utility.

## 14.4    Value of Information

*Value of information* quantifies the value associated with the action of gathering additional information about a state $X$ in order to reduce the epistemic uncertainty associated with its knowledge. In this section, we will cover two cases: *perfect* and *imperfect* information.

### 14.4.1    Value of Perfect Information

In cases where the value of a state $x \in \mathcal{X} = \{1, 2, \cdots, \mathtt{X}\}$ is imperfectly known, one possible action is to collect additional information about $X$. With the current knowledge of $X \sim p(x)$, the expected utility of the optimal action $a^*$ is

$$
\begin{aligned}
\overline{\mathbb{U}}(a^*) \equiv \mathbb{E}[\mathbb{U}(a^*, X)] &= \max_a \sum_{x \in \mathcal{X}} \mathbb{U}(a, x) \cdot p(x) \\
&= \sum_{x \in \mathcal{X}} \mathbb{U}(a^*, x) \cdot p(x).
\end{aligned}
\tag{14.2}
$$

If we gather perfect information so that $y = x \in \mathcal{X}$, we can then directly observe the true state variable, and the utility of the optimal action becomes

$$
\mathbb{U}(a^*, y) = \max_a \mathbb{U}(a, y).
$$

However, because $y$ has not been observed yet, we must consider all possible observations $y = x$ weighted by their probability of occurrence so the expected utility conditional on perfect information is

$$
\overline{\mathbb{U}}(\tilde{a}^*) \equiv \mathbb{E}[\mathbb{U}(\tilde{a}^*, X)] = \sum_{x \in \mathcal{X}} \max_a \mathbb{U}(a, x) \cdot p(x).
\tag{14.3}
$$

Notice how in equation 14.2, the max operation was outside of the sum, whereas in equation 14.3, the max is inside. In equation 14.2, we compute the expected utility where the state is a random variable and for an optimal action that is common for all states. In equation 14.3, we assume that we will know the true state once the observation becomes available, so we will be able to take the optimal action for each state. Consequently, the expected utility is calculated by weighting the utility corresponding to the optimal action for each state, times the probability of occurrence of that state. The *value of perfect information* (VPI) is defined as the difference between the expected utility conditional on perfect information and the expected utility for the optimal action,

$$
\mathrm{VPI}(y) = \overline{\mathbb{U}}(\tilde{a}^*) - \overline{\mathbb{U}}(a^*) \geq 0.
$$

Because the expected utility estimated using perfect information $\mathbb{E}[\mathbb{U}(\tilde{a}^*, X)]$ is greater than $\mathbb{E}[\mathbb{U}(a^*, X)]$, the VPI($y$) must be greater or equal to zero. The VPI quantifies the amount of money you should be willing to pay to obtain perfect information. Note that the concepts presented for discrete random variables can be extended for continuous ones by replacing the sums by integrals.

*Soil contamination example*   We apply the concept of the value of perfect information to the soil contamination example. The current expected utility conditional on the optimal action is

$$\mathbb{E}[\mathbb{U}(\text{▥}, X)] = (\$0 \times 0.9) + (-\$10K \times 0.1) = -\$1K$$

$$\mathbb{E}[\mathbb{U}(\text{♻}, X)] = (-\$100 \times 0.9) + (-\$100 \times 0.1) = \boxed{-\$100} = \overline{\mathbb{U}}(a^*),$$

so the current optimal action is to send the soil to a recycling plant with an associated expected utility of -$100. Figure 14.6 presents the decision tree illustrating the calculation for the value of perfect information. The expected utility conditional on perfect information is

$$\overline{\mathbb{U}}(\tilde{a}^*) = \sum_{x \in \mathcal{X}} \max_a \mathbb{U}(a, x) \cdot p(x)$$
$$= \underbrace{\$0 \times 0.9}_{y=x=\triangle} + \underbrace{-\$100 \times 0.1}_{y=x=\triangle} = \boxed{-\$10}.$$

Having access to perfect information reduces the expected cost because there is now a probability of 0.9 that the observation will indicate that no treatment is required and only a probability of 0.10 that treatment will be required. Moreover, the possibility of the worst case associated with a false negative where $\{a = \text{▥}, x = \text{⚠}\}$ is now removed from the possibilities. The value of perfect information is thus

$$\text{VPI}(y) = \overline{\mathbb{U}}(\tilde{a}^*) - \overline{\mathbb{U}}(a^*) = \boxed{\$90}.$$

It means that we should be willing to pay up to $90 for perfect information capable of indicating whether not the m$^3$ of soil is contaminated.

### 14.4.2   Value of Imperfect Information

It is common that observations of state variables are imperfect, so that we want to compute the value of imperfect information. For the case of a discrete state variable $x$ that belongs to the set $\mathcal{X} = \{x_1, x_2, \cdots, x_{\mathbf{x}}\}$, the expected costs conditional on imperfect

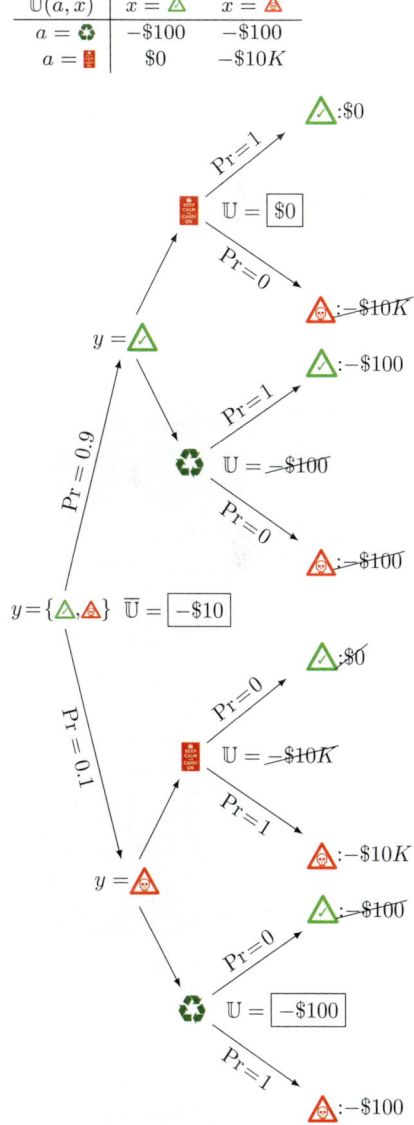

| $\mathbb{U}(a, x)$ | $x = \triangle$ | $x = \triangle$ |
|---|---|---|
| $a = \text{♻}$ | $-\$100$ | $-\$100$ |
| $a = \text{▥}$ | $\$0$ | $-\$10K$ |

Figure 14.6: Decision tree illustrating the value of perfect information.

| $p(x,y)$ | $y = \triangle$ | $y = \triangle$ |
|---|---|---|
| $x = \triangle$ | $0.9\cdot 1 = 0.9$ | $0.9\cdot 0 = 0$ |
| $x = \triangle$ | $0.1\cdot 0.05 = 0.005$ | $0.1\cdot 0.95 = 0.095$ |
| $p(y)$ | $0.905$ | $0.095$ |

| $p(x\|y)$ | $y = \triangle$ | $y = \triangle$ |
|---|---|---|
| $x = \triangle$ | $\frac{0.9}{0.905}$ | $\frac{0}{0.095}$ |
| $x = \triangle$ | $\frac{0.005}{0.905}$ | $\frac{0.095}{0.095}$ |

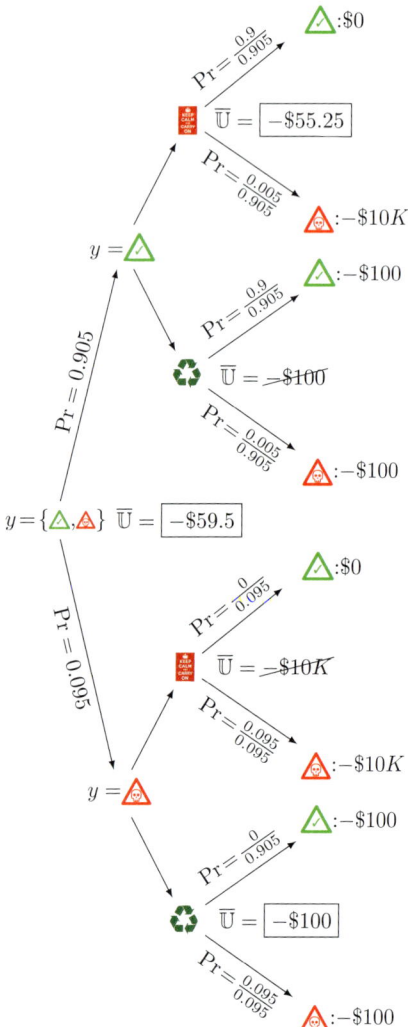

Figure 14.7: Decision tree illustrating the value of imperfect information.

[3] Russell, S. and P. Norvig (2009). *Artificial Intelligence: A Modern Approach* (3rd ed.). Prentice-Hall; and Bertsekas, D. P. (2007). *Dynamic programming and optimal control* (4th ed.), Volume 1. Athena Scientific.

information is obtained by marginalizing over both the possible states $x \in \mathcal{X}$ and the possible observations $y \in \mathcal{X}$, so that

$$\overline{\mathbb{U}}(\tilde{a}^*) = \sum_{y \in \mathcal{X}} \max_a \left( \sum_{x \in \mathcal{X}} \mathbb{U}(a,x) \cdot \underbrace{p(y|x) \cdot p(x)}_{p(y,x)} \right).$$

*Soil contamination example*   We apply the concept of the value of imperfect information to the soil contamination example. The conditional probability of an observation $y$ conditional on the state $x$ is

$$p(y|x) \begin{cases} \Pr(y = \triangle | x = \triangle) & = & 1 \\ \Pr(y = \triangle | x = \triangle) & = & 0.95, \end{cases}$$

where the observation is perfect if the soil is not contaminated, and the probability of a correct classification is 0.95 if the soil is contaminated.

Figure 14.7 depicts the decision tree for calculating the value of information for this example, where the probability of each observation is obtained by marginalizing the joint probability for both observations and states,

$$p(y) = \sum_{x \in \mathcal{X}} p(y|x) \cdot p(x).$$

The expected utility conditional on imperfect information is then

$$\overline{\mathbb{U}}(\tilde{a}^*) = \sum_{y \in \mathcal{X}} \max_a \left( \sum_{x \in \mathcal{X}} \mathbb{U}(a,x) \cdot \underbrace{p(y|x) \cdot p(x)}_{p(x,y)} \right)$$

$$= \underbrace{\underbrace{\$0 \times 0.9}_{x=\triangle} + \underbrace{-\$10K \times 0.005}_{x=\triangle}}_{y=\triangle} + \underbrace{\underbrace{-\$100 \times 0}_{x=\triangle} + \underbrace{-\$100 \times 0.095}_{x=\triangle}}_{y=\triangle}$$

$$= \boxed{-\$59.5}.$$

The *value of information* is now defined by the difference between the expected utility conditional on imperfect information and the expected utility for the optimal action, so that

$$\begin{aligned} VOI(y) &= \overline{\mathbb{U}}(\tilde{a}^*) - \overline{\mathbb{U}}(a^*) \geq 0 \\ &= -\$59.5 - (-\$100) \\ &= \boxed{\$40.5}. \end{aligned}$$

By comparing this result with the value of perfect information that was equal to $90, we see that having observation uncertainty reduces the value of the information.

The reader interested in advanced notions related to rational decisions and utility theory should consult specialized textbooks[3] such as those by Russell and Norvig or Bertsekas.

*Exercises*

P14.1  Explain what quantitative criterion defines a rational decision.

P14.2  What are the units of the utility?

P14.3  Explain the concepts associated with risk-neutral, risk-averse, and risk-seeking behaviors.

P14.4  What is the consequence associated with having a nonlinear utility function?

P14.5  Explain the value of perfect and imperfect information.

P14.6  Why is the value of perfect and imperfect information greater than or equal to 0?

P14.7  A structural health monitoring (SHM) system is employed to monitor the state of a structure and can lead to one of the two following actions: either *alert an engineer* (e) or *do nothing* ($\neg$e), $a \in \{e, \neg e\}$. The utility associated with each action and state is

$$\begin{aligned}
\mathbb{U}(e, n) &= -\$500 \\
\mathbb{U}(e, \neg n) &= -\$10{,}000 \\
\mathbb{U}(\neg e, n) &= -\$50 \\
\mathbb{U}(\neg e, \neg n) &= -\$100{,}000,
\end{aligned}$$

where the state of the structure can be either *normal* (n) or *abnormal* ($\neg$n), $s \in \{n, \neg n\}$. If the SHM system indicates that the probability of being in a normal state is 0.73:

a.  Compute the expected utility of each action $\mathbb{E}[\mathbb{U}(a, s)]$ and identify the optimal action to take.

b.  Compute the *value of perfect information*, that is, the maximal amount you would be willing to pay in order to know with certainty if the state is either n or $\neg$n.

# 15
## Sequential Decisions

In this chapter, we explore how to make optimal decisions in a sequential context. Sequential decisions differ from the decision context presented in chapter 14 because here, the goal is to select the optimal action to be performed in order to maximize the current reward as well as future ones. In that process, we must take into account that our current action will affect future states as well as future decisions. Let us consider a system that can be in a state $s \in \mathcal{S} = \{s_1, s_2, \cdots, s_S\}$. The set of actions $a \in \mathcal{A} = \{a_1, a_2, \cdots, a_A\}$ that should be performed given each state of the system is called a *policy*,

$$\pi = \{a(s_1), a(s_2), \cdots, a(s_S)\}.$$

In the context of sequential decisions, the goal is to identify the optimal policy $\pi^*$ describing for each possible state of the system the optimal action to take. This task can be formulated as a *reinforcement learning* (RL) problem, where the optimal policy is learned by choosing the actions $a^*(s)$ that maximize the *expected utility* over a given *planning horizon*. In order to compute the expected utility, we need a function describing for each set of current state $s$, action $a$, and next state $s'$ the associated *reward* $r(s, a, s') \in \mathbb{R}$. In practical cases, it can happen that the reward simplifies to a special case such as $r(s, a, s') = r(s, a)$, $r(s, a, s') = r(a, s')$, or even $r(s, a, s') = r(s)$. The last part we need to define is the *Markovian transition model* $p(s'|s_j, a_i) = \Pr(S' = s_k|s = s_j, a_i)$, $\forall k \in \{1 : S\}$, that is,

$$p(s'|s_j, a_i) \equiv \begin{cases} \Pr(S_{t+1} = s_1|s_t = s_j, a_i) \\ \Pr(S_{t+1} = s_2|s_t = s_j, a_i) \\ \quad\vdots \\ \Pr(S_{t+1} = s_S|s_t = s_j, a_i). \end{cases}$$

The *transition model* describes, given the current state of the system $s_j$ and an action taken $a_i$, the probability of ending in any

| | |
|---|---|
| States: | $s \in \mathcal{S} = \{s_1, s_2, \cdots, s_S\}$ |
| Actions: | $a \in \mathcal{A} = \{a_1, a_2, \cdots, a_A\}$ |
| Policy: | $\pi = \{a(s_1), a(s_2), \cdots, a(s_S)\}$ |
| Reward: | $r(s, a, s') \in \mathbb{R}$ |
| Transition model: | $p(s'|s, a)$ |

**Note:** Reinforcement learning differs from supervised learning because the optimal policy is not learned using other examples of policies. RL also differs from unsupervised learning because the goal is not to discover a structure or patterns in the data.

**Note:** A transition model is analogous to the concept of a conditional probability table presented in chapter 11.

state $s' \in \mathcal{S}$ at time $t + 1$. Here, the Markov property (see chapter 7) allows formulating a model where the probability of each state $s' \in \mathcal{S}$ at time $t + 1$ depends on the current state $\mathsf{s}_j$ at time $t$ and is independent of any previous states. Although sequential decision problems could also be formulated for continuous state variables, $s \in \mathbb{R}$, in this chapter we restrict ourselves to the discrete case.

*Example: Infrastructure maintenance planning*   The notions associated with reinforcement learning and the sequential decision process are illustrated through an application to infrastructure maintenance planning. Let us consider a population of bridges (see figure 15.1) for which at a given time $t$, each bridge can be in a state

$$s \in \mathcal{S} = \{ \underbrace{100\,\%}_{\mathsf{s}_1}, \underbrace{80\,\%}_{\mathsf{s}_2}, \underbrace{60\,\%}_{\mathsf{s}_3}, \underbrace{40\,\%}_{\mathsf{s}_4}, \underbrace{20\,\%}_{\mathsf{s}_5}, \underbrace{0\,\%}_{\mathsf{s}_6} \},$$

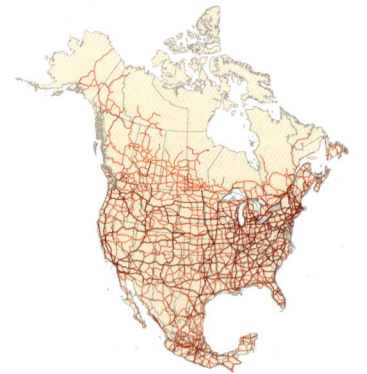

(a) Representation of North America's road network, which includes hundreds of thousands of bridges. (Photo: Commission for Environmental Cooperation)

(b) Example of a given structure within the network. (Photo: Michel Goulet)

Figure 15.1: Contextual illustration of a sequential decision process applied to the problem of infrastructure maintenance planning.

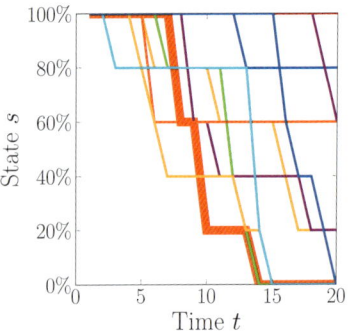

Figure 15.2: Example of bridge condition time histories.

where 100 percent indicates a perfect condition, and 0 percent corresponds to a state where the damage is so extensive that the structure must be closed. The objective of sequential decision making is to learn the optimal policy $\pi^*(s)$ that identifies among a set of possible actions,

$$a \in \mathcal{A} = \{\text{do nothing}\,(\mathsf{N}), \text{maintain}\,(\mathsf{M}), \text{replace}\,(\mathsf{R})\},$$

what is the optimal one $a^*(s)$ to take at each time, depending on the current condition $s$.

In this example, a bridge condition $s$ is rated every year from the results of visual inspections. We assume we have a database $\mathcal{D}$ containing the conditions obtained from inspection results for $\mathsf{N}$ structures each inspected $\mathsf{T}$ times,

$$\mathcal{D} = \left\{ \begin{array}{c} \{s_1, s_2, \cdots, s_\mathsf{T}\}_1 \\ \{s_1, s_2, \cdots, s_\mathsf{T}\}_2 \\ \vdots \\ \{s_1, s_2, \cdots, s_\mathsf{T}\}_\mathsf{N} \end{array} \right\}_{\mathsf{N} \times \mathsf{T}}.$$

Each line in this database corresponds to a given structure and each column corresponds to a given time $t \in \{1, 2, \cdots, \mathsf{T}\}$. Figure 15.2 presents an example for $\mathsf{N} = 15$ bridges and time histories of length $\mathsf{T} = 20$. One of them has been highlighted with a thicker red line to better visualize how a structure transits from state to state through time. In this data set, the action taking place at each time $t$ is $a =$ do nothing. We can use this data set to estimate the parameters of the Markovian transition model $p(s_{t+1}|s_t, a =$ do nothing) by employing the maximum likelihood estimate for each transition probability (see §11.4). The maximum likelihood estimate (MLE) corresponds to the number of structures that, from a state $i$ at time $t$, transitioned to a state $j$ at $t+1$, divided by the number of structures that started from a state $i$ at time $t$,

$$\hat{\Pr}(S_{t+1} = \mathsf{s}_j | s_t = \mathsf{s}_i, a_t = \text{do nothing}) = \frac{\#\{s_t = \mathsf{s}_i, s_{t+1} = \mathsf{s}_j\}}{\#\{s_t = \mathsf{s}_i\}},$$

where $\{\mathsf{s}_i, \mathsf{s}_j\} \in \mathcal{S}^2$ and the $\#$ symbol describes the number of elements in a set. These results are stored in a *transition matrix*,

$$p(s_{t+1} | s_t, a_t = \underbrace{\text{do nothing}}_{N}) = \begin{bmatrix} 0.95 & 0.03 & 0.02 & 0 & 0 & 0 \\ 0 & 0.9 & 0.05 & 0.03 & 0.02 & 0 \\ 0 & 0 & 0.8 & 0.12 & 0.05 & 0.03 \\ 0 & 0 & 0 & 0.7 & 0.25 & 0.05 \\ 0 & 0 & 0 & 0 & 0.6 & 0.4 \\ 0 & 0 & 0 & 0 & 0 & 1 \end{bmatrix}_{\mathsf{S} \times \mathsf{S}},$$

where each element $[\ ]_{ij}$ corresponds to the probability of landing in state $s_{t+1} = \mathsf{s}_j$ given that the system is currently in state $s_t = \mathsf{s}_i$. This matrix is analogous to the transition matrix we have seen in §12.2. In the case of infrastructure degradation, the transition matrix has only zeros below its main diagonal because, without intervention, a bridge's condition can only decrease over time, so a transition from $s_t = \mathsf{s}_i \rightarrow s_{t+1} = \mathsf{s}_j$ can only have a nonzero probability for $j \geq i$.

For maintenance and replacement actions, the transition matrix can be defined deterministically as

$$p(s_{t+1} | s_t, a_t = \underbrace{\text{maintain}}_{M}) = \begin{bmatrix} 1 & 0 & 0 & 0 & 0 & 0 \\ 1 & 0 & 0 & 0 & 0 & 0 \\ 0 & 1 & 0 & 0 & 0 & 0 \\ 0 & 0 & 1 & 0 & 0 & 0 \\ 0 & 0 & 0 & 1 & 0 & 0 \\ 0 & 0 & 0 & 0 & 1 & 0 \end{bmatrix}_{\mathsf{S} \times \mathsf{S}},$$

where from a current state $s_t = \mathsf{s}_i$, the state becomes $s_{t+1} = \mathsf{s}_{i-1}$. For example, if at a time $t$ the state is $s_t = \mathsf{s}_3 = 60\%$, the maintenance action will result in the state $s_{t+1} = \mathsf{s}_2 = 80\%$. Note that if the state is $s_t = \mathsf{s}_1 = 100\%$, it remains the same at $t+1$. For a replacement action, no matter the initial state at $t$, the next state becomes $s_{t+1} = \mathsf{s}_1 = 100\%$ so,

$$p(s_{t+1} | s_t, a_t = \underbrace{\text{replace}}_{R}) = \begin{bmatrix} 1 & 0 & 0 & 0 & 0 & 0 \\ 1 & 0 & 0 & 0 & 0 & 0 \\ 1 & 0 & 0 & 0 & 0 & 0 \\ 1 & 0 & 0 & 0 & 0 & 0 \\ 1 & 0 & 0 & 0 & 0 & 0 \\ 1 & 0 & 0 & 0 & 0 & 0 \end{bmatrix}_{\mathsf{S} \times \mathsf{S}}.$$

In order to identify the optimal policy $\pi^*$, we need to define the *reward* $r(s, a, s')$. Here, we assume that the reward simplifies to $r(s, a) = r(s) + r(a)$, so it is only a function of the state $s$ and the action $a$. For a given structure, the reward $r(s)$ is estimated as a function of the number of users quantified through the *annual average daily traffic flow* (AADTF = $10^5$ users/day), times a value of $3/user, times the capacity of the bridge as a function of its condition. The capacity is defined as $c(\mathcal{S}) = \{1, 1, 1, 0.90, 0.75, 0\}$, and the associated rewards are

$$
\begin{aligned}
r(\mathcal{S}) &= 10^5 \text{ users/day} \cdot 365 \text{ day} \cdot \$3/\text{user} \cdot c(\mathcal{S}) \\
&= \{109.5, 109.5, 109.5, 98.6, 82.1, 0\}\$M.
\end{aligned}
$$

The rewards for actions $r(a)$ correspond to a cost, so their values are lesser than or equal to zero,

$$
r(\mathcal{A}) = \{0, -5, -20\}\$M.
$$

For the context of reinforcement learning, we can generalize the example above by introducing the concept of an *agent* interacting with its *environment*, as depicted in figure 15.3. In the context of the previous example, the environment is the population of structures that are degrading over time, and the agent is the hypothetical entity acting with the intent of maximizing the long-term accumulation of rewards. The environment interacts with the agent by defining, for each time $t$, the state $s_t$ of the system and the reward for being in that state $s$, taking a given action $a$, and ending in the next state $s'$. The agent perceives the environment's state and selects an action $a$, which in turn can affect the environment and thus the state at time $t + 1$.

This chapter explores the task of identifying the optimal policy $\pi^*(s)$ describing the optimal actions $a^*$ to be taken for each state $s$. This task is first formulated in §15.1 using the model-based method known as the *Markov decision process*. Building on this method, §15.2 then presents two model-free reinforcement learning methods: *temporal difference* and *Q-learning*.

## 15.1   Markov Decision Process

In order to formulate a *Markov decision process* (MDP), we need to define a *planning horizon* over which the utility is estimated. Planning horizons may be either finite or infinite. With a *finite planning horizon*, the rewards are considered over a fixed period of time. In such a case, the optimal policy $\pi^*(s, t)$ is nonstationary because it depends on the time $t$. For the infrastructure maintenance example,

**Note:** Here the value of $3/user does not necessarily represent the direct cost per user as collected by a toll booth. It instead corresponds to the indirect value of the infrastructure for the society.

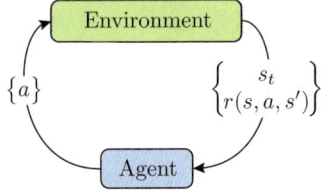

Figure 15.3: Schematic representation of the interaction between an agent and its environment.

a finite planning horizon could correspond to the case where we want to identify the optimal actions to take over the next $\mathtt{T} = 20$ years, after which we know a structure is going to be demolished. The optimal action for a state $s_t = 40\,\%$ would then not be the same whether we are at $t = 1$ or at $t = 19$. For an *infinite planning horizon*, the rewards are considered over an infinite period of time. In such a case, the optimal policy $\pi^*(s)$ is *stationary* as it does not depend on the time $t$. It means that at any time $t$, the optimal action to take is only a function of the current state $s$ we are in. In this chapter, we only study problems associated with an infinite planning horizon, as they allow identifying a stationary optimal policy $\pi^*(s)$ rather than one which depends on time $\pi^*(s, t)$, as is the case with a finite planning horizon.

**Finite planning horizon**
  Nonstationary policy

**Infinite planning horizon**
  Stationary policy

### 15.1.1   Utility for an Infinite Planning Horizon

The *utility* associated with an infinite planning horizon is defined as the reward for being in the current state plus the sum of the discounted rewards for all future states,

**Note:** Some references employ the term *utility*, others, the term *value* to refer to the sum of discounted rewards.

**Note:** In this subsection, the notation is simplified by assuming that $r(s, a, s') = r(s)$. Nevertheless, the same reasoning holds for the general case where the reward is $r(s, a, s')$.

$$
\begin{aligned}
\mathbb{U}(s_{t=0}, s_{t=1}, \cdots, s_{t=\infty}) &= r(s_{t=0}) + \gamma r(s_{t=1}) + \gamma^2 r(s_{t=2}) + \cdots \\
&= \sum_{t=0}^{\infty} \gamma^t r(s_t) \leq \frac{\max\limits_{s \in \mathcal{S}} r(s)}{1 - \gamma}.
\end{aligned}
$$
(15.1)

For a *discount factor* $\gamma \in (0, 1]$, the discounted sum of rewards over an infinite planning horizon is a finite quantity. This is an essential property because without a discount factor, we could not compare the performance of actions that would each have infinite utilities. In the special case where there is a *terminal state* for which the problem stops, the discount factor can be taken as $\gamma = 1$ because it will be possible to compute a finite estimate for the utility. Note that for the infrastructure maintenance planning example, there is no terminal state and the discount factor can be interpreted as an interest rate using the transformation $\frac{1}{\gamma} - 1$. The issue with our planning problem is that we do not know yet what will be the future states $s_{t=1}, s_{t=2}, \cdots$, so we cannot compute the utility using equation 15.1. Instead, the *expected utility* for being in the state $s_0$ is computed as

$$
\overline{\mathbb{U}}(s_0, \pi) \equiv \mathbb{E}[\mathbb{U}(s_0, \pi)] = r(s_0) + \mathbb{E}\left[\sum_{t=1}^{\infty} \gamma^t r(S_t)\right].
$$
(15.2)

In equation 15.2, the probability of each state $s$ at each time $t > 0$ is estimated recursively using the transition model $p(s_{t+1}|s_t, a)$, where at each step, the action taken is the one defined in the policy $\pi(s)$. Here, we choose to simplify the notation by writing the expected utility as $\mathbb{E}[\mathbb{U}(s,\pi)] \equiv \overline{\mathbb{U}}(s,\pi)$. Moreover, the notation $\overline{\mathbb{U}}(s,\pi) \equiv \overline{\mathbb{U}}(s)$ is employed later in this chapter in order to further simplify the notation by making the dependency on the policy $\pi$ implicit.

For a state $s$, the optimal policy $\pi^*(s)$ is defined as the action $a$ that maximizes the expected utility,

$$\pi^*(s) = \arg\max_{a \in \mathcal{A}} \sum_{s' \in \mathcal{S}} p(s'|s,a) \cdot \left(r(s,a,s') + \gamma \overline{\mathbb{U}}(s',\pi^*)\right).$$

The difficulty is that, in this definition, the optimal policy $\pi^*(s)$ appears on both sides of the equality. The next two sections show how this difficulty can be tackled using *value* and *policy iteration*.

### 15.1.2   Value Iteration

Equation 15.2 described the expected utility for being in a state $s$, where the dependence on the actions $a$ was implicit. The definition of this relationship with the explicit dependence on the actions is described by the *Bellman*[1] equation,

[1] Bellman, R. (1957). A Markovian decision process. *Journal of Mathematics and Mechanics* 6(5), 679–684.

$$\overline{\mathbb{U}}(s) = \max_{a \in \mathcal{A}} \sum_{s' \in \mathcal{S}} p(s'|s,a) \cdot \left(r(s,a,s') + \gamma \overline{\mathbb{U}}(s')\right), \tag{15.3}$$

where the optimal action $a^*(s)$ is selected for each state through the max operation. Again, the difficulty with this equation is that the expected utility of a state $\overline{\mathbb{U}}(s)$ appears on both sides of the equality. One solution to this problem is to employ the *value iteration* algorithm where we go from an iteration $i$ to an iteration $i+1$ using the *Bellman update step*,

**Note:** The symbol $\leftarrow$ indicates that the quantity on the left-hand side is recursively updated using the terms on the right-hand side, which are themselves depending on the updated term.

$$\overline{\mathbb{U}}^{(i+1)}(s) \leftarrow \max_{a \in \mathcal{A}} \sum_{s' \in \mathcal{S}} p(s'|s,a) \cdot \left(r(s,a,s') + \gamma \overline{\mathbb{U}}^{(i)}(s')\right). \tag{15.4}$$

In order to use the Bellman update, we start from the initial values at iteration $i = 0$, for example, $\overline{\mathbb{U}}^{(0)}(s) = 0$, and then iterate until we converge to a steady state. The value iteration algorithm is guaranteed to converge to the exact expected utilities $\overline{\mathbb{U}}^{(\infty)}(s)$ if an infinite number of iterations is employed. The optimal actions $a^*(s)$ identified in the process for each state $s$ define the optimal policy $\pi^*$. The sequence of steps for value iteration are detailed in algorithm 10.

*Example: Infrastructure maintenance planning*   We now apply the value iteration algorithm to the infrastructure maintenance example

---

**Algorithm 10:** Value iteration

---

1 define: $s \in \mathcal{S}$                (states)

2 define: $a \in \mathcal{A}$               (actions)

3 define: $r(s, a, s')$             (rewards)

4 define: $\gamma$            (discount factor)

5 define: $p(s'|s, a)$         (transition model)

6 define: $\eta$       (convergence criterion)

7 initialize: $\overline{\mathbb{U}}'(s)$     (expected utilities)

8 **while** $|\overline{\mathbb{U}}'(s) - \overline{\mathbb{U}}(s)| \geq \eta$ **do**

9     $\overline{\mathbb{U}}(s) \leftarrow \overline{\mathbb{U}}'(s)$

10     **for** $s \in \mathcal{S}$ **do**

11         **for** $a \in \mathcal{A}$ **do**

12             $\overline{\mathbb{U}}(s, a) = \sum_{s' \in \mathcal{S}} p(s'|s, a) \cdot \left( r(s, a, s') + \gamma \overline{\mathbb{U}}(s') \right)$

13         $\pi^*(s) \leftarrow a^* = \arg\max_{a \in \mathcal{A}} \overline{\mathbb{U}}(s, a);$

14         $\overline{\mathbb{U}}'(s) \leftarrow \overline{\mathbb{U}}(s, a^*)$

15 return: $\overline{\mathbb{U}}(s) = \overline{\mathbb{U}}'(s),\ \pi^*(s)$

---

defined at the beginning of this chapter. The last parameters that need to be defined are the convergence criterion $\eta = 10^{-3}$ and the discount factor taken as $\gamma = 0.97$, which corresponds to an interest rate of 3 percent. Starting from $\overline{\mathbb{U}}^{(0)}(\mathcal{S}) = \{0, 0, 0, 0, 0, 0\}\ \$M$, we perform iteratively the Bellman update step for each state $s$,

$$\overline{\mathbb{U}}^{(i+1)}(s) \leftarrow \max_{a \in \mathcal{A}} \begin{cases} \sum_{s' \in \mathcal{S}} p(s'|s, \mathsf{N}) \cdot \left( r(s, \mathsf{N}) + \gamma \overline{\mathbb{U}}^{(i)}(s') \right) \\ \sum_{s' \in \mathcal{S}} p(s'|s, \mathsf{M}) \cdot \left( r(s, \mathsf{M}) + \gamma \overline{\mathbb{U}}^{(i)}(s') \right) \\ \sum_{s' \in \mathcal{S}} p(s'|s, \mathsf{R}) \cdot \left( r(s, \mathsf{R}) + \gamma \overline{\mathbb{U}}^{(i)}(s') \right), \end{cases}$$

where we choose the optimal action $a^*$ leading to the maximal expected utility. The expected utilities for two first and last iterations are

$$\overline{\mathbb{U}}^{(1)}(\mathcal{S}) = \{\overbrace{109.5}^{s=100\%}, \overbrace{211}^{80\%}, \overbrace{309}^{60\%}, \overbrace{393}^{40\%}, \overbrace{459}^{20\%}, \overbrace{440}^{0\%}\}\ \$M$$

$$\overline{\mathbb{U}}^{(2)}(\mathcal{S}) = \{222.5, 329, 430, 511, 572, 550\}\ \$M$$

$$\vdots$$

$$\overline{\mathbb{U}}^{(356)}(\mathcal{S}) = \{3640, 3635, 3630, 3615, 3592, 3510\}\ \$M$$

$$\overline{\mathbb{U}}^{(357)}(\mathcal{S}) = \{3640, 3635, 3630, 3615, 3592, 3510\}\ \$M.$$

**Example setup**

$\mathcal{S} = \{100, 80, 60, 40, 20, 0\}\ \%$

$\mathcal{A} = \{\underbrace{\text{do nothing}}_{\mathsf{N}}, \underbrace{\text{maintain}}_{\mathsf{M}}, \underbrace{\text{replace}}_{\mathsf{R}}\}$

$r(\mathcal{S}) = \{109.5, 109.5, 109.5, 98.6, 82.1, 0\}\$M$

$r(\mathcal{A}) = \{0, -5, -20\}\$M$

$r(s, a, s') = r(s, a) = r(s) + r(a)$

$p(s'|s, \mathsf{N}) =$

$$\begin{bmatrix} 0.95 & 0.03 & 0.02 & 0 & 0 & 0 \\ 0 & 0.9 & 0.05 & 0.03 & 0.02 & 0 \\ 0 & 0 & 0.8 & 0.12 & 0.05 & 0.03 \\ 0 & 0 & 0 & 0.7 & 0.25 & 0.05 \\ 0 & 0 & 0 & 0 & 0.6 & 0.4 \\ 0 & 0 & 0 & 0 & 0 & 1 \end{bmatrix}$$

$p(s'|s, \mathsf{M}) = \begin{bmatrix} 1 & 0 & 0 & 0 & 0 & 0 \\ 1 & 0 & 0 & 0 & 0 & 0 \\ 0 & 1 & 0 & 0 & 0 & 0 \\ 0 & 0 & 1 & 0 & 0 & 0 \\ 0 & 0 & 0 & 1 & 0 & 0 \\ 0 & 0 & 0 & 0 & 1 & 0 \end{bmatrix}$

$p(s'|s, \mathsf{R}) = \begin{bmatrix} 1 & 0 & 0 & 0 & 0 & 0 \\ 1 & 0 & 0 & 0 & 0 & 0 \\ 1 & 0 & 0 & 0 & 0 & 0 \\ 1 & 0 & 0 & 0 & 0 & 0 \\ 1 & 0 & 0 & 0 & 0 & 0 \\ 1 & 0 & 0 & 0 & 0 & 0 \end{bmatrix}$

Here, we explicitly show the calculations made to go from $\overline{\mathsf{U}}^{(1)}(100\%) = \$109.5M \rightarrow \overline{\mathsf{U}}^{(2)}(100\%) = \$222.5M$,

$$\overline{\mathsf{U}}^{(2)}(100\%) \leftarrow \max_{a \in \mathcal{A}} \begin{cases} 0.95 \cdot (\$109.5M + 0.97 \cdot \$109.5M) \cdots \\ \quad + 0.03 \cdot (\$109.5M + 0.97 \cdot \$211M) \cdots \\ \quad + 0.02 \cdot (\$109.5M + 0.97 \cdot \$309M) = \boxed{\$222.5M} \\ 1 \cdot (109.5 - \$5M + 0.97 \cdot \$109.5M) = \$210.7M \\ 1 \cdot (109.5 - \$20M + 0.97 \cdot \$109.5M) = \$195.7M \end{cases}$$
$$= \$222.5M,$$

where the optimal action to perform if you are in state $s = 100\%$ is $\pi^{*(2)}(100\%) = a^* =$ do nothing. In order to complete a full iteration, this operation needs to be repeated for each states $s \in \mathcal{S}$. Figure 15.4 presents the evolution of the expected utility of each state as a function of the iteration number. Note how the expected utilities converge to stationary values for all state variables. The corresponding optimal policy is

$$\pi^*(\mathcal{S}) = \{\underbrace{\text{do nothing}}_{s=100\%}, \underbrace{\text{maintain}}_{80\%}, \underbrace{\text{maintain}}_{60\%}, \underbrace{\text{maintain}}_{40\%}, \underbrace{\text{replace}}_{20\%}, \underbrace{\text{replace}}_{0\%}\}.$$

Figure 15.4: Value iteration algorithm converging to stationary expected utilities.

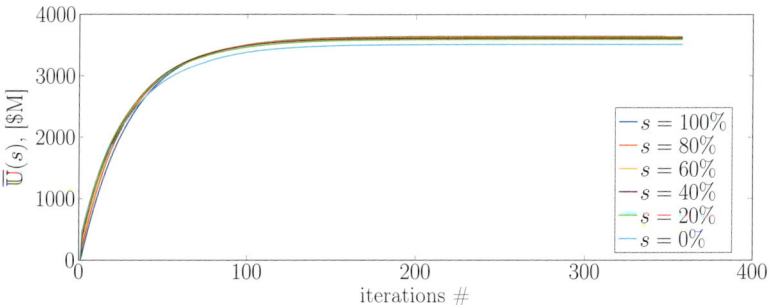

Figure 15.5 presents the expected utility for each action and state obtained using the value iteration algorithm. For each state, the expected utility corresponding to the optimal action is highlighted with a red border.

In the Bellman update step of the value iteration algorithm, the max operation is computationally expensive if the number of possible actions $\mathsf{A}$ is large. The difficulty is solved by the *policy iteration* algorithm.

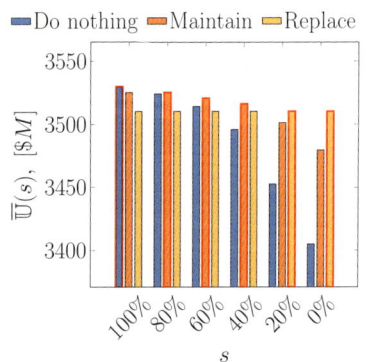

Figure 15.5: Expected utility for each action and state obtained using the value iteration algorithm. The expected utilities of the optimal action is highlighted with a red border.

### 15.1.3    Policy Iteration

Remember that a policy $\pi = \{a(\mathsf{s}_1), a(\mathsf{s}_2), \cdots, a(\mathsf{s}_\mathsf{S})\}$ defines an action to be taken for each state $s \in \mathcal{S}$. For the calculation of the expected utility of each state $s$, we can reduce the computational burden of the Bellman update step in equation 15.4 by replacing the max operation with the action $a$ defined by the policy $\pi^{(i)}(s)$,

$$\overline{\mathbb{U}}^{(i+1)}(s) \leftarrow \sum_{s' \in \mathcal{S}} p(s'|s, \underbrace{\pi^{(i)}(s)}_{a^*}) \cdot \left(r(s, \underbrace{\pi^{(i)}(s)}_{a^*}, s') + \gamma \overline{\mathbb{U}}^{(i)}(s')\right). \quad (15.5)$$

If we employ equation 15.5 instead of equation 15.4 to update the expected utility, we are required to perform only one sum rather than A sums, where A corresponds to the number of possible actions. Once the expected utility has been calculated for each state, the policy can be optimized using

$$\pi^{(i+1)}(s) \leftarrow a^* = \arg\max_{a \in \mathcal{A}} \sum_{s' \in \mathcal{S}} p(s'|s, a) \cdot \left(r(s, a, s') + \gamma \overline{\mathbb{U}}^{(i+1)}(s')\right).$$

We then repeat the successive steps of first calculating the expected utility with the value iteration algorithm using a fixed policy, and then update the policy. The details of the policy iteration procedure are presented in algorithm 11, where the *policy evaluation* step corresponds to the value iteration performed while employing equation 15.5 instead of equation 15.4. Note that identifying an optimal policy using the policy or the value iteration algorithm leads to the same results, except that policy iteration is computationally cheaper.

---

**Algorithm 11:** Policy iteration

---

1  define: $s \in \mathcal{S}$                                (states)
2  define: $a \in \mathcal{A}$                                (actions)
3  define: $p(s'|s, a)$                          (transition model)
4  define: $r(s, a, s')$                                 (reward)
5  define: $\gamma$                              (discount factor)
6  initialize: $\overline{\mathbb{U}}(s)$            (expected utilities)
7  initialize: $\pi'(s) \in \mathcal{A}$                         (policy)

8  **while** $\pi'(s) \neq \pi(s)$ **do**
9  $\quad$ $\pi(s) = \pi'(s)$
10 $\quad$ $\overline{\mathbb{U}}(s) \leftarrow$ policy evaluation$(\overline{\mathbb{U}}(s), \pi(s), p(s'|s, a), r(s, a, s'), \gamma)$
11 $\quad$ **for** $s \in \mathcal{S}$ **do**
12 $\quad\quad$ $\pi'(s) \leftarrow \arg\max_{a \in \mathcal{A}} \sum_{s' \in \mathcal{S}} p(s'|s, a) \cdot \left(r(s, a, s') + \gamma \overline{\mathbb{U}}(s')\right)$

13 return: $\overline{\mathbb{U}}(s)$, $\pi^*(s) = \pi'(s)$

---

*Example: Infrastructure maintenance planning*   Starting from $\overline{\mathbb{U}}^{(0)}(\mathcal{S}) = \{0, 0, 0, 0, 0, 0\}\$M$, we perform the policy iteration algorithm where the Bellman update step in the policy evaluation is defined by equation 15.5, where the optimal action $a = \pi^{(i)}(s)$ is defined by the optimal policy at iteration $i$. The expected utilities for the two first and last iterations are

$$
\begin{aligned}
\overline{\mathbb{U}}^{(1)}(\mathcal{S}) &= \{\overset{s=100\%}{2063}, \overset{80\%}{1290}, \overset{60\%}{768}, \overset{40\%}{455}, \overset{20\%}{197}, \overset{0\%}{0}\}\,\$M \\
\overline{\mathbb{U}}^{(2)}(\mathcal{S}) &= \{3483, 3483, 3468, 3457, 3441, 3359\}\,\$M \\
&\vdots \\
\overline{\mathbb{U}}^{(5)}(\mathcal{S}) &= \{3640, 3635, 3630, 3615, 3592, 3510\}\,\$M,
\end{aligned}
$$

and their corresponding policies are

$\mathcal{A} = \{\underbrace{\text{do nothing}}_{\text{N}}, \underbrace{\text{maintain}}_{\text{M}}, \underbrace{\text{replace}}_{\text{R}}\}$

$$
\begin{aligned}
\pi^{(0)}(\mathcal{S}) &= \{\text{N, N, N, N, N, N}\} \\
\pi^{(1)}(\mathcal{S}) &= \{\text{M, M, R, R, R, R}\} \\
\pi^{(2)}(\mathcal{S}) &= \{\text{N, N, M, M, R, R}\} \\
&\vdots \\
\pi^*(\mathcal{S}) = \pi^{(6)} = \pi^{(5)}(\mathcal{S}) &= \{\text{N, M, M, M, R, R}\}.
\end{aligned}
$$

Here, we explicitly show the calculations made to go from the policy iteration loop $1 \to 2$, $\overline{\mathbb{U}}^{(1)}(100\%) = \$2063M \to \overline{\mathbb{U}}^{(2)}(100\%) = \$3483M$, within which 351 policy evaluation calls are required,

$$
\begin{aligned}
\overline{\mathbb{U}}^{(1)}(100\%) \equiv \overline{\mathbb{U}}^{(2)(1)}(100\%) &= \overbrace{1 \cdot (109.5 - \$5M + 0.97 \cdot \$2063M)}^{\pi^{(1)}(100\%)\,=\,\text{M, maintain}} &= \$2106M \\
\overline{\mathbb{U}}^{(2)(2)}(100\%) &= 1 \cdot (109.5 - \$5M + 0.97 \cdot \$2106M) &= \$2147M \\
&\vdots \\
\overline{\mathbb{U}}^{(2)}(100\%) = \overline{\mathbb{U}}^{(2)(351)}(100\%) &= 1 \cdot (109.5 - \$5M + 0.97 \cdot \$3483M) &= \$3483M.
\end{aligned}
$$

Once the policy evaluation loop is completed, the policy must be updated. Here, we explicitly look at the specific iteration $\pi^{(1)}(100\%) = \text{M} \to \pi^{(2)}(100\%) = \text{N}$, so that

$$
\pi^{(2)}(100\%) = \arg\max_{a \in \mathcal{A}} \begin{cases} \sum_{s' \in \mathcal{S}} p(s'|s, \text{N}) \cdot \left( r(s, \text{N}) + \gamma \overline{\mathbb{U}}^{(2)}(s') \right) \\ \sum_{s' \in \mathcal{S}} p(s'|s, \text{M}) \cdot \left( r(s, \text{M}) + \gamma \overline{\mathbb{U}}^{(2)}(s') \right) \\ \sum_{s' \in \mathcal{S}} p(s'|s, \text{R}) \cdot \left( r(s, \text{R}) + \gamma \overline{\mathbb{U}}^{(2)}(s') \right) \end{cases}
$$

$$
= \arg\max_{a \in \mathcal{A}} \begin{cases} 0.95 \cdot (109.5 - \$0M + 0.97 \cdot \$3483M) \cdots \\ \quad + 0.03 \cdot (109.5 - \$0M + 0.97 \cdot \$3483M) \cdots \\ \quad + 0.02 \cdot (109.5 - \$0M + 0.97 \cdot \$3468M) = \boxed{\$3488M} \\ 1 \cdot (109.5 - \$5M + 0.97 \cdot \$3483M) = \$3483M \\ 1 \cdot (109.5 - \$20M + 0.97 \cdot \$3483M) = \$3468M \end{cases}
$$

$= \text{N (do nothing)}.$

The policy iteration converges in five loops and it leads to the same optimal policy as the value iteration,

$$\pi^*(\mathcal{S}) = \{\underbrace{\text{do nothing}}_{s=100\%}, \underbrace{\text{maintain}}_{80\%}, \underbrace{\text{maintain}}_{60\%}, \underbrace{\text{maintain}}_{40\%}, \underbrace{\text{replace}}_{20\%}, \underbrace{\text{replace}}_{0\%}\}.$$

Further details as well as advanced concepts regarding Markov decision processes can be found in the textbooks by Russell and Norvig,[2] and by Bertsekas.[3]

[2] Russell, S. and P. Norvig (2009). *Artificial Intelligence: A Modern Approach* (3rd ed.). Prentice-Hall.
[3] Bertsekas, D. P. (2007). *Dynamic programming and optimal control* (4th ed.), Volume 1. Athena Scientific.

### 15.1.4    Partially Observable Markov Decision Process

One hypothesis with the Markov decision process presented in the previous section is that state variables are exactly observed, so that at the current time $t$, one knows in what state $s$ the system is in. In such a case, the optimal action defined by the policy $\pi(s)$ can be directly selected. For the case we are now interested in, $s$ is a hidden-state variable, and the observed state $y$ is defined through the conditional probability $p(y|s)$. This problem configuration is called a *partially observable MDP* (POMDP). The challenge is that at any time $t$, you do not observe exactly what is the true state of the system; consequently, you cannot simply select the optimal action from the policy $\pi(s)$.

Because the state is observed with uncertainty, at each time $t$, we need to describe our knowledge of the state using a probability mass function $p(s) = \{\Pr(S = \mathsf{s}_1), \Pr(S = \mathsf{s}_2), \cdots, \Pr(S = \mathsf{s}_\mathsf{S})\}$. Given the PMF $p(s)$ describing our current knowledge of $s$, an action taken $a$, and an observation $y \in \mathcal{S}$, then the probability of ending in any state $s' \in \mathcal{S}$ at time $t + 1$ is given by

$$p(s'|a,y) = \frac{p(y|s') \overbrace{\sum_{s \in \mathcal{S}} \underbrace{\overbrace{p(s'|s,a) \cdot p(s)}^{p(s',s|a)}}_{}}^{\begin{array}{c} p(y,s'|a) \\ p(s'|a) \end{array}}}{p(y)}.$$

With the MDP, the policy $\pi(s)$ was a function of the state we are in; with the POMDP we need to redefine the policy as being a function of the probability of the state $\pi(p(s))$ we are in. In the context of an MDP, the policy can be seen as a function defined over discrete variables. With a POMDP, if there are $\mathsf{S}$ possible states, $\pi(p(s))$ is a function of a continuous domain with $\mathsf{S} - 1$ dimensions. This domain is continuous because $\Pr(S = s) \in (0, 1)$, and there are $\mathsf{S} - 1$ dimensions because of the constraint requiring that $\sum_{s \in \mathcal{S}} p(s) = 1$.

**MDP**
$\pi(s), \ s \in \underbrace{\mathcal{S}}_{\text{discrete}}$

**POMDP**
$\pi(p(s)), \ p(s) \in \underbrace{(0,1)^\mathsf{S}}_{\text{continuous}}$

Because the policy is now a function of probabilities defined in a continuous domain, the value and policy iteration algorithms presented for an MDP are not directly applicable. One particularity of the POMDP is that the set of actions now includes additional information gathering through observation. The reader interested in the details of value and policy-iteration algorithms for POMDP should refer to specialized literature such as the paper by Kaelbling, Littman, and Cassandra.[4] Note that solving a POMDP is significantly more demanding than solving an MDP. For practical applications, the exact solution of a POMDP is often intractable and we have to resort to approximate methods.[5]

## 15.2   Model-Free Reinforcement Learning

The Markov decision process presented in the previous section is categorized as a *model-based* method because it requires knowing the model of the environment that takes the form of the transition probabilities $p(s'|s, a)$. In some applications, it is not possible or practical to define such a model because the number of states can be too large, or can be continuous so that the definition of transition probabilities is not suited. Model-free reinforcement learning methods learn from the interaction between the agent and the environment, as depicted in figure 15.3. This learning process is typically done by subjecting the agent to multiple *episodes*. In this section, we will present two *model-free* methods suited for such a case: *temporal difference learning* and *Q-learning*.

### 15.2.1   Temporal Difference Learning

The quantity we are interested in estimating is $\mathbb{E}[\mathbb{U}(s, \pi)] \equiv \overline{\mathbb{U}}(s, \pi) \equiv \overline{\mathbb{U}}(s)$, that is, the expected utility for being in a state $s$. We defined in §15.1.1 that the utility is the sum of the discounted rewards,

$$\mathbb{U}(s, \pi) = \sum_{t=0}^{\infty} \gamma^t r(s_t, a_t, s_{t+1}),$$

obtained while following a policy $\pi$ so that $a_t = \pi(s_t)$. Moreover, we saw in §6.5 that we can estimate an expected value using the Monte Carlo method. Here, in the context where our agent can interact with the environment over multiple episodes, we can estimate $\overline{\mathbb{U}}(s, \pi)$ by taking the average over a set of realizations of $\mathbb{U}(s, \pi)$.

In a general manner, the average $\overline{x}_\mathrm{T}$ of a set $\{x_1, x_2, \cdots, x_\mathrm{T}\}$ can be calculated incrementally by following

$$\overline{x}_t = \overline{x}_{t-1} + \tfrac{1}{t}(x_t - \overline{x}_{t-1}).$$

[4] Kaelbling, L. P., M. L. Littman, and A. R. Cassandra (1998). Planning and acting in partially observable stochastic domains. *Artificial Intelligence 101*(1), 99–134.

[5] Hauskrecht, M. (2000). Value-function approximations for partially observable Markov decision processes. *Journal of Artificial Intelligence Research 13*(1), 33–94.

Figure 15.6 presents an example of an application of the incremental average estimation for a set of $\mathtt{T} = 500$ realizations of $x_t : X \sim \mathcal{N}(x; 3, 1)$. We can apply the same principle for the incremental estimation of the expected utility so that

$$\overline{\mathbb{U}}^{(i+1)}(s) \leftarrow \overline{\mathbb{U}}^{(i)}(s) + \frac{1}{\mathtt{N}(s)}\left(\mathbb{U}^{(i)}(s) - \overline{\mathbb{U}}^{(i)}(s)\right), \tag{15.6}$$

where $\mathtt{N}(s)$ is the number of times a state has been visited. Estimating $\mathbb{U}(s)$ through Monte Carlo sampling is limited to problems having a terminal state and requires evaluating the utilities over entire episodes. *Temporal difference* (TD) learning,[6] allows going around these requirements by replacing the explicit utility calculation $\mathbb{U}^{(i)}(s)$ by the *TD-target* defined as $r(s, a, s') + \gamma\overline{\mathbb{U}}^{(i)}(s')$, so that equation 15.6 can be rewritten as

$$\overline{\mathbb{U}}^{(i+1)}(s) \leftarrow \overline{\mathbb{U}}^{(i)}(s) + \alpha\Big(\underbrace{r(s, a, s') + \gamma\overline{\mathbb{U}}^{(i)}(s')}_{\text{TD-target}} - \overline{\mathbb{U}}^{(i)}(s)\Big). \tag{15.7}$$

Temporal difference learns by updating recursively the *state-utility* function $\overline{\mathbb{U}}^{(i+1)}(s)$ by taking the difference in expected utilities estimated at consecutive times while the agent interacts with the environment.

In equation 15.7, $\alpha$ is the *learning rate* defining how much is learned from the TD update step. Typically, we want to employ a learning rate that is initially large so our poor initial values for $\overline{\mathbb{U}}(s)$ are strongly influenced by the TD updates and then decrease $\alpha$ as a function of the number of times a particular set of state and action has been seen by the agent. In order to ensure the convergence of the state-utility function, the learning rate should satisfy the two following conditions,

$$\sum_{\mathtt{N}(s,a)=1}^{\infty} \alpha(\mathtt{N}(s, a)) = \infty, \quad \sum_{\mathtt{N}(s,a)=1}^{\infty} \alpha^2(\mathtt{N}(s, a)) < \infty.$$

A common learning rate function that satisfies both criteria is

$$\alpha(\mathtt{N}(s, a)) = \frac{c}{c + \mathtt{N}(s, a)},$$

where $\mathtt{N}(s, a)$ is the number of times a set of state and action has been visited and $c \in \mathbb{R}^+$ is a positive constant.

The TD-learning update step presented in equation 15.7 defines the behavior of a *passive agent*, that is, an agent, who is only evaluating the *state-utility function* for a predefined policy $\pi$. In other words, in its current form, TD-learning is suited to estimate

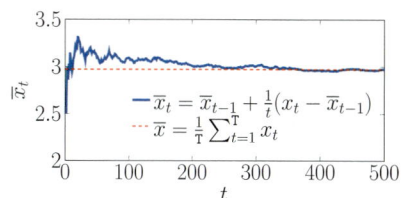

Figure 15.6: The incremental calculation of an average for a set $\{x_1, x_2, \cdots, x_{500}\}, x_t : X \sim \mathcal{N}(x; 3, 1)$.

[6] Sutton, R. S. (1988). Learning to predict by the methods of temporal differences. *Machine Learning 3*(1), 9–44.

$\overline{\mathbb{U}}(s)$ given that we already know the policy $\pi(s)$ we want the agent to follow. In practice, this is seldom the case because instead, we want an *active agent*, who will learn both the expected utilities and the optimal policy. The next section presents *temporal difference Q-learning*, a common active reinforcement learning method capable of doing both tasks simultaneously.

### 15.2.2  Temporal Difference Q-Learning

[7] Watkins, C. J. and P. Dayan (1992). Q-learning. *Machine Learning 8*(3–4), 279–292.

*Temporal difference Q-learning*[7] (often referred to as simply Q-learning) is an *active reinforcement learning* method that consists in evaluating the *action-utility function* $\mathbb{Q}(s,a)$, from which both the optimal policy

$$\pi^*(s) = \arg\max_{a\in\mathcal{A}} \mathbb{Q}(s,a)$$

and the resulting expected utilities

$$\overline{\mathbb{U}}(s) \equiv \overline{\mathbb{U}}(s,\pi^*) = \max_{a\in\mathcal{A}} \mathbb{Q}(s,a)$$

can be extracted. Analogously to equation 15.7, the TD update step for the *action-utility function* $\mathbb{Q}(s,a)$ is defined as

$$\mathbb{Q}^{(i+1)}(s,a) \leftarrow \mathbb{Q}^{(i)}(s,a) + \alpha\big(r(s,a,s') + \gamma \max_{a'\in\mathcal{A}} \mathbb{Q}^{(i)}(s',a') - \mathbb{Q}^{(i)}(s,a)\big). \tag{15.8}$$

Equation 15.8 describes the behavior of an *active agent*, which updates $\mathbb{Q}(s,a)$ by looking one step ahead and employing the action $a'$ that maximizes $\mathbb{Q}(s',a')$. Q-learning is categorized as an *off-policy* reinforcement learning method because the optimal one-step look-ahead action $a'$ employed to update $\mathbb{Q}(s,a)$ is not necessarily the action that will be taken by the agent to transit to the next state. This is because in order to learn efficiently, it is necessary to find a tradeoff between *exploiting* the current best policy and *exploring* new policies.

While learning the action-utility function using Q-learning, one cannot only rely on the currently optimal action $a^* = \pi^*(s)$ in order to transit between successive states, $s$ and $s'$. This is because such a *greedy* agent that always selects the optimal action is likely to get stuck in a local maxima because of the poor initial estimates for $\mathbb{Q}(s,a)$. One simple solution is to employ the $\epsilon$-greedy strategy, which consists in selecting the action randomly with probability $\epsilon\big(\mathtt{N}(s)\big)$ and otherwise selecting the action from the optimal policy $\pi^*(s)$. Here, $\mathtt{N}(s)$ is again the number of times a state has been visited. Therefore, an $\epsilon$-greedy agent selects an action following

$$a = \begin{cases} \mathsf{a}_i : p(a) = \frac{1}{\mathsf{A}}, \forall a, & \text{if } u < \epsilon\big(\mathsf{N}(s)\big) & \text{Random action} \\ \arg\max_{a \in \mathcal{A}} \mathbb{Q}(s, a), & \text{if } u \geq \epsilon\big(\mathsf{N}(s)\big) & \text{Greedy action,} \end{cases}$$

where $u$ is a sample taken from a uniform distribution $\mathcal{U}(0, 1)$. $\epsilon\big(\mathsf{N}(s)\big)$ should be defined so that it tends to zero when the number of times a state has been visited $\mathsf{N}(s) \to \infty$. Algorithm 12 details the sequence of steps for temporal difference Q-learning.

---

**Algorithm 12:** Temporal difference Q-learning

1   define: $s \in \mathcal{S}$             (states)
2   define: $a \in \mathcal{A}$             (actions)
3   define: $r(s, a, s')$          (rewards)
4   define: $\gamma$           (discount factor)
5   define: $\epsilon$       (epsilon-greedy function)
6   define: $\alpha$       (learning rate function)
7   initialize: $\mathbb{Q}^{(0)}(s, a)$     (action-utility function)
8   initialize: $\mathsf{N}(s, a) = 0, \forall\{s \in \mathcal{S}, a \in \mathcal{A}\}$   (action-state counts)

9   $i = 0$
10   **for** *episodes* $e \in \{1 : \mathsf{E}\}$ **do**
11     initialize: $s$
12     **for** *time* $t \in \{1 : \mathsf{T}\}$ **do**
13       $u : U \sim \mathcal{U}(0, 1)$
14       $a = \begin{cases} \mathsf{a}_i : p(a) = \frac{1}{\mathsf{A}}, \forall a, & \text{if } u < \epsilon\big(\mathsf{N}(s)\big) & \text{Random} \\ \arg\max_{a \in \mathcal{A}} \mathbb{Q}^{(i)}(s, a), & \text{if } u \geq \epsilon\big(\mathsf{N}(s)\big) & \text{Greedy} \end{cases}$
15       observe: $r(s, a, s'), s'$
16       $\mathbb{Q}^{(i+1)}(s, a) \leftarrow \mathbb{Q}^{(i)}(s, a) \cdots$
17         $+\alpha\big(\mathsf{N}(s, a)\big)\big(r(s, a, s') + \gamma \max_{a' \in \mathcal{A}} \mathbb{Q}^{(i)}(s', a') - \mathbb{Q}^{(i)}(s, a)\big)$
18       $\mathsf{N}(s, a) = \mathsf{N}(s, a) + 1$
19       $s = s'$
20       $i = i + 1$

21   return: $\overline{\mathsf{U}}(s) = \max_{a \in \mathcal{A}} \mathbb{Q}^{(i)}(s, a)$
22       $\pi^*(s) = \arg\max_{a \in \mathcal{A}} \mathbb{Q}^{(i)}(s, a)$

---

*Example: Infrastructure maintenance planning*   We revisit the infrastructure maintenance example, this time using the temporal difference Q-learning algorithm. The problem definition is identical to that in §15.1, except that we now initialize the action-utility

functions as $\mathbb{Q}^{(0)}(s,a) = \$0\,M,\ \forall\{s \in \mathcal{S}, a \in \mathcal{A}\}$. Also, we define the iteration-dependent learning schedule as

$$\alpha\big(\mathtt{N}(s,a)\big) = \frac{c}{c + \mathtt{N}(s,a)}$$

and the exploration schedule by

$$\epsilon\big(\mathtt{N}(s)\big) = \frac{c}{c + \mathtt{N}(s)},$$

where $c = 70$. We choose to learn over a total of 500 episodes, each consisting in 100 time steps. For each new episode, the state is initialized randomly. Figure 15.7 presents the evolution of the expected utility $\overline{\mathbb{U}}(s)$ for each state as a function of the number of episodes. Given that we employed 500 episodes, each comprising

Figure 15.7: Expected utility convergence for the Q-learning algorithm.

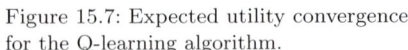
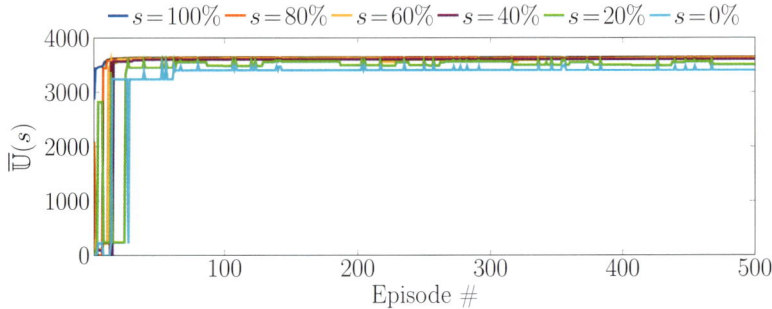

100 steps, the total number of iterations is 50,000. For the last iteration, the expected utilities are

$$\overline{\mathbb{U}}^{(50\,000)}(\mathcal{S}) = \{3640, 3635, 3630, 3615, 3592, 3510\}\,\$M.$$

Those values are identical to the ones obtained using either the policy- or value-iteration algorithm in §15.1. Analogously, the optimal policy derived from the action-utility function is also identical and corresponds to

$$\pi^*(\mathcal{S}) = \{\underbrace{\text{do nothing}}_{s=100\%}, \underbrace{\text{maintain}}_{80\%}, \underbrace{\text{maintain}}_{60\%}, \underbrace{\text{maintain}}_{40\%}, \underbrace{\text{replace}}_{20\%}, \underbrace{\text{replace}}_{0\%}\}.$$

Figure 15.8 present $\epsilon\big(\mathtt{N}(s)\big)$ as a function of the number of episodes, that is, for each state, the probability that the agent selects a random action. For the learning rate $\alpha\big(\mathtt{N}(s,a)\big)$, which is not displayed here, there are three times as many curves because it also depends on the action $a$.

For this simple infrastructure maintenance example, using an active reinforcement learning method such as Q-learning is not

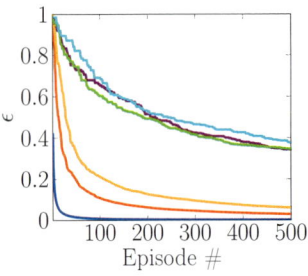

Figure 15.8: Evolution of $\epsilon\big(\mathtt{N}(s)\big)$ for each state $s$ as a function of the number of episodes.

beneficial in comparison with passive methods such as the policy- or value-iteration algorithm in §15.1. Nevertheless, active reinforcement learning becomes necessary for more advanced contexts where the number of states is large and often continuous. The reader interested in learning more about topics related to *on-policy* reinforcement learning methods such as *sarsa*, advanced exploration strategies, efficient action-utility function initialization, as well as continuous *function approximation* for Q-learning, should consult dedicated textbooks such as the one by Sutton and Barto.[8]

[8] Sutton, R. S. and A. G. Barto (2018). *Reinforcement learning: An introduction* (2nd ed.). MIT Press.

*Exercises*

P15.1  What is a policy, and why is it stationary for an infinite planning horizon and nonstationary for a finite one?

P15.2  Explain the Markov decision process (MDP).

P15.3  What does the Markov hypothesis imply in the context of an MDP?

P15.4  Explain the discount factor and what happens if $\gamma = 1$.

P15.5  How is an optimal policy defined in the context of an MDP?

P15.6  Why do we need to employ iterative algorithms to evaluate the Bellman equation?

P15.7  What is the advantage of the policy-iteration algorithm in comparison with value iteration?

P15.8  Explain the differences between model-based and model-free reinforcement learning.

P15.9  Using the same setup as the bridge maintenance example presented in this chapter:

    a.  Employ the data set in `MDP_bridge.csv` to estimate the transition matrix for the Markovian degradation model of structures for the action $a = $ do nothing.

    b.  Implement a policy-iteration algorithm and, using the discount factor $\gamma = 0.95$, identify the optimal action for each state $s \in \mathcal{S}$ and its corresponding expected utility.

    c.  Implement a temporal difference Q-learning algorithm to reproduce the results you obtained in (b).

# Bibliography

Abramowitz, M. and I. A. Stegun (1972). *Handbook of mathematical functions with formulas, graphs, and mathematical table*. National Bureau of Standards, Applied Mathematics.

Adelchi, A. and A. W. Bowman (1990). A look at some data on the Old Faithful geyser. *Journal of the Royal Statistical Society. Series C (Applied Statistics) 39*(3), 357–365.

Ang, A. H.-S. and W. H. Tang (1975). *Probability concepts in engineering planning and decision*, Volume 1—Basic Principles. John Wiley.

Armijo, L. (1966). Minimization of functions having Lipschitz continuous first partial derivatives. *Pacific Journal of Mathematics 16*(1), 1–3.

Bar-Shalom, Y., X. Li, and T. Kirubarajan (2001). *Estimation with applications to tracking and navigation: Theory algorithms and software*. John Wiley & Sons.

Bellman, R. (1957). A Markovian decision process. *Journal of Mathematics and Mechanics 6*(5), 679–684.

Ben-Akiva, M. E. and S. R. Lerman (1985). *Discrete choice analysis: Theory and application to travel demand*. MIT Press.

Bertsekas, D. P. (2007). *Dynamic programming and optimal control* (4th ed.), Volume 1. Athena Scientific.

Bertsekas, D. P., A. Nedi, and A. E. Ozdaglar (2003). *Convex analysis and optimization*. Athena Scientific.

Bishop, C. M. (2006). *Pattern recognition and machine learning*. Springer.

Blom, H. A. P. and Y. Bar-Shalom (1988). The interacting multiple model algorithm for systems with Markovian switching coefficients. *IEEE Transactions on Automatic Control 33*(8), 780–783.

Box, G. E. P. and G. C. Tiao (1992). *Bayesian inference in statistical analysis*. Wiley.

Brooks, S., A. Gelman, G. Jones, and X.-L. Meng (2011). *Handbook of Markov Chain Monte Carlo*. CRC Press.

Brynjarsdóttir, J. and A. O'Hagan (2014). Learning about physical parameters: The importance of model discrepancy. *Inverse Problems 30*(11), 114007.

Burnham, K. P. and D. R. Anderson (2002). *Model selection and multimodel inference: A practical information-theoretic approach* (2nd ed.). Springer.

Chong, E. K. P. and S. H. Zak (2013). *An introduction to optimization* (4th ed.). Wiley.

Dauphin, Y. N., R. Pascanu, C. Gulcehre, K. Cho, S. Ganguli, and Y. Bengio (2014). Identifying and attacking the saddle point problem in high-dimensional non-convex optimization. In *Advances in Neural Information Processing Systems*, 27, 2933–2941.

Del Moral, P. (1997). Non-linear filtering: interacting particle resolution. *Comptes Rendus de l'Académie des Sciences—Series I—Mathematics 325*(6), 653–658.

Dempster, A. P., N. M. Laird, and D. B. Rubin (1977). Maximum likelihood from incomplete data via the EM algorithm. *Journal of the Royal Statistical Society. Series B (methodological) 39*(1), 1–38.

Diard, J., P. Bessiere, and E. Mazer (2003). A survey of probabilistic models using the Bayesian programming methodology as a unifying framework. In *International Conference on Computational Intelligence, Robotics and Autonomous Systems (IEEE-CIRAS)*, Singapore.

Duvenaud, D. (2014). *Automatic model construction with Gaussian processes*. PhD thesis, University of Cambridge.

Ebden, M. (2008, August). Gaussian processes: A quick introduction. *arXiv preprint (1505.02965)*.

Friedman, J., T. Hastie, H. Höfling, and R. Tibshirani (2007). Pathwise coordinate optimization. *The Annals of Applied Statistics 1*(2), 302–332.

Gaudot, I., L. H. Nguyen, S. Khazaeli, and J.-A. Goulet (2019, May). OpenBDLM, an open-source software for structural health monitoring using bayesian dynamic linear models. In *13th Proceedings from the 13th International Conference on Applications of Statistics and Probability in Civil Engineering (ICASP)*.

Gelman, A., J. B. Carlin, H. S. Stern, and D. B. Rubin (2014). *Bayesian data analysis* (3rd ed.). CRC Press.

Gelman, A. and D. B. Rubin (1992). Inference from iterative simulation using multiple sequences. *Statistical Science 7*(4), 457–472.

Ghahramani, Z. (2001). An introduction to hidden Markov models and Bayesian networks. *International Journal of Pattern Recognition and Artificial Intelligence 15*(1), 9–42.

Ghahramani, Z. (2004). Unsupervised learning. In *Advanced lectures on machine learning*, Volume 3176, pp. 72–112. Springer.

Ghahramani, Z. and G. E. Hinton (1996). Parameter estimation for linear dynamical systems. Technical Report CRG-TR-96-2, University of Toronto, Dept. of Computer Science.

Gibbs, B. P. (2011). *Advanced Kalman filtering, least-squares and modeling: A practical handbook*. Wiley.

Goodfellow, I., Y. Bengio, and A. Courville (2016). *Deep learning*. MIT Press.

Goulet, J.-A. (2017). Bayesian dynamic linear models for structural health monitoring. *Structural Control and Health Monitoring 24*(12), 1545–2263.

Goulet, J.-A., C. Michel, and A. Der Kiureghian (2015). Data-driven post-earthquake rapid structural safety assessment. *Earthquake Engineering & Structural Dynamics 44*(4), 549–562.

Harrison, P. J. and C. F. Stevens (1976). Bayesian forecasting. *Journal of the Royal Statistical Society: Series B (Methodological) 38*(3), 205–228.

Hastings, W. K. (1970). Monte Carlo sampling methods using Markov chains and their applications. *Biometrika 57*(1), 97–109.

Hauskrecht, M. (2000). Value-function approximations for partially observable Markov decision processes. *Journal of Artificial Intelligence Research 13*(1), 33–94.

Hoerl, A. E. and R. W. Kennard (1970). Ridge regression: Biased estimation for nonorthogonal problems. *Technometrics 12*(1), 55–67.

Hotelling, H. (1933). Analysis of a complex of statistical variables into principal components. *Journal of Educational Psychology 24*(6), 417.

Jermyn, I. H. (2005). Invariant Bayesian estimation on manifolds. *The Annals of Statistics 33*(2), 583–605.

Kaelbling, L. P., M. L. Littman, and A. R. Cassandra (1998). Planning and acting in partially observable stochastic domains. *Artificial Intelligence 101*(1), 99–134.

Kalman, R. E. (1960). A new approach to linear filtering and prediction problems. *Transactions of the ASME–Journal of Basic Engineering 82*(Series D), 35–45.

Kennedy, M. C. and A. O'Hagan (2001). Bayesian calibration of computer models. *Journal of the Royal Statistical Society: Series B (Statistical Methodology) 63*(3), 425–464.

Kim, C.-J. and C. R. Nelson (1999). *State-space models with regime switching: Classical and Gibbs-sampling approaches with applications*. MIT Press.

Kreyszig, E. (2011). *Advanced engineering mathematics* (10th ed.). Wiley.

Ljung, L. (1979). Asymptotic behavior of the extended Kalman filter as a parameter estimator for linear systems. *IEEE Transactions on Automatic Control 24*(1), 36–50.

Lloyd, S. (1982). Least squares quantization in PCM. *IEEE Transactions on Information Theory 28*(2), 129–137.

Lowen, A. C. and J. Steel (2014). Roles of humidity and temperature in shaping influenza seasonality. *Journal of Virology 88*(14), 7692–7695.

MacKay, D. J. C. (1998). Introduction to Monte Carlo methods. In *Learning in graphical models*, pp. 175–204. Sprigner.

MacKay, D. J. C. (2003). *Information theory, inference, and learning algorithms*. Cambridge University Press.

McFadden, D. (2001). Economic choices. *American Economic Review 91*(3), 351–378.

Metropolis, N., A. W. Rosenbluth, M. N. Rosenbluth, A. H. Teller, and E. Teller (1953). Equation of state calculations by fast computing machines. *The Journal of Chemical Physics 21*(6), 1087–1092.

Murphy, K. P. (1998). Switching Kalman filters. Citeseer.

Murphy, K. P. (2002). *Dynamic Bayesian networks: representation, inference and learning.* PhD thesis, University of California, Berkeley.

Murphy, K. P. (2007). Conjugate Bayesian analysis of the Gaussian distribution. Citeseer.

Murphy, K. P. (2012). *Machine learning: A probabilistic perspective.* MIT Press.

Neal, R. (2008). The harmonic mean of the likelihood: Worst Monte Carlo method ever. URL https://radfordneal.wordpress.com/ 2008/08/17/the-harmonic-mean-of-the-likelihood-worst -monte-carlo-method-ever/. Accessed November 8, 2019.

Neal, R. M. (2001). Annealed importance sampling. *Statistics and Computing 11*(2), 125–139.

Neumann, M., S. Huang, D. E. Marthaler, and K. Kersting (2015). pyGPs: A python library for Gaussian process regression and classification. *The Journal of Machine Learning Research 16*(1), 2611–2616.

Newton, M. A. and A. E. Raftery (1994). Approximate Bayesian inference with the weighted likelihood bootstrap. *Journal of the Royal Statistical Society. Series B (Methodological) 56*(1), 3–48.

Ng, A. and M. Jordan (2002). On discriminative vs. generative classifiers: A comparison of logistic regression and naive Bayes. In *Advances in neural information processing systems*, Volume 14, pp. 841–848. MIT Press.

Ng, A., M. Jordan, and Y. Weiss (2002). On spectral clustering: Analysis and an algorithm. In *Advances in neural information processing systems*, Volume 14, pp. 849–856. MIT Press.

Nguyen, L. H., I. Gaudot, and J.-A. Goulet (2019). Uncertainty quantification for model parameters and hidden state variables in Bayesian dynamic linear models. *Structural Control and Health Monitoring 26*(3), e2309.

Nguyen, L. H. and J.-A. Goulet (2018). Anomaly detection with the switching Kalman filter for structural health monitoring. *Structural Control and Health Monitoring 25*(4), e2136.

Nielsen, T. D. and F. V. Jensen (2007). *Bayesian networks and decision graphs.* Springer.

Nocedal, J. and S. Wright (2006). *Numerical optimization.* Springer Science & Business Media.

Pearl, J. (1988). *Probabilistic reasoning in intelligent systems: Networks of plausible inference.* Morgan Kaufmann.

Quach, A. N. O., D. Tabor, L. Dumont, B. Courcelles, and J.-A. Goulet (2017). A machine learning approach for characterizing soil contamination in the presence of physical site discontinuities and aggregated samples. *Advanced Engineering Informatics 33*, 60–67.

Quiñonero-Candela, J. and C. E. Rasmussen (2005). A unifying view of sparse approximate Gaussian process regression. *Journal of Machine Learning Research 6*, 1939–1959.

Rasmussen, C. E. and H. Nickisch (2010). Gaussian processes for machine learning (GPML) toolbox. *The Journal of Machine Learning Research 11*, 3011–3015.

Rasmussen, C. E. and C. K. Williams (2006). *Gaussian processes for machine learning.* MIT Press.

Rauch, H. E., C. T. Striebel, and F. Tung (1965). Maximum likelihood estimates of linear dynamic systems. *AIAA Journal 3*(8), 1445–1450.

Rosenblatt, F. (1958). The perceptron: A probabilistic model for information storage and organization in the brain. *Psychological Review 65*(6), 386.

Rosenthal, J. S. (2011). Optimal proposal distributions and adaptive MCMC. In *Handbook of Markov Chain Monte Carlo*, 93–112. CRC Press.

Rumelhart, D. E., G. E. Hinton, and R. J. Williams (1986). Learning representations by back-propagating errors. *Nature 323*, 533–536.

Runnalls, A. R. (2007). Kullback-Leibler approach to Gaussian mixture reduction. *IEEE Transactions on Aerospace and Electronic Systems 43*(3), 989–999.

Russell, S. and P. Norvig (2009). *Artificial Intelligence: A Modern Approach* (3rd ed.). Prentice-Hall.

Simon, D. (2006). *Optimal state estimation: Kalman, H infinity, and nonlinear approaches.* Wiley.

Spiegelhalter, D. J., N. G. Best, B. P. Carlin, and A. van der Linde (2002). Bayesian measures of model complexity and fit. *Journal of the Royal Statistical Society: Series B (Statistical Methodology) 64*(4), 583–639.

Sutton, R. S. (1988). Learning to predict by the methods of temporal differences. *Machine Learning 3*(1), 9–44.

Sutton, R. S. and A. G. Barto (2018). *Reinforcement learning: An introduction* (2nd ed.). MIT Press.

Tolvanen, V., P. Jylänki, and A. Vehtari (2014). Expectation propagation for nonstationary heteroscedastic Gaussian process regression. In *IEEE International Workshop on Machine Learning for Signal Processing (MLSP)*, 1–6. IEEE.

Van der Maaten, L. and G. Hinton (2008). Visualizing data using t-SNE. *Journal of Machine Learning Research 9*, 2579–2605.

Van Loan, C. F. and G. H. Golub (1983). *Matrix computations.* Johns Hopkins University Press.

Vanhatalo, J., J. Riihimäki, J. Hartikainen, P. Jylänki, V. Tolvanen, and A. Vehtari (2013). GPstuff: Bayesian modeling with Gaussian processes. *Journal of Machine Learning Research 14*, 1175–1179.

Von Neumann, J. and O. Morgenstern (1947). *The theory of games and economic behavior.* Princeton University Press.

Wan, E. A. and R. van der Merwe (2000). The unscented Kalman filter for nonlinear estimation. In *Adaptive Systems for Signal Processing, Communications, and Control Symposium*, 153–158. IEEE.

Watanabe, S. (2010). Asymptotic equivalence of Bayes cross validation and widely applicable information criterion in singular learning theory. *Journal of Machine Learning Research 11*, 3571–3594.

Watkins, C. J. and P. Dayan (1992). Q-learning. *Machine Learning 8*(3–4), 279–292.

West, M. (2013). Bayesian dynamic modelling. In *Bayesian theory and applications*, 145–166. Oxford University Press.

West, M. and J. Harrison (1999). *Bayesian forecasting and dynamic models*. Springer.

Westgate, R. J. and J. M. W. Brownjohn (2011). Development of a Tamar Bridge finite element model. In *Conference Proceedings of the Society for Experimental Mechanics*, Volume 5, pp. 13–20. Springer.

Wikipedia (2017). Conjugate prior. URL `https://en.wikipedia.org/wiki/conjugate_prior`. Accessed November 8, 2019.

Wright, S. J. (2015). Coordinate descent algorithms. *Mathematical Programming 151*(1), 3–34.

# Index